BIOLOGICAL NITROGEN FIXATION: TOWARDS POVERTY ALLEVIATION THROUGH SUSTAINABLE AGRICULTURE

Current Plant Science and Biotechnology in Agriculture

VOLUME 42

Aims and Scope
The book series is intended for readers ranging from advanced students to senior research scientists and corporate directors interested in acquiring in-depth, state-of-the-art knowledge about research findings and techniques related to all aspects of agricultural biotechnology. Although the previous volumes in the series dealt with plant science and biotechnology, the aim is now to also include volumes dealing with animals science, food science and microbiology. While the subject matter will relate more particularly to agricultural applications, timely topics in basic science and biotechnology will also be explored. Some volumes will report progress in rapidly advancing disciplines through proceedings of symposia and workshops while others will detail fundamental information of an enduring nature that will be referenced repeatedly.

For other titles published in this series, go to
www.springer.com/series/6444

Biological Nitrogen Fixation: Towards Poverty Alleviation through Sustainable Agriculture

Proceedings of the 15th International Nitrogen Fixation Congress and the 12th International Conference of the African Association for Biological Nitrogen Fixation

Edited by

Felix D. Dakora
Cape Town Peninsula University of Technology,
Cape Town, South Africa

Samson B. M. Chimphango
University of Cape Town, Rondebosch, South Africa

Alex J. Valentine
University of the Western Cape, South Africa

Claudine Elmerich
Institut Pasteur, Paris, France

and

William E. Newton
Virginia Polytechnic Institute and State University, Blacksburg, VA, USA

 Springer

Editors

Felix D. Dakora
Cape Town Peninsula
University of Technology
Cape Town, South Africa

Samson B. M. Chimphango
University of Cape Town
Rondebosch, South Africa

Alex J. Valentine
University of the Western Cape
South Africa

Claudine Elmerich
Institut Pasteur
Paris, France

William E. Newton
Virginia Polytechnic Institute
and State University
Blacksburg, VA, USA

ISBN: 978-1-4020-8251-1   e-ISBN: 978-1-4020-8252-8

Library of Congress Control Number: 2008930755

Cover Illustration: A field trial with lentil showing the response to inoculation, and
fields of clover in bloom. Photographs courtesy of John Howieson, Murdoch
University, Western Australia and reproduced with permission.

Printed on acid-free paper

9 8 7 6 5 4 3 2 1

springer.com

15th INTERNATIONAL NITROGEN FIXATION CONGRESS

was supported by

Cape Peninsula University of Technology

UNESCO

CTA

(Technical Centre for Agricultural and Rural Cooperation,
operating within the Cotonou Agreement between the
African, Caribbean and Pacific Group and the European Union)

National Research Foundation, Pretoria

Department of Science and Technology, Pretoria

PlantBio (South Africa)

Academy of Sciences of South Africa

South African Airlines

partageons les connaissances au profit des communautés rurales

sharing knowledge, improving rural livelihoods

15th INTERNATIONAL NITROGEN FIXATION CONGRESS

was organized by the following committees

Local Organizing Committee

F. D. Dakora, Cape Peninsula University of Technology, South Africa
S. B. M. Chimphango, University of Cape Town, South Africa
P. Ndakidemi, Cape Peninsula University of Technology, South Africa
A. J. Valentine, University of the Western Cape, South Africa
N. Brewin, John Innes Centre, England, UK
E. Danckwerts, A & E Consultants, South Africa

Africa Regional Organizing Committee

I. Ndoye (Senegal)
M. Gueye (Senegal)
S. Sylla (Senegal)
S. Mpepereki (Zimbabwe)
N. Karanja (Kenya)
I. Yattara (Mali)
L. Thiombiano (FAO, Ghana)

S. K. A. Danso (Ghana)
D. Lesueur (Kenya)
M. Neyra (Senegal)
R. Abaidoo (IITA, Nigeria)
A. Beki (Algeria)
D. Odee (Kenya)
M. M. Spencer (Senegal)

15th INTERNATIONAL NITROGEN FIXATION CONGRESS

International Steering Committee

W. E. Newton (USA)
C. Elmerich (France)
T. Finan (Canada)
H. Hennecke (Switzerland)

F. O. Pedrosa (Brazil)
A. Pühler (Germany)
I. Tikhonovich (Russia)
Y.-P. Wang (China)

International Program Advisory Committee

N. Boonkerd (Thailand)
W. Broughton (Switzerland)
M. Buck (UK)
E. C. Cocking (UK)
H. Das (India)
F. de Bruijn (France)
R. Dixon (UK)
Z. M. Dong (Canada)
D. Emerich (USA)
R. Fani (Italy)
P. Gresshoff (Australia)
R. Haselkorn (USA)
R. Holm (USA)
C. Kennedy (USA)
A. Kolb (France)
J. K. Ladha (The Philippines)
P. Ludden (USA)
K. A. Malik (Pakistan)

E. Martinez-Romero (Mexico)
K. Minamisawa (Japan)
M. O'Brien (USA)
F. O'Gara (Ireland)
K. Pawlowski (Sweden)
B. Rolfe (Australia)
M. Rahman (Bangladesh)
C. Ronson (New Zealand)
T. Ruiz-Argueso (Spain)
S.-J. Shen (China)
H. P. Spaink (The Netherlands)
G. Stacey (USA)
S. Tabata (Japan)
N.-T. Hein (Vietnam)
R. Thorneley (UK)
C. Vance (USA)
D. Werner (Germany)

science
& technology

Department
Science and Technology
REPUBLIC OF SOUTH AFRICA

Contents

Edgar DaSilva
(1941–2007)

This volume is dedicated to the memory of Edgar DaSilva, former Director of the Division of Life Sciences at UNESCO, in recognition of his long-term interest in biological nitrogen fixation research. Edgar joined UNESCO in 1974 and was responsible for the life-sciences programs. His name is associated with the UNESCO Microbiological Resource Centers (MIRCEN) program, which was initiated in 1975 in partnership with the United Nations Environment Program (UNEP), and the United Nations Development Program (UNDP). He was involved in the implementation of five Biotechnology Education and Training Centres (BETCEN), one on each continent. The MIRCENs and BETCENs proved to be very successful and offered training to young scientists from developing countries as well as opportunities for international cooperation. He developed fellowship programs for research and training both within UNESCO's biotechnology program and also with partners like the American Society for Microbiology (ASM) and the International Union of Microbiological Societies (IUMS). One of his ambitions was that research carried out in developing countries should be recognized and given the necessary support. To this end, in 1985, he was instrumental in the launch of a journal, which is now published by Springer as "World Journal of Microbiology and Biotechnology". Edgar wrote numerous articles and reviews for many journals, especially those in the fields of microbiology and biotechnology. After his retirement in 2001, he continued his visits to many countries and also served as co-editor of the "Biotechnology" theme for the UNESCO-sponsored "Encyclopaedia for Life Support Systems". We thank Lucy Hoareau, Division of Basic and Engineering Sciences at UNESCO, for supplying this dedication.

Preface

Poverty is a severe problem in Africa, Asia, South America and even in pockets of the developed world. Addressing poverty alleviation via the expanded use of biological nitrogen fixation in agriculture was the theme of the 15th International Congress on Nitrogen Fixation. Because nitrogen-fixation research is multidisciplinary, exploiting its benefits for agriculture and environmental protection has continued to attract research by diverse groups of scientists, including chemists, biochemists, plant physiologists, evolutionary biologists, ecologists, agricultural scientists, extension agents, and inoculant producers.

The 15th International Congress on Nitrogen Fixation was held jointly with the 12th International Conference of the African Association for Biological Nitrogen Fixation. This joint Congress was hosted in South Africa at the Cape Town International Convention Centre, 21–26 January 2007, and was attended by about 200 registered participants from 41 countries world-wide. During the Congress, some 100 oral and approximately 80 poster papers were presented. The wide range of topics covered and the theme of the Congress justifies this book's title, *Nitrogen Fixation: Applications to Poverty Alleviation*.

Crop yields depend on many factors, but primarily on three major inputs; the capture of the Sun's light energy as chemical energy through photosynthesis, a source of water, and on the availability of a fixed-nitrogen (either mineral or organic) source. This Congress dealt with the last of these three major inputs. An enormous reservoir of nitrogen resides in the atmosphere as nitrogen (N_2) gas, however, this atmospheric nitrogen is not directly usable. It only becomes available to the biosphere through biological nitrogen fixation (BNF), a process that only the simplest microorganisms have developed. Through associations with these nitrogen-fixing microorganisms, plants can, in turn, derive a significant proportion of their fixed-nitrogen requirement for growth from BNF. The most agriculturally important associations are those of legume crops (for example, soybeans, peas, and feed legumes, like alfalfa and clover) with *Rhizobium* bacteria, where a tight symbiotic relationship occurs within a specially developed organ, the nodule, usually on the roots of the plants. An ecologically significant association involves *Frankia* microbes with trees and shrubs, which help reclaim devastated soils, and through inter-cropping can enhance the growth of valuable lumber-producing trees. Other associations include cyanobacteria with plants and trees as well as the more informal associative and endophytic associations of microorganisms with grasses, most particularly with sugarcane, where possibly all of its fixed-nitrogen requirement can be supplied by BNF. These examples clearly illustrate why BNF is a key metabolic process for food production and the maintenance of life on Earth.

The Congress theme, the application of BNF to sustainable agriculture, poverty alleviation, and environmental concerns, was well covered and included the introduction of nitrogen-fixing legumes into local small holdings, appropriate use of both soil and water, the use of indigenous soil microbes to provide N and P, and good agricultural practices generally. In addition, the basic sciences that underpin these more applied aspects were also well represented through presentations and progress reports, among others, on the fundamentals of nitrogen fixation (including a nitrogenase – the enzyme responsible for the chemical conversion of N_2 gas to ammonia – that can fix N_2 at 92°C!), plant breeding, plant and microbial molecular biology, legume-*Rhizobium* genetics, genomics, gene expression, evolution of symbioses, and nodulation physiology, stress responses, bioremediation, and forestry.

Sustainable agriculture not only depends on appropriate agricultural practices but, to maintain high yields, it requires the use of plant cultivars that respond to environmental constraints. In addition to classical plant-breeding technology, the modern-day engineering of high-performance crops and symbiotic associations can now access and use the incredible insight acquired through plant and bacterial genomics. Both topics were well represented in the Congress. Indeed, refined gene-sequence maps of the model legumes, *Lotus japonicus* and *Medicago truncatula*, have led to exciting genomic work in other legumes. On the microbial side, research on *Sinorhizobium meliloti* and *Mesorhizobium loti* as well as the broad host range *Rhizobium* NGR234 has reached the post-genomics era. The recent and rapid progress made in sequencing the genomes of *Azotobacter vinelandii*, *Azospirillum brasilense*, *Herbaspirillum seropedicae*, *Azorhizobium caulinodans*, *Gluconactetobacter diazotrophicus* and *Frankia* was also reported.

Moreover, there were reports of photosynthetic bradyrhizobia that lack the *nodAC* genes, which are necessary for Nod-factor biosynthesis and, therefore, for initiation of symbiosis. This discovery opens new opportunities for research in the area of plant-microbe signaling during the early stages of symbiotic establishment. New exciting data showed that members of the ß-proteobacteria (only recently recognized as microsymbionts), isolated from Mimosoideae can infect and fix N_2 in Papillionoideae. All this plus a cyanobacterium that anticipates sunrise each morning!

Taken together, the excellent technical presentations and the lively discussions that ensued, both during and after the oral and poster sessions, are clear indicators of the success of the Congress. We hope that this volume will serve as a living reminder of those sweet moments in Cape Town. On a sad note, during the preparation of these Proceedings, we learned of the untimely death of Edgar DaSilva, who was an avid supporter of biological nitrogen-fixation research. We feel it only appropriate to dedicate these Proceedings to his memory.

Finally, we wish to thank all the individuals who helped to organize this Congress and all those who gave their support, including Ms. Helen Zille, the Mayor of Cape Town, Mr. Ben Durham, who represented the RSA Department of Science and

Technology, Professor Wieland Gevers, the CEO of the Academy of Sciences of South Africa, and Dr. Ibrahima Ndoye of the AABNF. We also thank the sponsoring organizations, listed elsewhere, without whose financial support the Congress would not have been possible. We especially thank Elizabeth Danckwerts (and her Mum) for their commitment and tireless effort in organizing the Congress. We sincerely appreciate Vicki Newton for her hard work, dedication, and guidance towards getting these Proceedings published. Lastly, we thank Edwina Felix for her voluntary service to the Congress and all the postgraduate students who assisted during the Congress. All protocol observed!

Felix Dakora
Samson Chimphango
Alex Valentine
Claudine Elmerich
William E. Newton
Cape Town
December 31, 2007

KEYNOTE ADDRESS

GENE DISCOVERY AND MARKER DEVELOPMENT IN CROP LEGUMES

T. H. N. Ellis

John Innes Centre, Colney Lane, Norwich NR4 7UH, England, UK

The Taxonomic Context

Of all species of agronomic importance, legume crops are probably the most diverse taxonomic group. This species richness poses a problem for genetic and genomic approaches to support breeding because effort and investment is necessarily divided. Despite this, many cultivated legumes fall into two main taxonomic sub-groups represented by the ILRC and Phaseoleae clades, but lupin and peanut are notable exceptions to this convenient grouping (Figure 1) as are the many tree species that provide, for example, wood, fuel or ornamentation. This taxonomic structure provides a strong justification for the use of model species to focus effort and serve a wide community of users.

The model legume species, *Lotus japonicus* and *Medicago truncatula*, are both in the ILRC clade, providing a rich source of genetic and genomic information especially for the 'cool season' legumes in that group. The Phaseoloids are the other major taxonomic group that includes many legume crops. This group contains soybean, the legume of the greatest economic importance, for which extensive genomic information is available and which is itself the subject of a genomic sequencing project (see http: //soybase.agron.iastate.edu/). In contrast to soybean, the development of the model legume systems in *Lotus* and *Medicago* was driven by the need to find a plant which could be used to study the molecular genetics of the two root symbiotic systems that are notably absent from *Arabidopsis*. This was driven by the discovery of mutants that are defective in both mycorrhizal and rhizobial symbiosis (Duc et al., 1989), so the utility of these model species for legume crops is a bonus and we now have the opportunity to find how to make best use of them for crop legumes: in part, this is the objective of the current EU framework 6 funded project 'Grain Legumes' (www.eugrainlegumes.org).

Genetic Diversity

Together with the development of model systems, molecular genetical approaches have been developed more generally and have been applied to taxonomic studies in the widest sense. A gene sequence based phylogeny of the legumes has been developed that

F. D. Dakora et al. (eds.), *Biological Nitrogen Fixation: Towards Poverty Alleviation through Sustainable Agriculture*.
© Springer Science + Business Media B.V. 2008

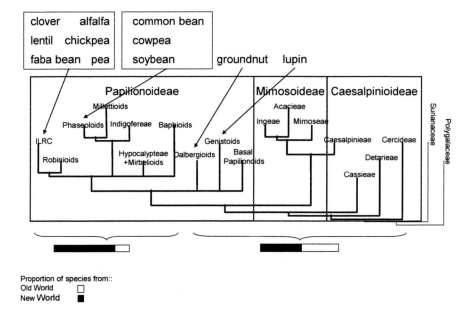

Figure 1. Taxonomic position of several agronomic legume species (Adapted from Lewis et al., 2005)

allows us to be precise in the formulation of evolutionary questions about biogeography or plant adaptation (Wojciechowski, 2003).

In addition, currently many studies of intra-species variation are underway, notably within the CGIAR (www.generationcp.org). A surprising observation for *Pisum* is that, for a predominantly in-breeding species, it displays an extensive history of introgression and recombination that segregates among lineages (Vershinin et al., 2003). Furthermore, estimates of allele ages propose that this species has maintained polymorphism for a very long period of time (Jing et al., 2005). Extension of these studies allows the understanding of the partitioning of genetic variation within germplasm collections and this can direct attention to important accessions or aid in the definition of subsets of accession either as core collections or for association studies.

Genome Organisation

The genome of *Pisum* is unusually large for a legume, but is typical of the Vicieae tribe (Figure 2). This tribe ranges widely in genome size and individual genera can vary ten-fold. For *Pisum,* the combination of the phylogenetic and taxonomic studies mentioned above, together with studies of repetitive DNA sequences in this genome, are consistent with a retrotransposon-driven increase in genome size within lineages that is averaged out between plants as a consequence of frequent introgression, recombination and segregation (Vershinin et al., 2003). The combination of information on the frequency distribution of alleles, together with estimates of allele ages, permitted the calculation of several population genetic parameters for *Pisum* (Jing et al., 2005) that suggest a relatively small effective population size. These results suggest that random replicative transposition events with little selective consequence can survive for extended periods within the *Pisum* genome and that, during this period, their frequency in the population of interbreeding lineages can wax and wane, whereas insertions are randomised with respect to each other through recombination. This model is sufficient to explain the observed pattern of marker diversity in *Pisum*, where loss of insertion sites does not appear to require any specific mechanism (Devos et al., 2002) distinct from the usual rules of neutral mutations in a population (Kimura, 1983).

Most legumes with large genomes are in the Vicieae.

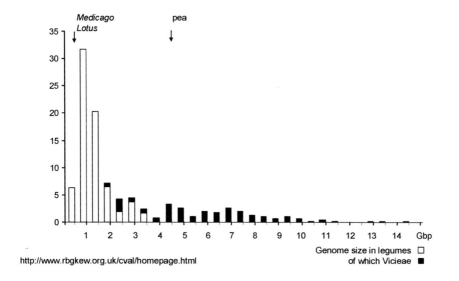

Figure 2. Genome size of legumes.

Conservation of Gene Order

The picture of extensive genomic diversity in the Vicieae contrasts with the observed conservation of the order of genes on the genetic maps of species within the ILRC clade (Kaló et al., 2004; Cannon et al., 2006; Phan et al., 2006) and more widely in legumes (Choi et al., 2004; Zhu et al., 2005), suggesting that the increase in genome size within the Vicieae is largely a consequence of the accumulation of intergenic repetitive elements rather than either segmental or whole-genome duplication as has been observed in soybean (Shoemaker et al., 2006).

Models and Crops

These observations suggest that model legume species should be useful for the understanding of traits and the isolation of corresponding genes in target crop species. This fine aspiration requires several conditions to be met in order to become a reality. First, the relationship between the model and target genomes needs to be established. While this has been achieved in general terms and to a moderate degree, any individual case needs to be verified. Second, for gene isolation in target species, it is necessary to order candidate genes with respect to trait determinants in segregating populations of the target species and, in addition to potentially difficult phenotype characterisation, this can require large population sizes for the collection of informative recombination events. Clearly if the trait determinant is either a cis-acting element or non-coding RNA, the problems of comparative genomics are challenging.

The presentation corresponding to this text discussed the relative merits of standard genetic mapping populations and populations of deletion mutants. In the former case, recombination events have to be collected one at a time and ordered with respect to the trait determinant, whereas in the second, paired deletion end-points are collected systematically and these are expected to be closer to one another than recombination events. However, the deletion strategy depends on having isolated the relevant deletion mutant. In turn, this raises the question of whether systematic approaches to collect deletion mutants in target species is a worthwhile investment of effort as compared to the generation and characterisation of segregating populations for selected traits as needed.

Acknowledgements

I thank the BBSRC, Defra and the European Commission for their funding for my lab, the Pulse Crop Genetic Improvement Network (www.pcgin.org) and the FP6 Integrated project 'Grain Legumes' FP6-2002-FOOD-1-506223, respectively. I also thank Maggie Knox, Carol Moreau, and Lynda Turner for their efforts in the projects I have described and Mike Ambrose, Andy Flavell, Julie Hofer and György Kiss for their unstinting support and encouragement.

References

Cannon SB et al. (2006) Proc. Natl. Acad. Sci. 103, 14959–14964.
Choi H-K et al. (2004) Proc. Natl. Acad. Sci. 101, 15289–15294.
Devos K et al. (2002) Genome Res. 12, 1075–1079.
Duc G et al. (1989) Plant Sci. 60, 215–222.
Jing R et al. (2005) Genetics 171, 741–752.
Kaló P et al. (2004) Mol. Genet. Genom. 272, 235–246.

Kimura M (1983) The Neutral Theory of Molecular Evolution, Cambridge University Press, Cambridge.
Lewis G et al. (eds) (2005) Legumes of the World. Royal Botanic Gardens, Kew.
Phan HTT et al. (2006) Theor. Appl. Genet. DOI 10.1007/s00122-006-0455-3.
Shoemaker RC et al. (2006) Curr. Opin. Plant Biol. 9, 104–109.
Vershinin AV et al. (2003) Mol. Biol. Evol. 20, 2067–2075.
Wojciechowski M (2003) Adv. Leg. Syst. 10, 5–35.
Zhu H et al. (2005) Plant Physiol. 137, 1189–1196.

SECTION 1

TOWARDS SUSTAINABLE AGRICULTURE

PART 1A

CROP IMPROVEMENT AND SUSTAINABLE AGRICULTURE FOR POVERTY ALLEVIATION

NUTRIENT MINING OR CARBON SEQUESTRATION? BNF INPUTS CAN MAKE THE DIFFERENCE

Robert M. Boddey[1], Claudia P. Jantalia[1], Lincoln Zotarelli[2], Aimée Okito[3], Bruno J. R. Alves[1], Segundo Urquiaga[1] and Telmo J. C. Amado[4]

[1]Embrapa Agrobiology, Km 47, BR 465, C.P: 74505, Seropédica, 23890-970, RJ, Brazil; Department of Crop Science, University of Florida, Gainsville, FL, USA; [3]Département Biologie, Faculté des Sciences, University of Kinshasa, B.P. 190, Kinshasa, Republique Democratique do Congo; [4]Department of Soil Science, Federal University Santa Maria, Santa Maria, RG, Brazil
(Email: bob@cnpab.embrapa.br)

Intensive mechanised grain production systems in Brazil occupy over 40 million hectares (Mha). Most of this area is dedicated to soybean and maize production in summer in rotation with oats, wheat, or green manure crops in the winter. In the 1960s and 1970s, it was common to grow soybean continuously, followed by wheat using conventional tillage. Apart from the problems of pest and disease build up owing to low crop diversity, the twice-yearly intensive tillage led to loss of soil organic matter (SOM) and increasing soil erosion, especially on more sandy soils. Because of this, at the start of the 1970s, several wealthy innovative farmers started to experiment with direct drilling of crops (zero tillage, ZT). With the help of the state and federal research institutes, university agricultural faculties and the pesticide companies, this technology gradually developed and approx. 1 Mha was under ZT by the start of the 1990s. However, as better practices and herbicides were developed, it became evident that as ploughing was abandoned, there were large savings to be made in terms of tractor fuel. Hence, even in areas where soil erosion was not an evident problem, adoption of ZT increased rapidly not only in the southern region, but also in the Cerrado (central tropical savanna) region of Brazil such that today over 24 Mha are cropped under this system.

There is a general consensus in North America that, where reduced or zero tillage has been adopted, SOM levels will increase with time. Recently, this consensus has been challenged by Baker et al. (2007), who suggested that the conclusion that the adoption of ZT stimulates soil C sequestration was probably an artefact of the fact that soil sampling in such studies was limited to only 20–30 cm depth. However, recent studies in Brazil, where soils were sampled to 80 or 100 cm depth (e.g., Sisti et al., 2004;

F. D. Dakora et al. (eds.), *Biological Nitrogen Fixation: Towards Poverty Alleviation through Sustainable Agriculture.*

Diekow et al., 2005), show conclusively that, in medium to high productivity systems when the cropping system showed a positive N balance, soil C accumulation can reach or exceed 1 mg ha^{-1} year^{-1} for at least 10 years. For the system to have a positive N balance, it is necessary that the external inputs of N to the system (fertiliser N and biological nitrogen fixation, BNF) exceed the N exported in grain or lost by leaching or in gaseous forms. In these studies, the N balances were positive because the crop rotations included N_2-fixing leguminous cover crops such as lupins, vetch or clover from which no grain was exported. Almost all of these rotations are based on soybean, which we have shown can contribute in excess of 200 kg N ha^{-1} crop^{-1}. However, the quantity of N exported in the grain (~80% of shoot N) usually exceeds the BNF contribution and the N balance of crop rotations comprised only of soybean and cereal crops rarely show a positive N balance and no detectable C sequestration occurs (Machado and Silva, 2001; Sisti et al., 2004).

The loss of SOM from tropical soils caused by frequent tillage is well documented and is particularly acute when agricultural products are removed from the field and there are no inputs of nutrients from fertilisers or other sources. In the countries of sub-Saharan Africa, the great majority of farming systems do export more from farm properties than is applied in external inputs (Stoorvogel and Smaling, 1998). This phenomenon, known as "nutrient mining", results in the gradual decrease in soil fertility with time. The questions are: can the technology of zero tillage and the introduction of leguminous cover crops be adapted for resource-poor farmers in such regions? And, will it both increase food crop yields and promote a gradual increase in SOM levels?

With regard to first question, the prerequisites appear to be some correction of soil fertility with modest applications of P and K fertilisers, and sufficient length of the rainy season (or a bimodal rainy season) to allow at least 90 days of legume cover crop growth prior to the main grain crop. Several reports from various regions have shown that crop yields can be increased with such technology (VanLauwe et al., 2002). In Swaziland, the FAO technical mission undertaken by Prof. Telmo Amado showed convincingly that yields of maize and sorghum on small-holder properties could be increased from less than 1 mg ha^{-1} to up to 4 mg ha^{-1} with the adoption of this package of technology (results shown at this Congress). The question of whether SOM levels will increase has not been answered satisfactorily as yet. Long-term monitoring of soil SOM levels in fields where this package of technologies has been adapted is highly desirable but nutrient balance studies can more rapidly, albeit less convincingly, demonstrate how SOM stocks are changing.

For nutrient balance studies, nitrogen is an excellent indicator because it is usually the element in the highest quantity in the crop, and is highly susceptible to losses (gaseous or via leaching). To compute an N balance, it is necessary to estimate the inputs of N from BNF and fertiliser (inorganic or organic) and the N exported in crop products or lost in gaseous forms or by leaching. With ZT, erosion losses should be minimal, and models may indicate the approximate magnitude of N losses.

In a study planted in an N-deficient soil at Embrapa Agrobiologia by Dr Aimée Okito, a simple N balance (BNF–N exported in grain; no N fertiliser was applied) was computed for the sequences groundnut, Mucuna or fallow (Okito et al., 2004). BNF was quantified using the ^{15}N natural abundance technique. The results of the N balance showed that, when both groundnut (1.0 mg ha^{-1}) and maize grain (2.57 mg ha^{-1}) were removed from the system, there was an overall removal from the system of at least (N losses were not quantified) 36.5 kg N ha^{-1} (Table 1), which showed that soil N was being mined from the system. In the sequence Mucuna-maize (Mucuna seeds were not harvested, maize yield 2.97 kg ha^{-1}), there was small positive balance of +8 kg N ha^{-1}.

Table 1. Simple N balance for the sequence of groundnut, velvet bean (*Mucuna pruriens*) or natural fallow, followed by maize variety Sol de Manhã (after Okito et al., 2004).

Legume crop before maize	N derived from BNF	N exported in legume grain	N exported in maize grain	Overall N balance[a]
		(kg N ha^{-1}) Maize variety: Sol de Manhã		
Groundnut	55.6	49.4	42.7	−36.5
Velvet bean	59.6	0.0	51.8	+7.8
Fallow[b]	30.9	0.0	33.6	−2.7

[a]N Balance calculated from export of N in grains minus input of N from BNF in aerial tissue
[b]The natural fallow was rich (70% of dry matter) in the legume *Indigofera hirsuta*

The benefit of the introduction of either legume on maize or total (maize + groundnut) grain yield was clearly demonstrated. The system that gave most income to the farmer was the groundnut followed by maize (US$520 ha^{-1} compared to Mucuna followed by maize, only US$276) would not be sustainable because SOM reserves were being depleted. This simple study sounds a note of warning. The introduction of legumes into farmers systems should be accompanied by nutrient balance (or equivalent) studies to investigate if SOM is being depleted or accumulated. If the balance is negative, as in the case of the sequence groundnut-maize in this study, it will be necessary to modify the package (e.g., addition of N fertiliser to the maize) until at least an approximate neutral nutrient balance is achieved.

Increased SOM levels bring very considerable direct benefits to the farmer and the removal of CO_2 from the atmosphere during this process pleases environmentalists everywhere. SOM build-up increases the capacity of the soil to store and exchange nutrients, and strengthens soil structure to resist erosive losses. It follows that to increase the productivity of resource-poor smallholders, it is essential to eliminate nutrient mining and promote SOM acquisition. This can only be viable using N_2-fixing legumes in a technological package which includes the elimination of soil tillage and, where necessary, liming and mitigation of P and K deficiency with fertiliser addition. The search for suitable genotypes of grain and cover crop legumes, which are drought tolerant, should be an urgent research priority. The vast biological diversity of such legumes has hardly been investigated, let alone exploited. For example, there are approximately 800 species of *Crotlaria*, but only very few genotypes of one or two species have been tested for use as cover crops and this resource could be of enormous benefit to building soil fertility on small-holdings all over the developing world. We as specialists in biological N_2 fixation should be encouraging funding agencies to finance projects which acquire and test such legumes in real on-farm situations and examine nutrient balances and SOM dynamics *in situ*.

Acknowledgements

The Brazilian authors would like to acknowledge financing of these nutrient balance and SOM dynamics studies by the Brazilian National Research Council (CNPq) and the Rio de Janeiro State Research Foundation through the program "Cientista de Nosso Estado".

References

Baker JM et al. (2007) Agric. Ecosyst. Environ. 118, 1–5.
Diekow J et al. (2005) Soil Till. Res. 81, 87–95.
Machado PLO de A and Silva CA (2001) Nutr. Cycl. Agroecosyst. 61, 119–130.
Okito A et al. (2004) Pesq. Agropec. Bras. 39, 1183–1190.
Sisti CPJ et al. (2004) Soil Till. Res. 76, 39–58.
Stoorvogel JJ and Smaling EMA (1998) Nutr. Cycl. Agroecosyst. 50, 151–158.
VanLauwe B et al. (2002) Integrated Plant Nutrient Management in Sub-Saharan Africa: From Concept to Practice. CABI, Wallingford/Oxon, UK.

TRIPARTITE SYMBIOTIC SYSTEM OF PEA (*PISUM SATIVUM L.*): APPLICATIONS IN SUSTAINABLE AGRICULTURE

A. Y. Borisov[1], T. N. Danilova[1], O. Y. Shtark[1], I. I. Solovov[2], A. E. Kazakov[1], T. S. Naumkina[2], A. G. Vasilchikov[2], V. K. Chebotar[1] and I. A. Tikhonovich[1]

[1]All-Russia Research Institute for Agricultural Microbiology, Podbelsky chaussee 3, Pushkin 8, St. Petersburg, 196608, Russia; [2]Institute of Grain Legumes and Groat Crops, Orel, p/b Streletskoe, 303112, Russia (Email: Alexey_Borisov@ARRIAM.spb.ru or AYBorisov@yandex.ru)

"…we enter the era of biotechnology knowing more and more about the growth of legumes at the gene level, but except for some producers in developed countries, unable to effectively translate these into major gains in productivity…"
G. Catroux et al. (*Plant and Soil*, 2001, 230: 21–30).

"The world is on the brink of a new agriculture, one that involves the marriage of plant biology and agroecology under the umbrella of biotechnology and germplasm improvement." C. P. Vance (*Plant Physiology*, 2001, 127: 390–397).

The striking increase in the use of nitrogen (N) and phosphorus (P) fertilizers between 1960 and 2000 by intensive agricultural practices has led to degradation of air and water quality (Tilman et al., 2001). However, the existence of common legume plant genes that are implicated in both nitrogen-fixing and arbuscular mycorrhizal symbioses – the two associations that improve mineral nutrition, water supply and tolerance to biotic and abiotic stresses of plants – poses the question of how such a tripartite symbiosis (legume plant + arbuscular mycorrhizal fungi + nodule bacteria) can be exploited in sustainable agriculture (Borisov et al., 2004). The great genetic variability shown in the effectiveness of such a system for pea (and, obviously, for other legumes) clearly demonstrates that we can and should breed legumes to increase the symbiotic potential of the tripartite symbioses (Jacobi et al., 2000).

This, in turn, raises the question of how best to develop new types of complex inoculants to match the highly symbiotically effective plants during the breeding process. Field trials (performed during the years 2000–04) demonstrated the highly beneficial effect of inoculation on plant biomass production and protein content. The initial material for breeding was selected and the only parameter to estimate symbiotic effectiveness,

F. D. Dakora et al. (eds.), *Biological Nitrogen Fixation: Towards Poverty Alleviation through Sustainable Agriculture*.
© Springer Science + Business Media B.V. 2008

plant biomass production, was determined (Shtark et al., 2006). The methodology of breeding and the production of the complex inoculants (BisolbiMix) were developed in cooperation with the Institute for Grain Legumes and Groats Crops (Orel, Russia) and the Innovation Company, Bisolbi-Inter (St. Petersburg, Russia). Two parameters were suggested to be included into the description of new legume commercial cultivars: "symbiotic effectiveness" (average percentage of additional yield on the score of tripartite symbiosis) and "symbiotic potential" (maximum percentage of additional yield on the score of tripartite symbiosis registered during trials).

For the first time in the history of legume breeding, the pea cultivar "Triumph" (Figure 1), which has increased potential for forming a tripartite symbiosis, has been purposefully created with a "symbiotic effectiveness" of 10% (Table 1) and a "symbiotic potential" of 30%.

Figure 1. Plant architecture of the cultivar "Triumph".

Table 1. Changes in agriculturally important traits for the cultivar "Triumph" after inoculation with BisolbiMix (2004–2006 г.г.).

Traits	Control	BisolbiMix
Plant height (cm)	92	88
Number of pods per productive nod	1.6	2.0
Number of seeds in the pod	6.72	7.80
Seed weight per plant (g)	28	32
Weight of 1,000 seeds (g)	240	250
Life cycle (days)	76	78
Protein content in the seeds (%)	22.3	24.0
Nitrogenase activity (μg of N(plant x hour)$^{-1}$)	10.2	15.7
Number of nodules per plant	73	112

This work was supported by the following grants: RFBR-INNO 02-04-08025; RFBR 06-04-01856, 07-04-01171, 07-04-01558; Grants of the President of Russia НШ-1103.2003.4, НШ-9744.2006.4, MD-350.2003.04; NATO-Russia JSTC.RCLG.979133; CRDF (ST-012-0); GLIP TTC FOOD-CT-2004-506223.

References

Borisov AY et al. (2004) Biologia 59, 137–144.
Jacobi LM et al. (2000) Agricultural Biology (Selskohoziastvennaia Biologia) 3, 94–102 (in Russian).
Shtark OY et al. (2006) Ecological Genetics (Ecologicheskaia Genetika) 4, 23–28 (in Russian).
Tilman D et al. (2001) Science 292, 281–284.

PLANT GROWTH-PROMOTING DIAZOTROPHS: OPTIMISING THEIR ROLE AS KEY AGENTS IN ACHIEVING MORE EFFICIENT NUTRIENT USE BY FIELD CROPS

Ivan R. Kennedy[1], Mihály Kecskés[1], Rosalind Deaker[1], Rodney J. Roughley[1], Sally Marsh[2], Michael Rose[1,4], Abu Choudhury[1], Nguyen Thanh Hien[3], Phan Thi Cong[4], Pham Van Toan[5] and Tran Thanh Be[6]

[1]Faculty of Agriculture, Food and Natural Resources, University of Sydney, NSW, Australia; [2]Faculty of Natural and Agricultural Sciences, University of Western Australia; [3]Hanoi University of Science, Vietnam; [4]Institute of Agricultural Science of South Vietnam, Vietnam; [5]National Institute of Soils and Fertilizers, Vietnam; [6]Mekong Delta Development Research Institute, Cantho University, Vietnam

There are widespread attempts by scientists and commercial producers to improve yields of crop plants with plant growth-promoting rhizobacteria (PGPR). Importantly, there is now statistically significant evidence (e.g., Nguyen et al., 2003) that benefits from inoculation can be obtained under field conditions, so reducing the need for high inputs of nutrients particularly nitrogen and providing significant economic benefits to farmers. We have advanced the hypothesis that diazotrophs are specially adapted to this PGPR role (Kennedy et al., 2004), because their ability to grow in conditions of high C:N ratios. However, such diazotrophic PGPRs do not yet provide a technology recognised as reliable enough to justify their application to crops such as rice and wheat on a global scale.

Bowen and Rovira (1999) included a discussion on possible mechanisms for the PGPR response that included plant growth-regulating effects (phytohormones) – both positive and negative, induced systemic resistance to microbial pathogens, siderophore production aiding plant nutrition by chelation, P solubilisation, and root-associated N_2 fixation. Dobbelaere et al. (2003) reviewed the diazotrophic PGPR in detail, highlighting their mechanisms of action, which include biological nitrogen fixation, plant growth promotion by production of auxins, cytokinins, gibberellins and ethylene, P-solubilisation, increased nutrient uptake, enhanced stress resistance, vitamin production, and biocontrol. We have advanced the thesis that PGPR may promote more sustainable crop yield increases by modifying soil-plant processes so that N and other nutrients are more completely retained in the plant-soil system (Kennedy et al., 2004).

F. D. Dakora et al. (eds.), *Biological Nitrogen Fixation: Towards Poverty Alleviation through Sustainable Agriculture.*
© Springer Science + Business Media B.V. 2008

In both laboratory and field work with rice in Vietnam and Australia, we have sought to analyse the factors controlling reliable expression of these PGPR effects in the field as vectors in a conceptual model known as *the yield polygon* (see Figure 1). The maximum yield or optimal economic yield involves a complex genotype x environment interaction that involves a set of non-linear yield vectors, which contribute to yield. By assessing the effectiveness of each of these vectors, using where possible a set of rapid tests (e.g., soil analyses and fertiliser potential), it should be possible to estimate which vectors are sub-optimal and thereby estimate the vectors most limiting yield. This procedure would provide a means of deciding the conditions under which application of management options, such as application of biofertilisers, would most likely be effective. If all the yield vectors are operating effectively, the extra yield potential would be minimal.

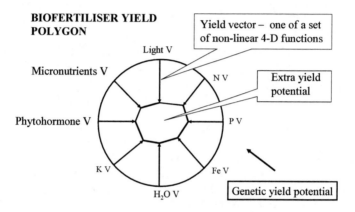

Figure 1. The yield polygon provides a framework for analysis of the factors controlling yield. Yield vectors (e.g., PV for phosphate supply) refer to the non-linear forces that mobilise interactions needing optimisation for either maximum yield or most profitable gross margins. The status of individual yield vectors may be assessed using rapid tests regarding their closeness to optimum levels.

However, this assumes that effective biofertiliser strains are already available or can be selected on demand. This is by no means assured and there is a need to generate these strains more efficiently if possible. Providing a reliable technology will require means of analysing the range of biodiversity available in both microbes and in cereals to enable the selection of superior genotypes of microbes and cereal cultivars for displaying of PGP effects. Up till now, PGPR organisms have been selected by direct approaches. For example, strains in the multi-strain biofertiliser (BioGro) designed for rice culture in the Hanoi area of Vietnam (Nguyen et al., 2003) were selected by methods such as ability to reduce C_2H_2 to C_2H_4 (as an indication of its potential for N_2

fixation), ability to solubilise precipitated $Ca_3(PO_4)_2$ in an agar medium, or production of toxic extra-cellular compounds similar to bacteriocins and inhibiting 50% of a test group of 100 rhizosphere organisms.

There is certainly potential already to apply such PGPR organisms for benefits in crop production. In Vietnam, the performance of a multi-strain biofertiliser (BioGro) designed for rice culture has been statistically assessed in field trials over three years. This project on rice in Vietnam has confirmed the need for repeated inoculation with each successive rice crop, the interaction of BioGro with rates of N and P application, and the capacity of the biofertiliser to provide needs of high quality rice in most cases on farms. Based on the results of these trials, an extension project has been conducted independently of the other researchers in the Mekong Delta rice-growing zone by Tran Thanh Be (Mekong Delta Development Research Institute). The results of the farmer trials, where half a farm was treated with BioGro with the remaining half receiving the normal application of chemical fertilisers (N as urea), showed significant increases in profits to the rice farmers. The data shown in Figure 2 take into account actual costs to farmers and include the use of insecticides, herbicides and related inputs.

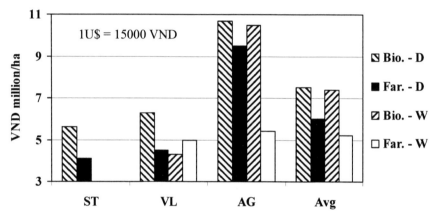

Figure 2. Extra profit per ha (Vietnamese Dong) obtained in three Mekong Delta sites (ST, VL, and AG) in wet (W) and dry (D) seasons in 2005–2006. Treatment with Bio-Gro (Bio.) and application of the normal rate of N-application (Far.) as urea is shown.

In these trials, the rice farmers verified an observation reported earlier related to the need for pest control. Rice grown with BioGro with a lower input of chemical nitrogen is apparently more resistant to insect attack, although the exact cause of this effect is unknown. Fewer insect sprays were required. These results have been obtained using strains obtained from the rhizosphere of rice. The strains used in BioGro have been modified over the past few years to further optimise its reliable effect. However, there are no doubt further opportunities of optimise the selection of strains.

To achieve this aim we suggest the following research strategy.

1. The use of bioinformatics to identify and confirm the regions of bacterial genomes responsible for PGP effects, which will enable continued improved selection of microbes from the natural biodiversity in the rhizosphere of crop plants.
2. Based on these results, develop novel molecular tools for the selection of the best PGPRs, e.g., sequence analysis of key genes or the use of immunoblotting techniques such as ELISA on imprinted membranes. These methods can be used to identify and estimate the numbers and viability (colony-forming units) of beneficial organisms and for selecting superior strains through hybridisation to specific sequences. These tools can also be used to confirm the function of PGP effects.
3. The use of proteomics to identify in plants the main gene products associated with effective PGP microbial activity, possibly allowing the selection of cultivars that can optimise yields.

It is anticipated that the regions of bacterial genomes, which code for IAA-synthesis enzymes (Dobbelaere et al., 1999; Xie et al., 1996) and for the repression of amino-cyclopropane carboxylate oxidase 1-amino cyclopropane-1-carboxylic acid deaminase (Glick et al., 1998; Gray and Smith, 2005) will form a basis for the development of these tools. But the use of bioinformatics to seek homologous regions in the genomes of PGPR organisms is also likely to provide information useful in isolating better strains, even where the function of such genetic information is not resolved.

The benefits from the application of such PGPR strains to crops will include economic gains such as increased income from higher yields and reduced fertiliser costs. However, it will also include environmental benefits, such as lower emission of greenhouse gas in fertiliser manufacture, decreased N_2O (with more than 300 times the global warming effect of CO_2) emissions, as well as decreased leaching of nitrate-N to ground water. Obtaining maximum benefits on farms from diazotrophic plant growth-promoting biofertilisers will require a systematic strategy of quality control designed to fully utilise all these beneficial factors, allowing crop yields to be maintained or even increased while fertiliser applications are decreased.

References

Bowen GD and Rovira AD (1999) Ad. Agron. 66, 1–102.
Dobbelaere S et al. (1999) Plant Soil 212, 155–164.
Dobbelaere S et al. (2003) Crit. Rev. Plant Sci. 22, 107–149.
Glick BR et al. (1998) J. Theor. Biol. 190, 63–68.
Gray EJ and Smith DL (2005) Soil Biol. Biochem. 37, 395–412.
Kennedy et al. (2004) Soil Biol. Biochem. 36, 1229–1244.
Nguyen TH et al. (2003) Symbiosis 35, 231–245.
Xie II et al. (1996) Curr. Microbiol. 32, 67–71.

MULTI-FACTOR APPROACH TO IDENTIFYING COWPEA GENOTYPES WITH SUPERIOR SYMBIOTIC TRAITS AND HIGHER YIELD FOR AFRICA

Samson B. M. Chimphango[1], Alphonsus Belane[2], Jesse B. Naab[3], Paul B. Tanzubil[4], Joseph A. N. Asiwe[5], Patrick A. Ndakidemi[2], Riekert P. D. van Heerden[6] and Felix D. Dakora[2]

[1]Botany Department, University of Cape Town, Rondebosch 7701, South Africa; [2]Cape Peninsula University of Technology, Cape Town 8000, South Africa; [3]Savanna Agricultural Research Institute, Wa Exp. Station, Ghana; [4]Savanna Agricultural Research Institute, Manga Exp. Station, Ghana; [5]ARC-Grain Crops Research Institute, Pochefstroom 2522, South Africa; [6]North West University, Pochefstroom 2522, South Africa

Cowpea (*Vigna unguiculata* L. Walp) is an important food crop in Africa and other developing countries. All the parts used as food are nutritious and provide proteins, vitamins (notably vitamin B), and minerals. The cowpea haulm is also a good source of livestock feed. In the last three decades, some efforts have been put into research aimed at improving the yield of cowpea. For example, since 1970, the International Institute of Tropical Agriculture (IITA) has worked on developing and distributing improved cowpea materials. Improved cultivars often with a single trait, such as early pmaturing, pest tolerance, disease tolerance, or high yielding, have been reported (Hall et al., 1997). However, the yields of cowpea at farmer level (240 kg ha^{-1}) are still very low (Quin, 1997) compared with the indicated yield potential of about 3,000 kg ha^{-1}. The aim of this study was to identify cowpea genotypes with superior multiple traits, including N$_2$-fixation, pest tolerance, and high grain yield.

A total of 126 cowpea landraces and cultivars were collected from Ghana, South Africa, and Tanzania on the basis of seed yield, growth habit, maturity dates, seed coat color, seed size, seed shape, and a known level of pest resistance. In the first year, each country planted seed material from its own collection for seed multiplication as well as for preliminary observation of the germplasm for symbiotic performance, pest resistance, and seed yield. About 27 cowpea genotypes were then selected and further screened in replicated field trials at Wa and Manga in Ghana, and Taung in South Africa. Plants were assessed at early pod-filling stage for biomass accumulation, nodule mass, and pest damage as well as for photosynthetic rates, and stomatal conductance and respiration. The amount of N-fixed was determined using the ^{15}N natural abundance technique.

F. D. Dakora et al. (eds.), *Biological Nitrogen Fixation: Towards Poverty Alleviation through Sustainable Agriculture*.
© Springer Science + Business Media B.V. 2008

23

At physiological maturity, yield components, such as number of pod-bearing peduncles, number of pods, seeds per plant, seed size and seed yield, were assessed. The density of trace elements in seeds was analyzed to evaluate the dietary importance of the selected genotypes. All measured parameters for the 27 genotypes at each site were ranked in descending order of high performance. The genotype with the lowest total ranking value was considered the best.

Photosynthetic rates, measured at flowering, showed significant differences between genotypes. An assessment of plants at pod-filling stage, and/or at physiological maturity, also showed significant differences in plant growth, nodule mass, pest damage, seed yield, and $\delta^{15}N$ values among the cowpea cultivars. Reproductive parameters revealed marked differences in the number of pod-bearing peduncles as well as number of pods and seeds per plant, which together resulted in significant yield differences among the genotypes. Analysis of edible cowpea grain showed that, irrespective of location, some genotypes have an intrinsic ability to accumulate significantly greater concentrations of trace elements, such as Fe, Zn, Mn, Cu and Se, which are important for human health.

Through the genotype ranking, a total of nine genotypes were selected at each site. These included the five overall best performing genotypes that had the lowest total rankings. The remaining four genotypes had at least a low ranking in one of the parameters. Because there was a significant site x cultivar interaction for most parameters measured, it is suggested that the identification of a superior cowpea genotype from the assessed traits would only be possible for a single location.

Acknowledgement

The authors are grateful to the McKnight Foundation for funding this project.

Reference

Hall EA et al. (1997) Cowpea Breeding. In: J Janick (ed.). Plant Breeding Reviews, Volume 15. Wiley, Hoboken, NJ.
Quin M (1997) Introduction. In: BB Singh, DR Mohan, KE Dasheil and L Jackai (eds.). Advances in Cowpea Research. IITA, Ibadan, Nigeria.

BNF APPLICATIONS FOR POVERTY ALLEVIATION

G. W. O'Hara, J. G. Howieson, R. J. Yates, D. Real and C. Revell

Centre for *Rhizobium* Studies, Murdoch University, Murdoch, Western Australia 6150

The causes of poverty worldwide are very complex, but solutions to poverty can be found. Although it is possible to easily identify many roles for biological nitrogen fixation (BNF) in poverty alleviation, it can be difficult for subsistence farmers to achieve sustainable change incorporating new developments that involve BNF. Moderate and extreme poverty occur in many countries located predominantly in a band both north and south of the equator in the tropics and sub-tropics across North and South America, Africa, and Asia. Significantly, many of these countries, in which more than 25% of the population lives on less than US$2 per day, have negative economic growth. The most important determinant in the declining economic growth of these countries is their poor food productivity. The poverty trap for individuals and communities in these countries is mainly a rural phenomenon of peasant farmers caught in a spiral of rising populations and stagnant or decreasing food production per person. Important factors in the poverty-trap situation that these farmers are unable to escape include the poor infrastructure for distribution and communication, the expense of products, such as essential fertilizers, which results in very little use of fertilizers, and under-funded scientific research and extension.

The clear role for BNF scientists to contribute to on-the-ground solutions for overcoming poverty is to expand the use and benefits of biologically-fixed nitrogen in agricultural systems and to increase use of legumes as secondary crops in fallows and rotations, as green manures and cover crops in fields, as well as to improve opportunities for training. But there are a range of identifiable constraints to adoption of BNF in agriculture for poverty alleviation. As recently outlined by Herridge (2006), these constraints include: a decline in many countries in the land areas being planted to legumes; the widespread use of N-fertilizers on legumes; lack of farmer knowledge about inoculation; difficulties in developing markets for inoculants in low-input systems; lack of expertise to make inoculants; lack of private sector involvement in inoculants; and a significant lack of support for research on BNF.

With an understanding of these constraints, Herridge (2006) identified the following critical research gaps and investment options for increasing broad-scale adoption of *Rhizobium*-based technology. These were: to increase legume plantings and inoculant

F. D. Dakora et al. (eds.), *Biological Nitrogen Fixation: Towards Poverty Alleviation through Sustainable Agriculture.*
© Springer Science + Business Media B.V. 2008

demand; to develop simple rapid "need to inoculate" tests; to enhance farmer knowledge of inoculation, inoculant manufacture, quality control, and distribution; to support *Rhizobium* R&D capacity; and to improve understanding of the role of legume- in farming systems.

The Centre for *Rhizobium* Studies (CRS) at Murdoch University, through ACIAR, aims to introduce legumes and their inoculants into the abandoned arable lands of the Eastern Cape of South Africa (acronym ECCAL). The project will address two of the major limitations to improving livestock production in the region; these are the quality and quantity of forages, and effective communal management of feed resources (from a social, economic and biological sense). To achieve success, two key elements have been identified. Firstly, the need for a "social incentive", i.e., a system of resource management that needs community ownership and secondly, a "financial incentive", i.e., the outcome must be an income-generating enterprise for the community. Strong in-country support comes from ARC and the ECDA. A key component is researching the microbiology of herbaceous legumes native to the southern and western parts of South Africa. Root-nodule bacteria from species of *Lotononis* and *Lessertia* have proven to be diverse and unusual nodule occupants, both taxonomically and physiologically. The Department of Agriculture and Food, Western Australia, is also evaluating species from these legume genera for their potential in Australian agriculture.

The scientific challenges for improving the application of BNF for poverty alleviation are often rudimentary. These can often be to do with determining which legumes will grow and overcoming constraints to adoption. It is clear that for N_2-fixing forages and pastures, the rhizobiology must be addressed first, so that plant selection and breeding can be undertaken using effective N_2-fixing symbiotic plants. In addition, it is important that simple-to-use robust inoculation technologies are available for subsistence-level farmers to adopt. By contrast, the social and cultural challenges for achieving sustainable use of BNF can be significant. In essence, community ownership and management is essential for poverty alleviation to be successful. One important aspect is that often an investment of funding is required to implement change, such as the need to introduce a new technique or a new crop, the need to provide necessary P and K fertilizers, or the need to overcome loss of a planting season. Implementation of programs aimed at alleviating poverty by application of BNF in subsistence-level agricultural systems requires a multi-faceted approach that must be built on foundations where the rhizobiology of effective symbioses is integrated with plant selection, breeding and evaluation.

Reference

Herridge (2006) Report on a review of ACIAR-funded projects on *Rhizobium* during 1983–2004. ACIAR Working Paper, No. 62.

BIOLOGICAL NITROGEN FIXATION IN RESOURCE-POOR AGRICULTURE IN SOUTH AFRICA

Jacomina F. Bloem

Agricultural Research Council, Plant Protection Research Institute, Private Bag X134, Queenswood, Pretoria 0121, South Africa
(Email: BloemJ@arc.agric.za)

The majority of rural farmers produce maize in mono-cropping systems. Despite the relatively high rainfall, problems with acidity, low soil fertility, erosion, and poor management hamper crop production. Cost-effective strategies are required to improve the yields of resource-poor farming communities.

A Participatory Research and Development approach was followed in the projects at Mlondozi, Bergville, Lusikisiki, Bizana, Qunu, and Thohoyandou. Information regarding the agricultural and socio-economic status of communities was gathered using Rural Appraisals. The aim was to determine the needs, problems, fears, and aspirations of the people living in a community. The tribal authority and community were informed of the results, after which a planning session was held with representatives in the community on agricultural intervention options. Leader farmers were selected and trained in conservation agriculture. The technologies included reduced tillage, retention of crop residues, increased crop diversification, and integrated soil fertility management as well as plant pest and disease management. 'Best practice' manuals were compiled in the specific language of an area and translation was done during training sessions. Due to the incidence of illiteracy, the use of pictograms and clear associations clarified concepts, such as the process of nitrogen fixation, inoculation, and nodulation. Capacity building is an important aspect in agricultural development, and we have found that project sustainability largely depends upon the success of farmer-to-farmer training and the interest of the second-generation farmers. The process, benefits, and factors that influence biological nitrogen fixation (BNF) are well studied, but its significance needs to be conveyed to farmers continually. Since 1999, BNF was promoted in the crop diversification strategy of the Landcare. Traditional and alternative production methods were compared in researcher- and farmer-managed demonstration trials.

The versatility of legumes in farming systems was emphasized as a protein-rich food source and/or animal feed, while improving soil fertility. Crop rotation, intercropping, and multiple cropping were evaluated, using cowpeas, soybeans, dry beans, lablab, and

F. D. Dakora et al. (eds.), *Biological Nitrogen Fixation: Towards Poverty Alleviation through Sustainable Agriculture*.
© Springer Science + Business Media B.V. 2008

mung beans. Legumes were inoculated with appropriate inoculants and nodulation of trials was assessed in conjunction with farmers. Significant increases in yield were achieved when soil fertility management was combined with practices like the use of certified seed, timely planting, and in-season weeding. The importance of continuous technology transfer and monitoring and evaluation cannot be underestimated. On-farm visits by the research team are crucial for addressing crop-production concerns and encouraging farmers to recognize the benefits of the introduced technology. The Plan-Act-Observe-Reflect action-learning cycle was followed, and the approach proved to be feasible for improving crop production and subsequently improved livelihoods of the farmers.

Changing farmers' traditional way of thinking and doing things has been a challenge. As scientists with no previous training in social sciences, particular mind-set changes were necessary. Language, cultural differences, and gender issues had to be handled with care, but respect towards people and a passion for agriculture overcame all differences. Seed, fertiliser, and legume inoculant starter packs were supplied to leader farmers. The risk of crop failure during the learning process is carried by the project. Care should be taken not to create and enhance a dependency syndrome. Farmers were encouraged to expand newly acquired knowledge and technology at their own cost to larger areas of their fields. Availability, transport, and continuous escalating costs of agricultural input remain a problem in rural areas, not only in South Africa. Sustainable use of new technology is dependant upon the establishment of effective supply links. Small and affordable packaging may address some issues, but manufactures should be encouraged not to compromise on product quality and safety. The availability of good quality seed for the various legumes promoted in projects continues to hamper expansion to larger production areas. Seed companies could sub-contract resource-poor farmers in areas with suitable soil and climatic conditions to produce legume seed.

Farmers' participation in project development encouraged taking ownership and sustainability of projects. Farmers were able to choose from a 'basket of technologies' and it contributed towards recorded achievements. Resource-poor farmers included legumes and BNF technology as part of their cropping systems. Farmers adopted and continued with these methods following increased crop yields and improved food security and safety. Agro-processing or value adding of harvested crops encouraged, especially women, to continue with the newly introduced conservation agricultural practices. Despite the challenges faced, we were able to merge a scientific and a 'soft systems' approach for the benefit of these farmers.

The Congress' theme of addressing poverty alleviation could be expanded to include poverty eradication. Our responsibility as researchers remains to share the knowledge with farmers who need the cost-effective and applicable technology offered by BNF.

COMBATING FOOD INSECURITY ON SANDY SOILS IN ZIMBABWE: THE LEGUME DILEMMA

Florence Mtambanengwe[1] and Paul Mapfumo[1,2]

[1]Department of Soil Science and Agricultural Engineering, University of Zimbabwe, P.O. Box MP 167, Mount Pleasant, Harare, Zimbabwe; [2]SOFECSA Coordination Unit, CIMMYT Southern Africa Regional Office, P.O. Box MP 163, Mount Pleasant, Harare, Zimbabwe
(Email: fmtamba@agric.uz.ac.zw)

Maize, the staple food of Zimbabwe, is the most widespread crop grown under rainfed conditions in these systems but average maize yields remain low (<0.5 t ha^{-1}), threatening household food security. There are calls for diversification into high value crops, which may include both food and non-food leguminous crops as a means of sustaining household food security. However, several key questions with respect to diversification need to be answered. Firstly, what opportunities exist? Secondly, who has the capacity to diversify? Thirdly, which crops should be considered in this diversification process? The objectives of this study were to: (i) quantify farm-level maize and legume benefits generated through use of ISFM technologies; and (ii) promote appropriate targeting of ISFM technologies to different farmer-resource groups, paying particular attention to their capacity to adapt such technologies at field/farm levels.

The study was carried out on smallholder farms in Chinyika (32°20: 18°14' S), rainfall 650–750 mm year^{-1}. Farmer participatory research approaches were used in farmer and field selection. Three distinct farmer resource groups were identified on the basis of resource-endowment (Mtambanengwe and Mapfumo, 2005). These were: (i) resource-endowed farmers; (ii) intermediate farmers; and (iii) resource-constrained farmers. Four field sites were selected in this initial phase and each site was then sub-divided into six sub-plots. Three sub-plots were planted to maize; they received either (a) fertilizer alone, (b) cattle manure plus mineral N, or (c) woodland litter plus mineral N. One was planted to the soyabean (*Glycine max*) varieties, Solitaire and Storm. Sugarbean (*Phaseolus vulgaris*) and sunnhemp (*Crotalaria juncea*) occupied the last two sub-plots. The organic nutrient sources were used in combination with varying quantities of mineral-N fertilizer depending on what the different farmers could ideally afford.

F. D. Dakora et al. (eds.), *Biological Nitrogen Fixation: Towards Poverty Alleviation through Sustainable Agriculture*.
© Springer Science + Business Media B.V. 2008

The biomass and grain yield levels for all the four crops tested were consistent with farmer classification. Although the soyabean variety, Storm, had high biomass production (0.2–4 Mg ha^{-1}) compared to the variety Solitaire, grain yields were about 10–25% higher on farmer-managed fields (Figure 1). Sugarbean grain yield varied from 15 to 250 kg ha^{-1}, whereas sunnhemp yields were about 15 t ha^{-1} for resource-endowed, 9 t ha^{-1} for intermediate, and 0.8 t ha^{-1} for the resource-constrained farmer. Researcher-managed plots yielded about 6 t ha^{-1}. Manure generally gave superior maize yields on farmers' fields compared to woodland litter and fertilizer alone. Yields were generally far below potential levels, even on the resource-endowed farmer's fields. Farmers do have the intention to grow legumes within their cropping cycles, but continue to be frustrated by persistent low yields that are in turn linked to poor soil fertility.

Several studies have confirmed these low yields (e.g., Mpepereki and Pompi 2003; Nhamo et al., 2003). Results from on-station work showed >100% yield gain following incorporation of C. juncea and C. calothyrsus (both leguminous) compared to same quantities of the traditional cattle manure.

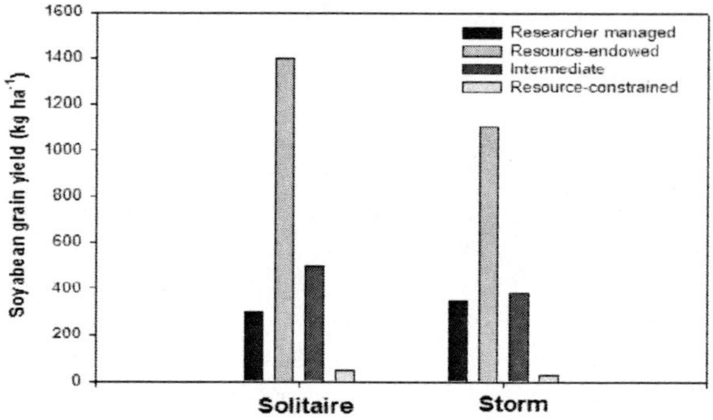

Figure 1. Soyabean grain yield on fields belonging to farmers differing in resource endowment.

The results show that there is scope for enhancing the contribution of legumes to both soil fertility and household nutrition within smallholder farming systems. The challenge is improving accessibility to quality and improved legume seed and promoting the use of mineral fertilizers to stimulate high biomass productivity. A significant part of the legume-BNF research should focus on promoting utilization aspects along specific legume value chains.

Acknowledgement

This work was conducted under the auspices of the Soil Fertility Consortium for Southern Africa (SOFECSA) and funded by the Rockefeller Foundation.

References

Mpepereki S and Pompi I (2003) In SR Waddington et al. (eds.), Soil Fert. Net, CIMMYT, Harare.
Mtambanengwe F and Mapfumo P (2005) Nutr. Cycl. Agroecosyst. 73, 227–243.
Nhamo et al. (2003) In SR Waddington et al. (eds.), Soil Fert. Net, CIMMYT, Harare, pp. 119–128.

NITROGEN FIXATION IN AGRICULTURE: FORAGE LEGUMES IN SWEDEN AS AN EXAMPLE

Kerstin Huss-Danell[1], Georg Carlsson[1], Eugenia Chaia[1,2] and Cecilia Palmborg[1]

[1]Department of Agricultural Research for Northern Sweden, Crop Science Section, Swedish University of Agricultural Sciences (SLU), Umeå, Sweden; [2]Universidad Nacional del Comahue, Centro Regional Universitario Bariloche, Quintral 1250, SC Bariloche, Argentina

Perennial grasslands, managed for forage production and leys, represent the majority of cultivated land in Sweden. In northern Sweden, the leys are typically comprised of the grasses timothy, *Phleum pratense*, and meadow fescue, *Festuca pratensis*, together with red clover, *Trifolium pratense*. It is common to establish the leys as an undersowing in barley and then keep them for about 3 years in a crop rotation. Leys are normally harvested twice per year for silage, but some leys are also grazed.

Red clover has several important functions in leys (Frame et al., 1998). Red clover increases palatability of the forage, which leads to increased forage consumption and thereby increased milk production. Red clover is deep-rooted and can thus improve nutrient uptake, especially of divalent cations, from deeper soil layers. Due to its symbiosis with N_2-fixing *Rhizobium leguminosarum* bv. *Trifolii*, red clover has a great potential to increase sustainability in ley-dominated crop rotations. These rhizobia occur in high numbers in agricultural soil in northern Scandinavia. We found a large genotypic variation but only minor differences in N_2-fixation capacity among rhizobia isolated from *T. pratense* in northern Scandinavia. The rhizobia are thus well adapted to the environmental conditions in the studied soils (Duodu et al., 2007).

In spite of the widespread use of red clover in Sweden, information about its N_2 fixation activity and consequent fixed-N additions below ground is scarce. We used the ^{15}N-based ID (isotope dilution) and NA (natural abundance) methods to quantify the proportion of clover N derived from air (pNdfa) and the amount of N_2 fixed per area in a first-year ley, a second-year ley, and a third-year ley. The leys were in neighbouring fields in Umeå, northern Sweden. The so-called B value needed in the NA method was established in red clover grown in the laboratory (Carlsson et al., 2006).

F. D. Dakora et al. (eds.), *Biological Nitrogen Fixation: Towards Poverty Alleviation through Sustainable Agriculture.*
© Springer Science + Business Media B.V. 2008

Measurements of pNdfa were made separately in leaves, stem, stubble, and in roots. pNdfa values were very high, usually ≥ 0.8, which is in agreement with previous results for forage legumes grown in mixture with grasses in other northern temperate/boreal areas (reviewed by Carlsson and Huss-Danell, 2003). Our measurements of pNdfa in leaves could provide useful indications of pNdfa in shoots and whole red clover plants, thus avoiding the need for time-consuming root analyses in field-grown plants (Huss-Danell and Chaia, 2005). When N_2 fixation was measured only in the herbage, (i.e., the plant parts removed at harvest), it amounted on average to 45% of whole plants at first harvest and 77% of whole plants at second harvest.

Biomass production, N content, and N_2 fixation were studied during three parts of the growing season; the start of the season until late June (first harvest), late June until mid-August (second harvest), and mid-August until early October, when first frost occurred at the site. In clover the biomass production, N content, and N_2 fixation per area was much higher in late summer than in early summer and autumn parts of the season. Grasses, on the other hand, had more similar biomass production and N content in early and late summer, but low biomass and N content in the autumn part of the season. There were none or only small effects of ley age in the studied parameters. Input of N from added fertilizer plus N_2 fixation in whole plants balanced the amount of N removed in clover and grasses in the two harvests. Nitrogen compounds in stubble and roots left in field are then used for regrowth of the plants after harvest and are decomposed in the soil. The decomposition is particularly important when the soil is ploughed for a next crop in the crop rotation. Our studies have thus highlighted that red clover consistently derives most of its N from N_2 fixation and that companion and subsequent plant species benefit from clovers.

In contrast, the influence of neighbouring species on pNdfa in clover was not known. We addressed this question in a long-term biodiversity experiment with three clover species and a varied number and composition of companion species (grasses, legumes and other forbs). Under non-fertilized conditions, the pNdfa (NA method) was always high irrespective of companion species. This is further evidence of an inherently high ability for N_2 fixation in our domesticated clovers.

Acknowledgements

Thanks to the Swedish Foundation for International Cooperation in Research and Higher Education, the Swedish Research Council for Environment, Agricultural Sciences and Spatial Planning, and the Carl Trygger Foundation for financial support (to KHD).

References

Carlsson G and Huss-Danell K (2003) Plant Soil 253, 353–372.
Carlsson G et al. (2006) Acta Agric. Scand. Sect. B., Soil Plant Sci. 56, 31–38.
Duodu S et al. (2007) J. Appl. Microbiol. doi:10.1111/j.1365-2672.2006.03196.x.
Frame J et al. (1998) Temperate Forage Legumes. CAB International, Wallingford, UK.
Huss-Danell K and Chaia F. (2005) Physiol. Plant. 125, 21–30.

BIOLOGICAL NITROGEN FIXATION WITH THE SOYBEAN AND COMMON BEAN CROPS IN THE TROPICS

Mariangela Hungria, Rubens J. Campo, Fernando G. Barcellos,
Ligia Maria O. Chueire, Pâmela Menna, Jesiane S. Batista,
Fabiana G. S. Pinto, Daisy R. Binde, Leandro P. Godoy and
Alan A. Pereira

Embrapa Soja, Cx. Postal 231, 86001-970, Londrina, Paraná, Brazil
(Email: hungria@cnpso.embrapa.br)

Brazil's economy is based firmly on agriculture, but the high costs of mostly imported fertilizers has resulted in a major research focus on biological N_2 fixation to enhance crop productivity and agricultural sustainability. One good example is the successful adoption of the soybean (*Glycine max* (L.) Merr.) crop in the country, in large part due to the approach of using biological N_2 fixation as a main component of the crop's production.

Nitrogen Fixation with Soybean and Common Bean in Brazil

The area under cultivation with soybean in Brazil (~22 million hectares) accounts for more than one third of world grain production and the great majority of the soils show populations of *Bradyrhizobium* established by previous inoculations and estimated at 10^3–10^6 cells g^{-1} soil. However, most farmers practice inoculation because the new strains and technologies available today result in grain yield increases averaging at 8%. Furthermore, N-fertilizers do not benefit the crop. Rates of N_2 fixation with the soybean under field conditions can exceed 300 kg of N ha^{-1}, providing up to 94% of total plant N and resulting in an economy to the country estimated at US\$3 billion per year. In addition, release of N (~30 kg of N ha^{-1}) to following crops has been reported. Those numbers indicate higher rates of N_2 fixation than in other producing countries, such as U.S.A., Australia, Argentina and China. Explanations for this success in Brazil include: (1) emphasis on selection of both soybean cultivars and rhizobia with higher N_2 fixation capacity; (2) development of technologies aimed at increasing the contribution of the biological process, e.g., studies on compatibility of inoculants with agrochemicals and micronutrients; (3) education of the farmers towards the importance of spending time and money with inoculation; and (4) establishment of a strong legislation controlling both the strains used and the quality of commercial inoculants. The selection programs for superior symbiotic performance have been underway since the late 1960s and still continue, aimed at supplying the increased demands for more productive cultivars.

F. D. Dakora et al. (eds.), *Biological Nitrogen Fixation: Towards Poverty Alleviation through Sustainable Agriculture.*
© Springer Science + Business Media B.V. 2008

Common bean (*Phaseolus vulgaris* L.) represents the most important source of protein for most countries in South America, with Brazil being the largest grower and consumer of the legume worldwide. Among the rhizobial species, *Rhizobium tropici* has proven to be outstanding in terms of genetic stability and tolerance to tropical environmental stresses. Strains very effective in fixing N_2 belonging to this species have been selected and, when used as inoculants, increase nodulation and N_2-fixation rates, allowing plants to produce up to 4,000 kg ha^{-1} without any N-fertilizer.

What About the Future?

Higher rates of N_2 fixation will be needed in future decades because of: (1) the release of more productive cultivars; (2) the use of legumes for biodiesel; (3) the need for recovering degraded areas; and (4) an increased demand for food. New and promising tools and approaches might help to speed the process. The selection of cultivars with a higher capacity of N_2 fixation and adaptability to a variety of environmental conditions might be faster if molecular markers, such as the simple sequence repeat (SSR) markers, are employed. New insights have been gained from genomic and proteomic research on very effective strains, like *Bradyrhizobium japonicum* CPAC 15 and *R. tropici* PRF 81, which are commercially recommended for the soybean and common bean crops, respectively, in Brazil. Comparative genomics, in addition to ecological studies on long-term field experiments, are revealing broad genome plasticity, high rates of horizontal gene transfer from inoculant strains to indigenous bacteria in the Brazilian soils, and significant shifts in microbial community structure subsequent to displacement of natural vegetation. It is also noteworthy that genetic characterization of dozens of bacteria collected from various Brazilian ecosystems are indicating the presence of many new species and may also represent a valuable reservoir of new genes. In the next decade, there will be an increased demand for higher N_2-fixation rates for many legumes and the results obtained in the next few years may be critical in convincing society and its leaders of the importance of the biological process for agricultural sustainability, environmental quality, energy supply, and food security.

Brazil, which is considered the largest reserve of productive land still available in the world, is emerging as a leading grain producer. However, a major challenge will be to devise models – combining basic and applied research – to improve crop yields by bringing degraded soils back to productivity and developing sustainable agricultural systems. N_2 fixation will play a key role in achieving these goals and may benefit large and small landholders, and subsistence and cash legume crops.

Acknowledgement

This work was partially supported by CNPq (Conselho Nacional de Desenvolvimento Científico e Tecnológico, Brazil): Instituto do Milênio, Edital Universal (471773/2004-2), CABBIO (400710/2004-8), Genome (505499/2004-2), and Coleção de Culturas (552393/2005-3).

BIODIVERSITY IN AMAZONIAN DARK EARTH: A CONTRIBUTION FOR THE SUSTAINABILITY OF TROPICAL SOILS FROM THE MICROBIAL SYMBIOSES

S. M. Tsai[1], B. O'Neill[2], F. Cannavan[1], D. Campos[1], R. Medau[1],
S. Fedrizzi[1], J. Grossmann[2] and J. Thies[2]

[1]Cell and Molecular Biology Laboratory, Center for Nuclear Energy in Agriculture, University of São Paulo, Piracicaba, SP, CEP 13.416-000, Brazil; [2]Department of Crop and Soil Science, College of Agriculture and Life Sciences, Cornell University, Ithaca, NY, USA
(Email: tsai@cena.usp.br)

The anthropogenic addition of organic amendments, including plant and animal material, pottery, and charcoal, into extant soil formed what are known as Terra Preta do Índio (TPI) or Amazonian Dark Earth (ADE). Whether or not these soil amendments were part of an agricultural management strategy by pre-Colombian cultures, today ADE soils are prized by farmers for their sustained fertility, in regions where chronic soil infertility has lead, in part, to on-going destruction of primary forest to create marginal cropland. However, although prehistoric practices have left a legacy of fertile soil for a few isolated farmers, expanding the benefits of ADE to other soils will require more than serendipity. The unusual chemistry resulting from organic amendments to ADE results in distinct microbial communities that are involved in nutrient cycling and ecological processes. The ongoing biogeochemical activity in ADE is the true legacy of prehistoric practices, and the microbial processes at work in these unique soils are crucial to these activities (O'Neill et al., 2006; Thies and Suzuki, 2003; Zhou et al., 1996).

In reviewing the current state of research on microbial communities in ADE, we were able to determine the substantial contribution from biological systems, including root-nodule bacteria and arbuscular mycorrhizal fungi, and decide future directions towards understanding the role of soil microbial communities in both soil biodiversity and conservation of deforested tropical soils. In ADE, the higher fertility at greater soil depth appears to be stabilized by the presence of pyrogenic carbon or Black Carbon (BC), leading to large and diverse microbial populations. Understanding the biogeochemical processes involved in maintaining the fertility of ADE soils may lead to new technologies for soil management in the tropics and also provide a novel strategy for mitigating atmospheric CO_2 by sequestering BC in soils, which may also serve as a nucleus for improved soil fertility and biological activity.

F. D. Dakora et al. (eds.), *Biological Nitrogen Fixation: Towards Poverty Alleviation through Sustainable Agriculture.*
© Springer Science + Business Media B.V. 2008

For unique soils, such as ADE, culturing techniques offer an excellent opportunity to screen for and reveal novel organisms that can be maintained and studied *in vivo*. Most probable number (MPN) estimations only reflect a small portion of soil organisms, but are a traditional means by which to compare microbial population levels. Bacterial populations have been shown to be high at four ADE sites and to remain at high levels with increasing soil depth compared to adjacent nutrient-poor soils (Figure 1).

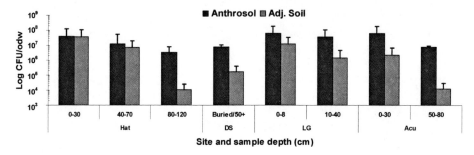

Figure 1. Bacterial populations at various depths in ADE (Anthrosol) and adjacent soils at a variety of sites (Hat, DS, LG and Acu).

A comparison of soil bacterial communities was made based on growth on selective media followed by screening of 16S rDNA amplicons. A phylogeny of sequenced 16S rDNA that was used to compare isolates across soil types, depths, and sites indicated a higher diversity in ADE. In one site in Balbina-AM, there was an obvious shift in bacterial communities in ADE versus the adjacent soil: Comamonadaceae (8% vs. 20%); Moraxellaceae (8% vs. 21%); Pseudomonadaceae (22% vs. 11%); Neisseriaceae (8% vs. 0%); Oxalobacteraceae (8% vs. 0%); Bacillaceae (8% vs. 16%); Rhizobiaceae (8% vs. 5%); Burkholderiacea (30% vs. 11%); and Micrococcaceae (0% vs. 16%). Leguminous roots collected from both ADE and adjacent soil showed high rates of mycorrhizal infection, ranging from 45–50%. A non-specific fluorescent stain was used to distinguish between living and dead cells and microscopy imaged living microorganisms on the surface of C-black.

Our group is developing T-RFLP, DGGE, and clone libraries to describe community-level differences across ADE sites and starting identification of microbial metabolites excreted by bacterial isolates from ADE as initial steps toward understanding the biotechnological potential of natural compounds from these communities.

References

O'Neill B et al. (2006) In: Proc. 17th World Soil Sci. Congress, July 2006.
Thies J and Suzuki K (2003) In: J Lehmann et al. (eds.), Amazonian Dark Earths: Origin, Properties, and Management, Kluwer, New York.
Zhou JZ et al. (1996) Appl. Environ. Microbiol. 62, 316–322.

TECHNIQUES FOR THE QUANTIFICATION OF PLANT-ASSOCIATED BIOLOGICAL NITROGEN FIXATION

Robert M. Boddey[1], Claudia P. Jantalia[1], Lincoln Zotarelli[2], Aimée Okito[3], Bruno J. R. Alves[1] and Segundo Urquiaga[1]

[1]Embrapa Agrobiologia, Km 47, BR 465, C.P: 74505, Seropédica, 23890-970, RJ, Brazil; [2]Department of Crop Science, University of Florida, Gainsville, FL, USA; [3]Département Biologie, Faculté des Sciences, Université Kinshasa, Kinshasa, Republica Democratique du Congo (Email: bob@cnpab.embrapa.br)

There are many techniques available for the quantification of plant-associated biological nitrogen fixation (BNF) and they can be divided roughly into those that are applicable in the field and those that are not. This short review will mainly consider the former and we cite key references that should be consulted for further details. The techniques that are rarely applicable in the field include the acetylene-reduction assay, exposure to $^{15}N_2$-enriched atmospheres and the total N-balance method.

Glasshouse and Laboratory Techniques

The acetylene-reduction assay: It relies on the fact that nitrogenase is able to use acetylene as an alternative substrate and reduce it to ethylene without immediate change in electron flux. These two gases can be detected at very high sensitivity using flame-ionisation gas chromatography; the assay is rapid and does not require sophisticated equipment. However, it suffers from many artefacts and disadvantages (Boddey, 1987) and, with work on legumes, exposure of the nodules to acetylene or any physical disturbance (e.g., removal from soil) quickly decreases their activity, so making the assay almost useless quantitatively. Despite these great disadvantages, it is still very useful for ascertaining if legume nodules and bacterial cultures are actively fixing N_2.

$^{15}N_2$ enrichment: The exposure of an intact soil/plant system to a $^{15}N_2$-enriched atmosphere can be considered definitive proof of a BNF contribution to a plant/diazotroph association (see, e.g., Eskew et al., 1981). However, because exposure to the gas requires complete control of the enclosed environment (pCO_2, pO_2, light intensity, temperature, etc.), it is virtually impossible to apply in the field.

F. D. Dakora et al. (eds.), *Biological Nitrogen Fixation: Towards Poverty Alleviation through Sustainable Agriculture.*
© Springer Science + Business Media B.V. 2008

Total N balance technique: The total N-balance technique requires the accurate analysis of N in the soil-plant system at the planting of a crop (usually in pots) and then, after one or more crops, the total N in the system is again quantified. It is advisable either to quantify N losses from the system or to minimise them. Although it can be applied in long-term experiments in the field, best results are obtained from pot experiments (see, e.g., App et al., 1980; Lima et al., 1987).

Field Techniques

There are basically three techniques that are applicable in the field. These are: (i) the two forms of the ^{15}N-isotope dilution technique, either the natural-abundance technique or that using ^{15}N-enriched soil; and (ii) the ureide-abundance technique.

^{15}N natural-abundance technique: The ^{15}N natural-abundance technique relies on the fact that most soils are generally found to be slightly different in ^{15}N abundance (usually slightly enriched in ^{15}N) compared to that of N_2 in the atmosphere. The N_2 in the atmosphere is such a large pool, and so well mixed, that the proportion of ^{15}N atoms in this gas (0.3663 atom % ^{15}N) is so uniform that even the most accurate mass spectrometers available cannot detect differences between samples taken from the air anywhere on the planet (Mariotti, 1983). To determine the proportion of N derived from the air via BNF (%Ndfa), it is necessary to measure: (a) the ^{15}N abundance of the "N_2-fixing" plant under study; (b) the ^{15}N abundance of the N in this plant derived from the soil; and (c) the ^{15}N abundance of the N in this plant derived from BNF. It is assumed that the "N_2-fixing" plant will accumulated N from the soil with the same ^{15}N abundance as non-N_2-fixing plants growing in its close proximity and these values may be used to estimate the ^{15}N abundance of the N in the legume plant derived from the soil. Although this assumption cannot be tested directly, it is reasonable to assume that if several different neighbouring non-N_2-fixing plants all have very similar values of ^{15}N abundance, then this value will be suitable to use for the N derived from the soil by the "N_2-fixing" plant. In agricultural soils, the neighbouring non-N_2-fixing plants generally present very similar values of ^{15}N abundance. To determine the possible range in ^{15}N abundance of soil-derived N, it is recommended to sample several non-N_2-fixing reference plants (Boddey et al., 2000). However, in natural ecosystems, due to the diversity of N sources (plant litter, and organic and inorganic forms of N, etc.) and the different N-acquisition strategies of the plants, non-N_2-fixing plants often have such large differences in ^{15}N abundance that it is impossible to either quantify BNF associated with nodulated legumes or even identify which are fixing N_2 from the δ^{15}N data (Pate et al., 1993; Gehring and Vlek, 2004).

The ^{15}N abundance of the N in the "N_2-fixing" plant derived from BNF is known as the 'B' value. To determine this value, it is usual to grow the plant under conditions where there can be no input of N from sources other than the atmosphere (e.g., in N-free hydroponic or sand culture in the greenhouse). Alternatively, it is possible to calculate the value of 'B' for a plant if the %Ndfa has been determined using another independent method (see, e.g., Doughton et al., 1992; Okito et al., 2004). Because it is almost impossible to remove all roots and nodules from the soil in the field, only shoot tissue is generally harvested and thus the relevant 'B' value is that for the shoot tissue, which is usually considerably lower than zero. There are several reports that the 'B' value can vary considerably depending on which rhizobial strains occupy the nodules (Okito et al., 2004; Carlsson et al., 2006) and, to a lesser extent, it can vary with plant variety and age (Boddey et al., 2000). Determining the exact value of the 'B' value becomes critical if the ^{15}N natural abundance of the soil is low and the proportion of N derived from BNF is high. In this case, it may even be necessary to determine nodule occupancy and the 'B' value of legumes infected with the different strains.

^{15}N-enrichment isotope-dilution technique: The principle of this technique is the same as the natural-abundance technique, except that the soil is enriched with ^{15}N, typically between 0.1 and 1.0 atom % above natural abundance (atom % ^{15}N excess). The soil is most commonly enriched with ^{15}N by adding ^{15}N-enriched fertiliser (e.g., ammonium sulphate, urea, calcium nitrate). However, when ^{15}N-labelled NO_3^- or NH_4^+ is added to soil, the labelled N enters the soil mineral N pool. This pool becomes depleted by plant N uptake and N immobilisation and is subsequently recharged with unlabelled N from mineralisation of soil organic matter. Hence, as time progresses, the ^{15}N enrichment of the plant-available N decreases. If the ^{15}N enrichment of soil mineral-N changes with depth and/or time, then plants with different N-uptake patterns will accumulate N with different ^{15}N enrichments. In other words, it will not be possible to decide which reference crop obtained N from the soil with the same ^{15}N enrichment as that of the "N_2-fixing" plant. To slow down the rate of change in ^{15}N enrichment of soil mineral N with time, it is possible either to use slow-release labelled fertiliser or to mix the soil with ^{15}N-enriched organic matter. To determine the possible range in ^{15}N enrichment, it is recommended that several non-N_2-fixing reference plants be sampled. These problems, and ways to mitigate them, are discussed at length by Boddey et al. (1995).

The ureide-abundance technique: Many legumes of tropical origin synthesise allantoin and allantoic acid (collectively known as ureides) as the principal products of N_2 fixation in nodules. The ureides are exported to the shoot tissue via the xylem along with NO_3^-, the principal form of N derived from the soil. Hence, at any one time, the ratio of ureide N to ureide + nitrate N in the plant xylem sap is approximately proportional to the proportion of N derived from BNF at that time. Legumes that originate from temperate regions, such as pea, clover, faba beans (*Vicia faba*) or lupins, as well as some important tropical legumes, such as groundnut (*Arachis* spp.), *Stylosanthes* spp., and almost all legume trees and shrubs, do not produce ureides, hence this method cannot be applied with them (for a more complete list of ureide and non-ureide producers, see Peoples et al., 1989). To apply this technique for ureide-producing legumes, it is necessary to collect sap from the cut stems of legumes. Techniques for doing this have been fully described (Peoples et al., 1989). Sometimes it is impossible to extract sap and then it may be necessary to analyse hot-water extracts of dried stems (Herridge, 1982; Alves et al., 2000a). The techniques for the analysis of xylem sap have also been fully described (Peoples et al., 1989), however, it may be desirable to sub-stitute the recommended method for nitrate analysis, which uses very corrosive chemi-cals, with the flow-injection technique developed by Alves et al. (2000b).

The ureide assay estimates the proportion of N derived from BNF by the legume at the time of sampling. Thus, to quantify the total BNF contribution throughout plant growth, it is necessary to make several (we suggest a minimum of four) samplings (destructive harvests) of the sap during plant ontogeny (up to maximum N accumu-lation). Not all nitrate reduction always occurs in roots and some biologically fixed N may be exported from the nodules as amino acids, thus, there is a need to calibrate the technique with another method. This can be accomplished in sand culture in the greenhouse with increasing levels of ^{15}N-labelled nitrate (Herridge, 1982; Alves et al., 2000a).

Conclusions

The ^{15}N-enrichment technique requires the establishment of ^{15}N-labelled plots (or pots), which may not be feasible on-farm. For both ^{15}N-isotope dilution techniques, it is nece-ssary to harvest and analyse several neighbouring non-fixing reference plants. Both ^{15}N-isotope dilution (natural abundance and ^{15}N enrichment) techniques estimate the total BNF contribution to the plants throughout growth until the time of harvest, whereas the ureide-abundance technique requires several harvests to estimate the overall BNF contribution to the plants during plant ontogeny. Finally, if more than one technique can be applied simultaneously, the resulting estimates can be compared and this greatly strengthens confidence in the results.

Acknowledgements

The Brazilian authors gratefully acknowledge financing of their recent studies on quan-tification of BNF by Embrapa, the Brazilian National Research Council (CNPq), and the Rio de Janeiro State Research Foundation (FAPERJ) through the program "Cientista de Nosso Estado".

References

Alves BJR et al. (2000a) Nutr. Cycl. Agroecosyst. 56, 165–176.
Alves BJR et al. (2000b) Comm. Soil Sci. Plant Anal. 31, 2739–2750.
App AA et al. (1980) Soil Sci. 130, 283–289.
Boddey RM (1987) CRC Crit. Rev. Plant Sci. 6, 209–266.
Boddey RM et al. (1995) Fert. Res. 42, 77–87.
Boddey RM et al. (2000) Nutr. Cycl. Agroecosyst. 57, 235–270.
Carlsson D et al. (2006) Acta Agric. Scand. B. 56, 31–38.
Doughton JA et al. (1992) Plant Soil 144, 23–29.
Eskew et al. (1981) Plant Physiol. 68, 48–52.
Gehring C and Vlek PLG (2004) Basic Appl. Ecol. 5, 567–580.

Herridge DF (1982) Plant Physiol. 70, 1–6.
Lima E et al. (1987) Soil Biol. Biochem. 19, 165–170.
Mariotti A (1983) Nature 303, 685–687.
Okito A et al. (2004) Soil Biol. Biochem. 36, 1179–1190.
Pate JS et al. (1993) Plant Cell Environ. 16, 365–373.
Peoples MB et al. (1989) Methods for Evaluating Nitrogen Fixation by Nodulated Legumes in the Field. ACIAR Monograph No. 11, Canberra, Australia.

MEASURING N_2 FIXATION IN LEGUMES USING ^{15}N NATURAL ABUNDANCE: SOME METHODOLOGICAL PROBLEMS ASSOCIATED WITH CHOICE OF REFERENCE PLANTS

Felix D. Dakora[1], Amy C. Spriggs[2], Patrick A. Ndakidemi[1] and Alphonsus Belane[1]

[1]Cape Peninsula University of Technology, Cape Town 8000, South Africa; [2]University of Cape Town, Private Bag, Rondebosch 7701, South Africa

Nodulated legumes contribute significantly to the N economy of soils and, in so doing, increase agricultural yields of food crops (Dakora and Keya, 1997). The amount of N fixed in symbiotic legumes by root-nodule bacteria (species of *Rhizobium*, *Bradyrhizobium*, *Allorhizobium*, *Agrorhizobium*, *Mesorhizobium and Sinorhizobium*) has generally been assessed under field conditions using the ^{15}N natural-abundance technique and/or the N difference method. Although the ^{15}N natural-abundance technique has been used successfully for field measurements of N_2 fixation in cultivated food grain and pasture legumes, the method is beset by problems associated with the choice of reference plants, especially in natural settings. When using this method, the N derived from atmospheric nitrogen fixation (%Ndfa or P_{fix}) is usually estimated (Shearer and Kohl, 1986) as:

$$P_{fix} = \frac{\delta^{15}N_{ref} - \delta^{15}N_{leg}}{\delta^{15}N_{ref} - B \text{ value}} \times 100$$

where $\delta^{15}N_{ref}$ is the mean $\delta^{15}N$ value of the reference plant, $\delta^{15}N_{leg}$ is the mean $\delta^{15}N$ value of the test legume, and the B value is the $\delta^{15}N$ of the legume wholly dependent on N_2 fixation for its N nutrition. The B value represents the value of atmospheric N_2 and usually incorporates the isotopic fractionation associated with N_2 reduction in root nodules (Shearer and Kohl, 1986).

Suitability of Reference Plants for Quantifying Legume N_2 Fixation

The choice of reference plant can affect the precision of the N_2-fixation measurement. The reference plant is usually a non-fixing species chosen for its ability to measure the ^{15}N abundance of soil-N taken up by the test legume (Shearer and Kohl, 1986). Thus, the reference plant is assumed to have similar rooting pattern, root depth, and the same level of N-isotope fractionation during N uptake by the test legume. However, where the $\delta^{15}N$ of soil is independent of soil depth as in the fynbos of South Africa (Spriggs et al., 2003), the root pattern and depth need not be the same as those for the legume.

F. D. Dakora et al. (eds.), *Biological Nitrogen Fixation: Towards Poverty Alleviation through Sustainable Agriculture*.

Several studies have assessed the suitability of various plant species as references for measuring N_2 fixation in field legumes, using the ^{15}N natural-abundance method (Spriggs et al., 2003; Nyemba and Dakora, 2005). Whether in on-station or on-farm studies, intercropped cereal plants in mixed culture with legumes are generally used as the reference plants. However, below-ground transfer of biologically-fixed N by the legume to the cereal has been demonstrated for intercropped systems (Eaglesham et al., 1981) and it appears that the closer the cereal roots are to those of the legume (i.e., the more interwoven), the more fixed-N is transferred to the cereal. As shown in Table 1, the N concentration was highest in the intra-hole maize (where legume and cereal were planted in the same hole), followed by intra-row planting (where cereal and legume alternated within a row). This increase in cereal N was due to the transfer of fixed-N from the legume partner. This is evidenced by the significantly lower $\delta^{15}N$ values of the intra-hole, intra-row and inter-row maize plants relative to the mono-cultured maize, clearly indicating maize acquisition of biologically-fixed N from symbiotic cowpea. These data therefore suggest that the use of intercropped cereals (whether inter-row, intra-row, or intra-hole planted non-legumes) as reference plants is likely to under-estimate N_2 fixation. Cereal plants from mono-cultures and/or weeds occurring outside legume stands are therefore recommended as reference plants.

Table 1. Effects of cropping system on $\delta^{15}N$ and N concentration in cowpea and maize plants grown in a mixed culture. Values followed by dissimilar letters in a column are significant at P ≤ 0.05 (Ndakidemi and Dakora, unpublished data).

Cropping system	Parameter			
	$\delta^{15}N$ (‰)		N (%)	
	Cowpea	Maize	Cowpea	Maize
Sole	2.6a	5.3a	1.9a	0.9b
Inter-row	2.2b	4.7ab	1.9a	1.0ab
Intra-row	2.0c	4.5b	2.0a	1.10a
Intra-hole	1.9c	4.0b	2.1a	1.15a

Mycorrhizal Effect on $\delta^{15}N$ Value of Reference Plants

Mycorrhizal infection of plants can alter the isotopic fractionation of N during uptake by roots (Spriggs et al., 2003). Mycorrhizally-infected non-legume plants were therefore found to be unsuitable for measuring N_2 fixation in natural stands of *Cyclopia*. As shown in another study (Table 2), the $\delta^{15}N$ values of the reference plants were found to be highly negative and lower than those of *Cyclopia* plants in the different sites, thus, making it impossible to use those values for estimating N_2 fixation. To measure N_2 fixation in such situations would require that both the legume and reference plants are mycorrhizally infected. More research is needed on the effect of N transfer and mycorrhizal infection on the measurement of N_2 fixation in field legumes.

Table 2. The δ^{15}N values (means $\pm$ SE‰) of *Cyclopia* and reference plants for different study sites (Spriggs and Dakora, unpublished data).

Site	*Cyclopia* species	δ^{15}N		B value (‰)
		Cyclopia	Reference plant	
W1	*C. subternata*	-1.79 ± 0.19	-5.35 ± 0.40	-0.24 ± 0.13
W3	*C. subternata*	-0.36 ± 0.06	-5.36 ± 1.19	-0.85 ± 0.32
W4	*C. genistoides*	-0.11 ± 0.18	2.87 ± 1.24	-0.53 ± 0.20
W5	*C. genistoides*	-0.38 ± 0.07	-5.06 ± 1.52	-0.86 ± 0.21
W8	*C. intermedia*	-0.63 ± 0.15	-4.65 ± 0.48	-1.29 ± 0.44
W9	*C. intermedia*	-0.75 ± 0.33	-5.31 ± 1.61	-0.54 ± 0.19
W10	*C. maculata*	-0.94 ± 0.06	0.48 ± 0.59	-0.50 ± 0.20
W11	*C. maculata*	-0.84 ± 0.09	-2.01 ± 0.52	-0.89 ± 0.14
W12	*C. sessiliflora*	-0.00 ± 0.13	-2.69 ± 1.09	-0.77 ± 0.14
W13	*C. sessiliflora*	-3.64 ± 0.23	-6.08 ± 0.34	-0.50 ± 0.09

References

Dakora FD and Keya SO (1997) Soil Biol. Biochem. 29, 809–817.
Eaglesham ARJ et al. (1981) Soil Biol. Biochem. 13, 169–171.
Nyemba RC and Dakora FD (2005) Symbiosis 40, 79–86.
Shearer G and Kohl DH (1986) Aust. J. Plant Physiol. 13, 699–756.
Spriggs AC et al. (2003) Plant Soil 258, 495–502.

HOW TO QUANTIFY BIOLOGICAL NITROGEN FIXATION IN FORAGE LEGUMES IN THE FIELD

Georg Carlsson and Kerstin Huss-Danell

Department of Agricultural Research for Northern Sweden, Crop Science Section, Swedish University of Agricultural Sciences (SLU), Umeå, Sweden

Reliable measurements of the input of fixed-N from forage legumes are essential in estimating the need for N fertilization in grasslands. Such measurements rely on methods that can be applied in the field for either the entire growing season or parts of it. High precision and reliability should always be aimed at when measuring N_2 fixation. However, depending on the aims and availability of resources (time and equipment), methods with varying precision can be chosen. Methods with high precision are needed if the aim is to reveal factors that affect the N_2-fixation rate, whereas N balances at the field or farm level may be based on N_2 fixation estimates obtained with lower precision. Of the methods listed in Table 1, we only recommend ^{15}N natural abundance (NA) and ^{15}N isotope dilution (ID) for measurements with high precision. In addition, we propose that simple formulas based on legume biomass can be used for rapid estimates with lower precision. Nitrogen difference is not a reliable method for forage legumes because forage grasses usually are more efficient than legumes in taking up soil N (Table 1; Carlsson and Huss-Danell, 2003 and references therein).

In addition, the following is important when comparing NA and ID methods. Although ID gives a measure of N_2 fixation during a defined period, from application of ^{15}N to harvest, NA gives a measure of N_2 fixation integrated over a longer time, up to the whole lifetime of a plant (Huss-Danell and Chaia, 2005). NA is a simpler method than ID because it does not require fertilization with ^{15}N, but NA is sensitive to the B value used (Carlsson et al., 2006; Högberg, 1997). The B value corresponds to the ^{15}N/^{14}N ratio of the symbiotic association when grown with N_2 in air as the only N source. Both NA and ID require a reference plant that only uses soil N, and the legume and the reference plant should take up soil N with the same ^{15}N/^{14}N ratio (Högberg, 1997).

Grasses may not be ideal reference plants because they have different rooting patterns from forage legumes (Huss-Danell and Chaia, 2005). We therefore recommend using the mean ^{15}N/^{14}N ratio of several reference plants, e.g., by using weeds.

F. D. Dakora et al. (eds.), *Biological Nitrogen Fixation: Towards Poverty Alleviation through Sustainable Agriculture.*
© Springer Science + Business Media B.V. 2008

Table 1. Methods to measure N_2 fixation in legume-rhizobia symbioses.

	N content	N difference	$^{15}N_2$ fixation	^{15}N natural abundance	^{15}N isotope dilution	H_2 evolution	C_2H_2 reduction
Substrate for N_2ase	N_2	N_2	N_2	N_2	N_2	H^+	C_2H_2
Study period	Weeks –years	Weeks– years	Hour(s)	Weeks– years	Weeks– years	Minutes– hours	Minutes– hours
Destroys sample?	Yes	Yes	Yes	Yes	Yes	No	No
Analyses needed	N content	N Content	^{15}N content	^{15}N content	^{15}N content	H_2 sensoror GC^a	GC^*
In field	No	Yes (?)	No	Yes	Yes	No	No (yes)
Require-ments, limits	Needs N-free sub-strate, which is rare in the field	Legume and non-legume must take up soil-N with same efficiency	$^{15}N_2$ is expen-sive; gas-tight incub-ation needed	Conc. of ^{15}N in soil and air must differ; B-value in legume should be known	Added ^{15}N must be evenly distri-buted in the soil profile	Uptake H_2ase (Hup) must be inactive; difficult to make gas-tight for H_2	C_2H_2 causes decline in N_2ase activity; conver-sion factor for C_2H_4 to NH_3 needed

[a]GC, gas chromatography

Because forage legumes grown together with grasses consistently derive most of their N from N_2 fixation, rough estimates of N_{fix} (kg N ha^{-1}) can be calculated as: N_{fix} = 0.026 × legume dry matter (DM; kg ha^{-1}) for *Trifolium pratense* L.; N_{fix} = 0.031 × DM for *T. repens* L.; and N_{fix} = 0.021 × DM for *Medicago sativa* L. (Modified from Carlsson and Huss-Danell, 2003.) Such rough but rapid estimates of N_2 fixation may be of great value for establishing on-farm N budgets or N fertilization plans.

Thanks to the Swedish Research Council for Environment, Agricultural Sciences and Spatial Planning and the Carl Trygger Foundation for financial support (to KHD).

References

Carlsson G and Huss-Danell K (2003) Plant Soil 253, 353–372.
Carlsson G et al. (2006) Acta Agric. Scand. B., Soil Plant Sci. 56, 31–38.
Högberg P (1997) New Phytol. 137, 179–203.
Huss-Danell K and Chaia E (2005) Physiol. Plant. 125, 21–30.

PART 1B

USE OF INOCULANTS

NODULAR DIAGNOSIS FOR INTEGRATED IMPROVEMENT OF SYMBIOTIC NITROGEN FIXATION IN CROPPING SYSTEMS

J. J. Drevon[1], S. Gugliemni[2], G. Boyer[3], E. Lafosse-Bernard[4], R. Métral[2], C. Pernot[1] and H. Vailhe[1]

[1]INRA-Montpellier-Supagro, France; [2]CT Montpellier-Supagro, France; [3]Chambre d'Agriculture, Catelnaudary, France; [4]CIVAM-Bio, Montpellier, France

Nodular diagnosis consists of measuring the nodulation of a legume in an area of production, and relating it to the growth and subsequent yield of the legume (Drevon, 2001; Drevon et al., 2001; 2003). The objective of this procedure is to respond to the following questions: (i) does symbiotic nitrogen fixation (SNF) cover the N requirement of the legume in the cropping system? and (ii) what are the environmental factors limiting SNF in the area?

The nodular diagnosis is based on sampling a site chosen from the fields of the bean farmers who agreed to participate in the agronomic survey on the nodulation of their plants. Each site was divided into two parts, one without N fertilization and the other receiving a non-limiting N fertilizer with the aim of establishing whether the N nutrition is indeed the major factor limiting the legume growth in that area. Once this was established, two practices (local practice versus an alternative) were assessed in an agronomic trial within the area. Thus, the sites of nodular diagnosis were multi-locational, so that tests could be shared with many extension agents. However, it was decided not to test more than one alternative bio-technique with farmers. A reliable number of sites for use was determined to be ten per area. At flowering, corresponding to the stage when the SNF potential starts to decline, 20 plants were dug out (20-cm depth) at four sampling points within a homogeneous site. Plants with roots and nodules were preserved in a cold room for subsequent measurement of their individual biomass of shoots and nodules.

Spatial Variation in Nodulation and Efficiency in Use of the Rhizobial Symbiosis

Figure 1 illustrates the variation in nodule number, from less than 5 up to more than 50, which was generally observed at each site with instances of the total absence of nodules on some or all plants surveyed in fields of common bean in rotation with wheat in Lauragais farming systems.

F. D. Dakora et al. (eds.), *Biological Nitrogen Fixation: Towards Poverty Alleviation through Sustainable Agriculture.*
© Springer Science + Business Media B.V. 2008

A simple regression was obtained from plots of the nodulation parameter (number or biomass) and the growth of plants as shown in Figure 1. This allowed for calculation of the ratio of additional shoot growth for each additional nodule, i.e., the slope of the regression was considered as an assessment of the efficiency of use of the rhizobial symbiosis (EURS) at that site. This EURS varied significantly between sites.

gsDW / pl (R6-7)

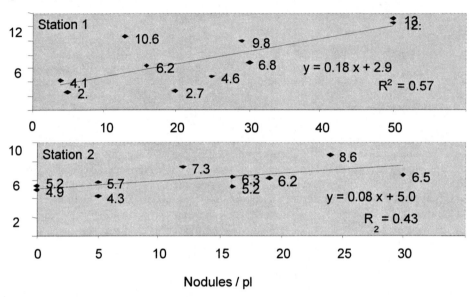

Figure 1. Efficiency in utilisation of the rhizobial symbiosis varies among fields in Lauragais.

The comparison of growth and yield of common bean with or without an optimal N fertilization within each site (Figure 2) made it possible to assess whether the nodulation or the shoot is the determining variable of the regression. As observed in site 5 and 6, the higher growth with N fertilization established that the low nodulation at those sites was the major factor that limited bean yield. These were fields where improving SNF could directly contribute to yield improvement. This may be obtained by inoculation with a native rhizobia isolated from sites with high EURS. This was not the case with sites 1 and 2 where it was found that SNF could complement soil N efficiently to support such a high bean grain yield of 3 t ha^{-1}. However, exceptionally higher grain yield than 4 t ha^{-1} were found at site 3, although the EURS was relatively high in the non-fertilized part of the site. This result suggested that the plant genotype did not have an effective capacity for SNF. This prompted the farmers to request genetic improvement of the lingot type beans, including (among other selection parameters) the height of the pods in order to decrease the losses at harvest.

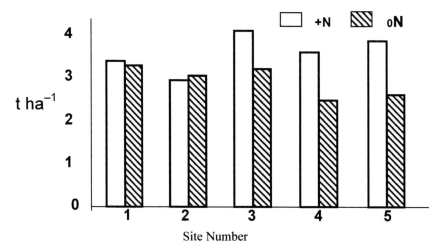

Figure 2. Yield of common bean with or without N in tons of bean per hectare.

Temporal Variation in Nodulation

Nodulation was found to vary extensively during a follow-up nodule analysis over a number of years of participation in the bean-cereal cropping systems in Lauragais (Figure 3). The low nodulation values found could be due to high residual soil mineral N that varied during and between years.

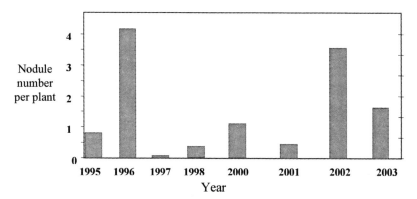

Figure 3. Temporal variations of nodulation in one season.

Participatory Assessment of Contrasting Line for EUP and SNF

The relationship of nodular diagnostics with participatory breeding was based on sowing 2-m rows of genotypes in the research plots within each site of the nodulation survey in that area. The data revealed the extent of variation in both nodulation and growth of snap bean in four fields, which belonged to horticultural organic farmers, where the mean nodulation value for the local cv Pongo was higher than 50 mg nod (DW plant)$^{-1}$, in contrast to six other sites in the area, where the nodule number per plant was less or even nil (Figure 4). The nodulation was generally higher for the recombinant inbred lines with higher efficiency in the use of phosphorus for symbiotic nitrogen fixation, namely 115, and this was associated with a significantly higher shoot growth than for 147.

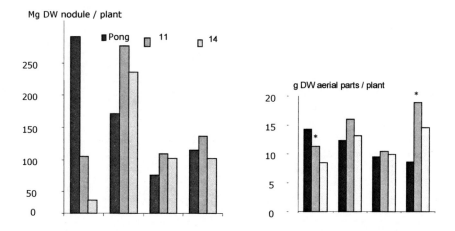

Figure 4. Growth and nodulation of snap-bean in stations where N_2 fixation contributes to growth in organic horticulture of Lauragais.

Conclusions

The nodule analysis for common bean in either the conventional or organic cropping systems of Languedoc in southern Mediterranean France showed large spatial and temporal variation in nodulation and efficiency in the use of the symbiotic nitrogen fixation for growth and subsequent grain yield. The higher nodulation and growth in the bean line selected for the efficient use of P suggest that P is a limiting factor of the symbiosis in the organic horticultural systems and that improving the use of P for N_2 fixation could contribute to an increase in the yield of bean and also benefit the N and P biogeochemical cycles.

References

Drevon JJ (2001) In: Nitrogen Assimilation by Plants (Morot-Gaudry JF, ed.), INRA, Paris, pp. 417–426.

Drevon JJ et al. (2001) J. Biotech. 91, 257–268.

Drevon JJ et al. (2003) In: Fixation Symbiotique de l'Azote et Développement Durable dans le Bassin Méditerranéen (Drevon JJ and Sifi B, eds), INRA Les Colloques, Paris, 100, 141–148.

EVALUATION OF SEED AND LIQUID INOCULATION ON BIOLOGICAL NITROGEN FIXATION AND GRAIN YIELD OF SOYBEAN

A. D. C. Chilimba

Chitedze Agricultural Research Station, P.O. Box 158, Lilongwe. Malawi
(Email: achilimba@hotmail.com or chitedze@malawi.net)

Soybean production is being encouraged in Malawi to supply protein to its population. Cultivation of soybean varieties currently recommended in Malawi requires seed inoculation with a highly effective rhizobium strain. Using rhizobial inoculants ensures that the correct rhizobial bacteria associate with the plant. Until the late 1980s and early 1990s, seed-applied peat-based inoculants dominated the commercial inoculant market. However, although peat was recognized as a very good carrier of rhizobia, there was interest in developing alternate formulations because it was considered time-consuming and impractical. As alternative liquid formulations were introduced, along with new packaging, this allowed farmers to treat seed directly from the packaged product as it was passing through the grain auger into the seeding equipment. Brockwell et al. (1980) reported that, when conditions were stressful and generally unfavourable, soil inoculation, such as granular soil implants, resulted in better nodulation and better yield than seed-applied inoculants. Hynes et al. (2001) reported that of eight lentil trials, six indicated that liquid and peat formulations were equally effective at enhancing final yields relative to the control. In contrast, Clayton et al. (2003) reported that seed yield by peat-based powder typically out-performed the liquid inoculant, which did not differ significantly from uninoculated. However, there were no data in the country on which to base rhizobia-inoculation recommendations. A trial was initiated in Malawi to determine the best method of inoculation and the rate of application of rhizobial inoculants for optimal biological nitrogen fixation and grain yield of soybean.

The inoculation methods used were either coated seed or liquid. The rates of application were 1 sachet, 1.5 sachets, 2 sachets, 2.5 sachets, or 3 sachets (each of 50 g) to 25 kg of seed and zero inoculation. The coated-seed inoculation used 200 mL of sugar solution mixed with the inoculant sachet(s) to make a paste, which was then poured onto 25 kg of seed and mixed thoroughly. The liquid inoculation used 5 L of water, which was thoroughly mixed and then spread in the split ridge where the seed was

F. D. Dakora et al. (eds.), *Biological Nitrogen Fixation: Towards Poverty Alleviation through Sustainable Agriculture.*
© Springer Science + Business Media B.V. 2008

planted. The trials were conducted at three sites, namely Lisasadzi, Biloti, and Tembwe in Malawi.

The results showed that liquid inoculation significantly outperformed the seed inoculation in pod numbers, pod dry weight, nodule numbers, nodule fresh weight, and top dry weight at Lisasadzi (Table 1). Good performance of the liquid application could be due the fact that the seed pelleting is dried under a shed before planting, whereas the liquid is applied directly to the seed in the ridge. Drying under the shed might have detrimental effects on the survival of the rhizobia, hence the poor performance. One 50-g sachet of inoculant significantly increased nitrogen fixation and grain yield compared to no inoculation and was not significantly different with the higher rates of application. The lower rate of application was adequate to improve nitrogen fixation, which is in agreement with the finding that, in the absence of soil rhizobia, just 100 cells/g soil is required (Beattie and Handelsman, 1993). However, the grain yield was not significantly different between the two methods of application (Table 1). Inoculation significantly increased grain yield in comparison to the control, but there were no significant differences among the rates of application in grain yield (Table 1). Similar results were recorded at the Biloti and Tembwe sites.

Table 1. Effect of inoculation methods and rate of inoculation on nitrogen fixationand grain yield of Gelduld soybean variety at Lisasadzi.

Treatments	Pod no.	Pod dry wt (g)	Nodule no.	Nodule fresh wt (g)	Top dry wt (g)	Grain yield (kg/ha)
Seed inoculation	146	42	502	16	63	1,890
Liquid inoculation	178	55	664	19	78	1,787
1 × 50 g	159	46	519	16	73	1,907
1.5 × 50 g	181	58	775	24	80	1,969
2 × 50 g	155	44	614	18	60	1,917
2.5 × 50 g	181	51	671	22	74	2,030
3 × 50 g	173	60	658	19	77	2,031
No inoculation	122	32	263	7	59	1,470
SE (methods)	6.15	2.16	50.19	1.34	3.65	85.49
SE (rates)	10.65	3.74	86.92	2.32	6.32	148.08
CV%	18.6	21.0	42.0	36.0	25.0	22.0

References

Beattie GA and Handelsman J (1993) J. Gen. Microbiol. 139, 529–538.
Brockwell J et al. (1980) Aust. J. Agric. Res. 31, 47–60.
Clayton G et al. (2003) Can. J. Plant Sci. 84, 89–96.
Hynes RK et al. (2001) Can. J. Microbiol. 47, 575–600.

INITIATION, LOCALIZATION, AND GROWTH OF NODULES WITHIN THE ROOT SYSTEM OF PEA AS AFFECTED BY ASSIMILATE AVAILABILITY

A.-S. Voisin[1], N. G. Munier-Jolain[1], L. Pagès[2] and C. Salon[1]

[1]INRA, Unité de Génétique et d'écophysiologie des légumineuses, BV 86510, 21065 Dijon, France; [2]INRA, Unité Plantes et Systèmes de Culture Horticoles, Site Agroparc, 84914 Avignon, France

In pea, symbiotic nitrogen fixation is highly related to nodule biomass (Voisin et al., 2003). Although many studies have focussed on nodule organogenesis at the cellular and organ level, little information has been gained concerning the dynamics of nodule biomass during the plant growth cycle. Therefore, our aim was to characterize nodule establishment and growth at the plant level as a function of internal (carbon and nitrogen fluxes) and external (temperature, radiation and nitrate) factors.

The effect of carbon nutrition on pea nodule growth was investigated in two experiments either by modifying C availability (using plant shading) or by modulating competition for C between nodules and other plant organs using contrasted nodulation conditions (produced by inoculating at weekly intervals from radicle emergence, which resulted in various shoot and root developments). Another treatment in which plants were grown with 14 mol m^{-3} as N in NO_3^- served as a control for maximal root development (as this nitrate dose totally inhibited nodulation) and for maximal N supply. Plants were grown in hydroponics in a greenhouse with a mean temperature of 20°C. They were harvested at two- or three-day intervals during the whole vegetative period. The taproot was divided into 2-cm long segments for morphological observations, which included the number and maximal length of lateral roots as well as the number and biomass of nodules on the taproot segment and its lateral roots. The plants were divided into several parts (sowed seed, shoot, roots and nodules) that were weighed separately and ground for determination of their N content.

The time course of nodule number was characterised by several nodulation waves. For all treatments, there was an initial increase in nodule number after inoculation and then nodule number stabilised. For some treatments, there was a second later nodulation wave that was parallel to the rapid increase in plant biomass. Seven weeks after germination, the number of nodules varied greatly among treatments, from 300 to over 2,000. To account for variations in the total nodule number during the first nodulation wave,

F. D. Dakora et al. (eds.), *Biological Nitrogen Fixation: Towards Poverty Alleviation through Sustainable Agriculture*.
© Springer Science + Business Media B.V. 2008

the number of nodules was first related to the number of lateral roots present at the time of nodulation. Indeed, these lateral roots represent the potential number of root tips that are likely to be infected by rhizobia. However, this did not explain the whole variability among treatments. Nodule number appeared to be more related to total plant biomass, which indicated the extent of plant N needs at the time of nodulation. The total number of nodules formed was probably a combination of both variables.

We localised nodules on both tap root and lateral roots for every 2 cm tap-root segment (Figure 1). With the tap root, there was always a zone without nodules, then a restricted zone where nodule number peaked, followed by a zone with no or very few nodules. The zone with numerous nodules corresponded to the deepest soil penetration and its position corresponded to tap-root length at inoculation. As such, nodulation of the tap root was limited in time to a short period subsequent to inoculation. Nodule number was far higher on lateral roots, accounting for more than 80% of the total, and mainly on lateral roots originating in the first 2–4 cm of the tap root. This could be related to root architecture, as shown by the distribution of maximal length of lateral roots, because nodules were found on the longest lateral roots, i.e., those with highest elongation rates.

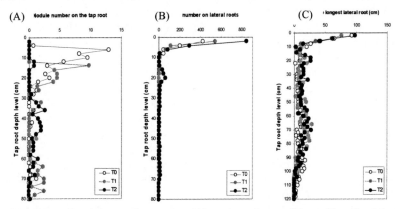

Figure 1. Number of nodules on the tap root (A), on lateral roots (B), and length of the longest lateral root (C), for each 2 cm segment of the tap root, for pea plants inoculated at germination (T0), one week later (T1) or two weeks later (T2).

The respective effects of C and N nutrition on nodule number, growth, and localisation will be further analysed, as related to N uptake by symbiotic fixation. The resulting model will combine trophic and architectural approaches (inspired by Thaler and Pagès, 1998) and will help test hypotheses concerning factors that regulate nodulated root establishment and N-uptake function. This model will also become a powerful tool for analysing and constructing genetic variability associated with legumes N nutrition.

References

Thaler P and Pagès L (1998) Plant Soil 201, 307–320.
Voisin AS et al. (2003) J. Exp. Bot. 54, 2733–2744.

ENHANCED EARLY NODULATION OF *MEDICAGO TRUNCATULA* CO-INOCULATED WITH *SINORHIZOBIUM MEDICAE* AND *ACHROMOBACTER XYLOSOXIDANS*

S. Fox[1], G. W. O'Hara[1], L. Brau[1], J. Howieson[1,2] and W. Reeve[1]

[1]Centre for *Rhizobium* Studies, Murdoch University, Western Australia;
[2]Agriculture Western Australia, South Perth, Western Australia
(Email: slfox@aapt.net.au)

Efforts to increase the productivity and sustainability of agro-ecosystems have resulted in increased research focused on generating improved microbiological inoculant technologies. Plant growth-promoting rhizobacteria (PGPR) are one area receiving increased attention due to their potential for use as biofertilizers, plant-growth promoters, and biocontrol agents for weed and disease control in farming systems (Siddiqui, 2006).

Although most PGPR research has focused on improving the yields of cereals and other broad acre crops, more recently, some research has indicated that there may be a role for using PGPR as co-inoculants with root-nodule bacteria to increase nodulation and, hence, the yields of pasture and grain legumes (Bai et al., 2003; Camacho et al., 2001; Marek-Kozaczuk and Skorupska, 2001; Zhang et al., 1996). Currently, the mechanisms of PGPR-mediated enhanced nodulation are largely speculative and there is a need for mechanistic studies to be completed in order to understand how these organisms influence nodulation. Understanding these interactions should allow us to improve the consistency and reproducibility of these responses and may allow the development of PGPR inoculants either for specific soil conditions or for specific stress responses.

Research was undertaken to characterize the mechanisms of enhanced nodulation and yields of *Medicago truncatula* when co-inoculated with *Sinorhizobium medicae* WSM419 and a PGPR. Initial glasshouse trials, using four PGPRs from the Centre for Rhizobium Studies (CRS) collection, resulted in the selection of *Achromobacter xylosoxidans* WSM3457 for further study. Subsequent glasshouse trials confirmed that this isolate increased nodule scores, plant nodule mass, and enhanced yields by 25–50% when *M. truncatula* was challenged with a low inoculum dose of *S. medicae* WSM419 (3×10^3 cfu/mL).

F. D. Dakora et al. (eds.), *Biological Nitrogen Fixation: Towards Poverty Alleviation through Sustainable Agriculture*.
© Springer Science + Business Media B.V. 2008

Nodule initiation studies were conducted to determine if co-inoculation of *M. truncatula* with WSM3457 and WSM419 affects the rate of nodule initiation and development. Plants grown in soil under glasshouse conditions were harvested at 5, 7, 9, 11, 14, 17 and 21 days after inoculation. Roots were stained with brilliant green and examined under a dissecting microscope for nodule initials. The data revealed that early nodule development was enhanced with the co-inoculation treatment. Nodule initials were first observed on day 5 and day 7 in the co-inoculation and rhizobium treatments, respectively. Nodules were first evident on the co-inoculated plants on day 7, whereas nodules were only evident on day 9 in plants inoculated with WSM419 alone. This trend continued throughout the sampling period with significantly higher nodule numbers on the co-inoculation treatment by day 17.

A further experiment was aimed at investigating if the enhanced nodulation on co-inoculated plants was a result of PGPR stimulation of *M. truncatula* root development. Four treatments were investigated: (i) inoculation with WSM419 alone; (ii) inoculation with WSM419 and WSM3457; (iii) inoculation with WSM3457 alone; and (iv) an un-inoculated control. There was no significant difference in root length or lateral root formation between *M. truncatula* co-inoculated with WSM419 and *A. xylosoxidans* WSM3457 or inoculated with WSM419 alone. There was a transient increase in root-hair density between days 7 and 9 on all inoculated treatments when compared to the un-inoculated treatment. Although this suggested that bacterial inoculation increased root-hair densities during this time, there was no difference between co-inoculated roots and those inoculated with rhizobium alone. The data suggest that WSM3457 does not increase root development in such a way as to explain the enhanced early nodulation of co-inoculated *M. truncatula*.

These results indicate that *A. xylosoxidans* WSM3457 is exerting its effect over the *M. truncatula/S. medicae* WSM419 symbiosis during the period of the early signalling processes between this legume and its micro-symbiont, resulting in enhanced early root infection by *S. medicae* WSM419. Future work will focus on investigating the mechanisms of this novel interaction.

References

Bai Y et al. (2003) Crop Sci. 43, 1774–1781
Camacho M et al. (2001) Can. J. Micro. 47, 1058–1062.
Marek-Kozaczuk M and Skorupska A (2001) Biol. Fert. Soils 33, 146–151.
Siddiqui, ZA (2006) PGPR: Biocontrol and Biofertilization. Springer, Dordrecht, The Netherlands.
Zhang F et al. (1996) Ann. Bot. 77, 453–460.

GENE TRANSFER IN THE ENVIRONMENT PROMOTES THE RAPID EVOLUTION OF A DIVERSITY OF SUBOPTIMAL AND COMPETITIVE RHIZOBIA FOR *BISERRULA PELECINUS* L.

K. G. Nandasena, G. W. O'Hara, R. P. Tiwari and J. G. Howieson

Centre for *Rhizobium* Studies, Murdoch University, Murdoch, Western Australia 6150

The emergence of biodiversity in rhizobia after the introduction of exotic legumes and their respective rhizobia to new regions is a challenge for contemporary rhizobiology. *Biserrula pelecinus* L. is a pasture legume species that was introduced to Australia from the Mediterranean basin and which is having a substantial impact on agricultural productivity on acidic and sandy soils of Western Australia and New South Wales (Howieson et al., 2000). This deep-rooted plant is also valuable in reducing the development of dryland salinity. This legume is nodulated by a specific group of root-nodule bacteria that belongs to *Mesorhizobium* (Nandasena et al., 2001, 2007).

We have recently shown the evolution of diverse opportunistic rhizobia able to nodulate *B. pelecinus* following *in situ* transfer of symbiotic genes, located on a mobile symbiosis island, from an inoculant strain to other soil bacteria (Nandasena et al., 2006). A symbiosis island was first described in *M. loti* strain ICMP3153 (Sullivan and Ronson, 1998). Genomic islands could become inert and stabilised due to genomic rearrangements. Therefore, the aim of our current research is to investigate whether the current commercial inoculant strain for *B. pelecinus* (WSM1497) potentiated the development of a diversity of strains able to nodulate this legume via lateral transfer of the symbiotic island from WSM1497 to other soil bacteria.

Nodules from commercially grown *B. pelecinus* were collected (in 8/2005; 5–6 years after introduction and inoculation) from four different sites in Western Australia. The 387 pure cultures from nodule crushes (Table 1) were fingerprinted with the RPO1 PCR primer (Richardson et al., 1995) to show that only 50.1% of the nodules were occupied by WSM1497. These 193 isolates were authenticated on *B. pelecinus* cv. Casbah in a glasshouse experiment, when 184 nodulated this legume. Nodule occupancy by WSM1497 varied significantly between field sites, from >80% of the nodules collected from Brookton and Kondinin to <30% of the nodules from Karlgarin and Wickepin (Table 1). Furthermore, there were only a few dominant strains at the latter two sites. At Karlgarin, 20%

F. D. Dakora et al. (eds.), *Biological Nitrogen Fixation: Towards Poverty Alleviation through Sustainable Agriculture.*
© Springer Science + Business Media B.V. 2008

of the nodules were occupied by a strain designated as type B, 25% of the nodules by another strain designated as type C, whereas only 22% of nodules were occupied by WSM1497, thus indicating that these recently evolved strains are highly competitive for nodulation of *B. pelecinus*.

Table 1. Nodule occupancy of commercially grown B. pelecinus collected from different fields in the WA wheat belt five or six years after inoculation with WSM1497.

Site	Total no of isolates	% WSM1497	% novel isolates	% non symbionts
Brookton	44	86.4	9.1	4.5
Kondinin	120	84.2	13.3	2.5
Karlgarin	120	21.7	77.5	0.8
Wickepin	103	28.2	68.9	2.9
Total	**387**	**50.1**	**47.5**	**2.3**

To maximize the value of *B. pelecinus* in farming systems, its N_2-fixing symbiosis must be maintained at the highest level of efficiency. Therefore, we tested the N_2-fixation efficiency of 53 randomly selected authenticated isolates (methods as Nandasena et al., 2004), all of which were found to be less effective than WSM1497. However, the N_2-fixation efficiency among these isolates ranged from no N_2 fixation (six isolates) to about 70% of that of the commercial inoculant. Fourteen of these 53 diverse isolates were randomly selected for the sequencing of *dnaK* to infer the phylogenetic relationships. All 14 strains clustered within *Mesorhizobium* and distantly to the Mediterranean *Biserrula* mesorhizobia. Furthermore, the symbiosis-island insertion regions of these 14 strains (Nandasena et al., 2006) had an identical sequences that was 100% similar to that of WSM1497, indicating a possible transfer from the commercial inoculant.

The results show the rapid evolution of competitive, yet suboptimal, strains for N_2 fixation on *B. pelecinus*, following the lateral transfer of a symbiosis island from the commercial inoculant to other soil bacteria. At present, we are constructing a genetically stable inoculant strain for *B. pelecinus* by inactivating the genes responsible for symbiosis-island transfer and thereby, reducing the chances of lateral gene transfer between the inoculant and soil bacteria in order to manage the future productivity of *B. pelecinus* at optimum levels.

References

Howieson et al. (2000) Field Crop. Res. 65, 107–122.
Nandasena et al. (2001) Int. J. Syst. Evol. Microbiol. 51, 1983–1986.
Nandasena et al. (2004) Soil Biol. Biochem. 36, 1309–1317.
Nandasena et al. (2006) Appl. Environ. Microbiol. 72, 7365–7367.
Nandasena et al. (2007) Int. J. Syst. Evol. Microbiol. 57, 1041–1045.
Richardson et al. (1995) Soil Biol. Biochem. 27, 515–524.
Sullivan and Ronson (1998) Proc. Natl. Acad. Sci. USA 95, 5145–5149.

EFFECT OF A GENETICALLY MODIFIED *RHIZOBIUM LEGUMINOSARUM* STRAIN ON BACTERIAL AND FUNGAL DIVERSITY IN THE RHIZOSPHERE OF *PISUM SATIVUM*

C. Lantin[1], B. Bulawa[1], A. Haselier[1], R. Defez[2] and U. B. Priefer[1]

[1]Aachen University, Institute for Biology I, Department of Soil Ecology, Aachen, Germany; [2]Institute for Genetics and Biophysics, Naples, Italy (Email: c.lantin@biol.rwth-aachen.de or priefer@biol.rwth-aachen.de)

At present, legumes account for one third of the total protein intake of the human diet. Therefore, any improvement of forage and grain legume yields achieved without using potentially harmful chemical fertilizers are of extreme importance, in particular to the developing countries. One promising approach is the inoculation with a genetically modified *Rhizobium* strain that releases the plant hormone indole-3-acetic acid (IAA) to the rhizosphere of its host plant. The main objective of our experiments was to investigate whether the inoculation with a genetically modified *Rhizobium* strain had any impact on the composition and diversity of the bacterial and fungal communities in the rhizosphere of *Pisum sativum*.

For this purpose, we used *Rhizobium leguminosarum* bv. *viciae* VF39, which was modified by inserting the *iaaM* (tryptophan monooxygenase) and *tms2* (indole-3-acetamide hydrolase) genes for IAA-synthesis into two separate plasmids pRD20 and pBBRIAA, each derived from the same origin. The expression of these two genes in pRD20 is under control of the strong constitutive *cat* promotor, whereas in pBBRIAA, the *rolA* promotor is stationary-phase and acid inducible. We also investigated the impact of the modified *Rhizobium* strain on the composition of the root exudates and determined the IAA concentration in the rhizosphere by means of GC-MS analyses.

We applied four different treatments:

1. Inoculation with the wild type strain (VF39)
2. Inoculation with the modified strain (VF39 pRD20 or pBBRIAA)
3. Addition of IAA with a concentration of 10^{-6} M (instead of inoculation)
4. Control (untreated plant)

F. D. Dakora et al. (eds.), *Biological Nitrogen Fixation: Towards Poverty Alleviation through Sustainable Agriculture*.
© Springer Science + Business Media B.V. 2008

The incubation time for the first analysis of each treatment was 9 weeks (sampling after week 1, 4 and 9). In addition, a second analysis was performed with an incubation time of 10 days (sampling after day 3 and 10).

We used denaturing gradient gel electrophoresis (DGGE) to compare the similarity of the total and active part of the microbial community based on the separation of PCR products obtained after amplification of 16S/18S ribosomal DNA (total) and cDNA of 16S/18S rRNA (active) extracted simultaneously from all experimental approaches. The DGGE gels were analysed with the GelCompar software by means of the DICE correlation matrix and the unweighted pair group method using arithmetic averages (UPGMA). In this analysis, the presence or absence of a band at a distinct position in the banding pattern was determined. In order to analyse the impact of the modified *Rhizobium* strain on the bacterial and fungal diversity, we used the Shannon–Weaver index.

The DGGE band patterns from the analysis of the bacterial community did not indicate any effect of the genetically modified VF39 (p*RD20*) strain on the total and active bacteria in the rhizosphere of the host plant (incubation: 9 weeks). In contrast, VF39 (p*BBRIAA*) showed an obvious effect on the bacterial composition of the rhizosphere as well as on the diversity and richness of the bacteria (incubation: 10 days).

The composition of the fungal community in the rhizosphere inoculated with VF39 (p*RD20*) and VF39 (p*BBRIAA*) showed little similarity with the other approaches regarding their band patterns (9 weeks/10 days). This did not only apply to the composition of the total part, but also to the active part of the fungal rhizosphere community. The effect of VF39 (p*RD20*) increased over time. We concluded that there obviously is an impact of the genetically modified *Rhizobium* strain on the active as well as total fungal community in soil. However, the diversity and richness of the fungal rhizosphere community was not affected.

Further investigation of the root exudate patterns by GC-MS analysis did not show any effect of the genetically modified *Rhizobium* strain releasing IAA on its composition. The IAA concentration in all approaches differed at the beginning of the experiment, but it decreased over time and was similar for all treatments at the end of the experiment. We concluded that the higher amount of IAA in the first week was due to the additional inoculation with the IAA-releasing *Rhizobium* strain at the beginning experiment.

In summary, the results of our experiments displayed only an effect of the genetically modified *Rhizobium* strain VF39 p*BBRIAA* (IAA synthesis under control of the inducible *rolA* promotor) on the total and active part of the bacterial community in the rhizosphere of *Pisum sativum*. In contrast, both modified *Rhizobium* strains VF39 p*RD20* and VF39 p*BBRIAA* had a large impact on the total as well as on the active fungal community in the rhizosphere of the host plant.

INOCULANTS FOR SUGAR CANE: THE SCIENTIFIC BASES FOR THE ADOPTION OF THE TECHNOLOGY FOR BIOFUEL PRODUCTION

V. M. Reis[1], A. L. M. de Oliveira[2], M. F. da Silva[3], F. L. Olivares[4], J. I. Baldani[1], R. M. Boddey[1] and S. Urquiaga[1]

[1]Embrapa Agrobiologia, BR 465, km 47 and km 07, 23890-970 Seropédica, RJ, Brazil; [2]Departamento do Biologia, Universidade Estadual de Londrina, Londrina, PR, Brazil; [3]Universidade Federal Rural do Rio de Janeiro, Instituto de Agronomia, doutorado em Ciência do Solo, km 47, Seropédica, RJ, Brazil; [4]Centro de Biociências e Biotecnologia, UENF, Campo dos Goytacazes, RJ, Brazil (Email: veronica@cnpab.embrapa.br)

Sugar cane is one of the graminaceous species that can obtain most of its nitrogen from biological N_2 fixation (BNF). Available evidence suggests that BNF contributions are dependent on cane variety and soil properties. Using the ^{15}N natural-abundance technique, the proportion of N derived from BNF may range from zero to 70%. The impact of saving just half of the nitrogen fertilizer used on the crop in Brazil (mean of 60 kg N ha^{-1}) would be approximately 150,000 mg of fertiliser N per year.

Sugarcane is propagated vegetatively via stem pieces (setts) and a considerable quantity of sucrose is present and starts to decompose as the sett germinates. This decomposition pathway liberates sub-products normally used by different microorganisms, including diazotrophs, present in the soil and plant. The selection for use of an inoculant containing a mixture of diazotrophic bacteria, based on the knowledge of the ecological behaviour of the species involved (Oliveira et al., 2006), is the key to obtaining high BNF contributions to this crop.

The experiments performed in Brazil used a mixture of five different diazotrophic species all isolated from sugarcane. These were: *Gluconacetobacter diazotrophicus* strain BR 11281^T; *Herbaspirillum seropedicae* strain BR 11335; *H. rubrisubalbicans* strain BR 11504; *Azospirillum amazonense* strain BR 11145, and *Burkholderia tropica* strain BR 11366. All strains were deposited in the diazotrophic bacterial collection of Embrapa Agrobiologia. This mixture was applied to two sugarcane varieties, SP 70-1143 and SP 81-3250, grown under commercial field conditions at three sites with contrasting soil types; an Alfisol, an Oxisol and an Ultisol, which are equivalent to low,

F. D. Dakora et al. (eds.), *Biological Nitrogen Fixation: Towards Poverty Alleviation through Sustainable Agriculture.*
© Springer Science + Business Media B.V. 2008

medium and high natural fertility, respectively. In the first trial, sterile micro-propagated plantlets were inoculated with the diazotrophs (as described by Reis et al., 1999). In the Alfisol, the response to inoculation of the stem yield, dry matter and N accumulation of the variety SP70-1143 (previously identified as high in BNF) was equivalent to the effect of the usual annual nitrogen fertilization. The plants were grown without N fertilizer for three consecutive years (a plant crop and two ratoons) and the ^{15}N data indicated that inoculation promoted a significant increase in the contribution of BNF.

Because sugarcane is normally propagated by setts, the next step was to perform a new procedure to introduce these five diazotrophs into stem pieces by immersion, after the heat treatment (50°C for 30 min) normally used to control ratoon-stunting disease. The strains were grown individually, then adjusted to the same cell density (10^8 cells mL^{-1}), and mixed before planting the setts. Two new sugarcane varieties, RB 72-454 and RB 86-7515, were used to test this inoculation procedure. The assay was performed using plastic trays containing a mixture of vermiculite + sand (2:1) and, after 50 days, plant colonization and biomass accumulation were evaluated. The bacterial population colonized plant tissue in numbers higher than 10^6 cells g^{-1} fresh weight. Root biomass accumulation was higher in the presence of bacterial application. This inoculation procedure can be used to select bacterial strains in a rapid trial and adapted to new varieties of sugarcane, which are regularly being introduced to the cropping systems.

Using this methodology, Medeiros et al. (2006) evaluated the effect of inoculation of *G. diazotrophicus* and the addition of increasing amounts of two sources of mineral nitrogen (ammonium sulphate and calcium nitrate) on the population of this diazotroph, its nitrogenase activity (acetylene reduction), and accumulation of N by two sugarcane hybrids, SP 701143 and SP 792312. The results showed that both varieties differed in the form of nitrogen they prefer to take up from the soil. In both varieties, the addition of increased doses of ammonium and nitrate decreased the population of *G. diazotrophicus* but, in the variety SP 70-1143, the inhibition was more pronounced in the presence of calcium nitrate. Acetylene-reduction activity was inhibited in both varieties, especially in variety SP 79-2312, in the presence of either of the nitrogen sources. These results demonstrate that the interaction of sugarcane and diazotrophic bacteria is sensitive to nitrogen application in a similar manner to legume symbioses.

Acknowledgements

This work was partially funded by Empresa Brasileira de Pesquisa Agropecuária – Embrapa Grant No. 02.02.5.13.00.04; PRONEX II Grant No. E-26/171.208/2003; PADCT III Grant No. CiBio 77.97.1138.00; and FAPERJ – "Bolsa Cientista do Nosso Estado".

References

Medeiros et al. (2006) Plant Soil 279, 141–152.
Oliveira et al. (2006) Plant Soil 284, 23–32.
Reis et al. (1999) Plant Soil 206, 205–211.

DETECTION AND ENUMERATION OF PLANT GROWTH-PROMOTING BACTERIA

M. L. Kecskés, E. Michel, B. Lauby, M. Rakotondrainibe, A. Palágyi and I. R. Kennedy

Faculty of Agriculture, Food, and Natural Resources, Room 307, Bldg A03, The University of Sydney, NSW 2006, Australia

Quantitative and specific monitoring of live microorganisms under laboratory and especially field conditions is still one of the unsolved basic problems facing agricultural (soil) microbiology. Without the use of genetic modification, tracing associative inoculant biofertiliser strains released into the environment is a demanding task because of the large number of species of native microorganisms present in the rhizosphere. "Colony immuno-blotting" is a promising tool for following the success or failure of the colonizing ability/survival of bacterial strains in the soil matrix because it detects live numbers of a target strain by combining the advantages of traditional dilution plating with specific immuno-detection. However until now, this technique has been mostly employed in molecular biology using pure cultures. Therefore, adjustments for testing samples originated from the complex soil environment are needed.

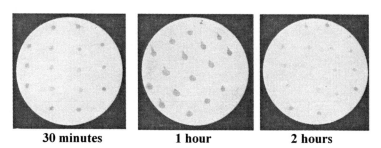

| 30 minutes | 1 hour | 2 hours |

Figure 1. Effect of different incubation times (0.5, 1 and 2 hours) on the immunoblots of *Citrobacter freundii* using anti-*Citrobacter freundii* polyclonal antibody.

Following the immuno-blotting procedure of Duez et al. (2000), we evaluated it first by even spreading of *Citrobacter freundii* mimicking colonies (see Figure 1 above). The following incubation parameters were varied: colony transfer time from the agar plate to

F. D. Dakora et al. (eds.), *Biological Nitrogen Fixation: Towards Poverty Alleviation through Sustainable Agriculture.*
© Springer Science + Business Media B.V. 2008

the nitrocellulose membrane, (polyclonal) antibody (Ab) concentration and incubation time, horse radish peroxidase (HRP) incubation time, and chromogen concentration and incubation time. A recurring problem, accidental (mechanical) removal of the chromogen product by the fixing solution (H_2SO_4), was overcome using a spray technique. Based on visual observations, an optimised protocol could be established, which mostly involved a shorter incubation time and lower concentrations of reagents and antibody, without compromising the results. In subsequent tests, colony immuno-blotting allowed us to detect and quantify the inoculant biofertiliser strain *Citrobacter freundii* in both artificial and "real" test soil samples (Figure 2). Our data indicated that, at higher to equal ratios of inoculant fertiliser to background population concentration, it was possible to determine the identity and number of strains within a few hours.

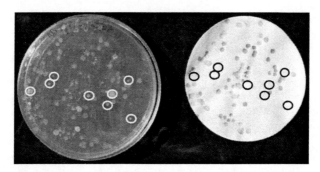

Figure 2. Dilution plate (left) and subsequent immuno-blot (right) carried out with anti-*C. freundii* Ab originated from a soil sample inoculated by *C. freundii* (dark dots). The empty white circles mark some of the colonies of the strains in the background population that are missing (empty black circles) on the immuno-blot.

However, the relative low number of target strains and the possible antigenic cross reactions to some of the native cultivable strains may lower the potential of the method. A combination of selective media, incorporation of antibiotics into the culture media, the use of monoclonal Abs, and the study of the polyclonal/monoclonal Ab recognition sites should enhance the sensitivity of this assay. Further, there is also an option for applying two or more selected monoclonal Abs with different target epitopes and coloured tagged dye in future protocols. DNA-colony blotting might also be a viable option for following population dynamics of an inoculant biofertiliser strain in soil.

Further improvements in the reliability of the application of immuno-blotting are expected. Once it has been fully developed, in addition to its relatively simple application, it provides a cheap and relatively rapid means of monitoring *Citrobacter freundii* or any other inoculant biofertiliser microbial strains in soils.

Reference

Duez H et al. (2000) J. Appl. Microbiol. 88, 1019–1027.

PART 1C

TREE LEGUMES AND FORESTRY

NODULATED TREE LEGUMES AND THEIR SYMBIOTIC *BRADYRHIZOBIUM* IN AFRICAN AND SOUTH-AMERICAN TROPICAL RAINFORESTS

Y. Prin, B. Dreyfus, C. Le Roux, G. Bena, M. Diabaté, P. de Lajudie, A. Bâ, S. M. de Faria, A. Munive and A. Galiana

Laboratoire des Symbioses Tropicales et Méditerranéennes (LSTM), Campus International de Baillarguet, 34398 Montpellier Cedex 5, France (Email: dreyfus@mpl.ird.fr)

Leguminosae, the third largest family of angiosperms, are of major agricultural, ecological and economic importance. Several recent studies have clarified the taxonomic and phylogenetic relationships among the 19,400 species that constitute this family and traced the "road map of legume diversity" (Doyle and Lückow, 2003).

Following the work of Doyle et al. (1997) on the *rbc*L gene, Wojciechowski et al. (2004) analysed the plastid *matK* genes of 330 legume species and brought new insight on clade organization within this family. Using the phylogenetic tree obtained by these latter authors, we superimposed on each plant species the taxonomic identity of its symbiotic N_2-fixing partner (from the databank http://www.ncbi.nlm.nih.gov/). We also checked the nodulation status of each plant species (from http://www.ars-grin.gov/~sbmljw/cgi-bin/taxnodul.pl). Surprisingly, the bacterial genera are randomly distributed among the plant clades and the genus *Bradyrhizobium* is increasingly represented from the Milletioid to the Caesalpinioid clades.

Although these analyses have to be confirmed on a larger number of plants and bacteria, with respect to each plant clade size, diversity, and natural geographic distribution, they reveal the need for plant and bacterial phylogenists to work together hand-in-hand and to seek the help of botanists and foresters. Care should be taken with the protocols used to obtain the bacterial isolates, i.e., directly from field collected nodules or through baiting with homologous or heterologous (i.e., *Macroptilium atropurpureum*) species. Molecular characterization of bacterial DNA from nodule extracts can complement the surveys when bacteria failed to be isolated.

F. D. Dakora et al. (eds.), *Biological Nitrogen Fixation: Towards Poverty Alleviation through Sustainable Agriculture.*
© Springer Science + Business Media B.V. 2008

Table 1. Growth response of eight leguminous tree species 4 months after rhizobial inoculation. (One-way ANOVA showed significant effects of inoculation on height at P < 0.01 in each tree species. Means followed by different letters (a, b or c) are significantly different according to the Newman & Keuls test at P = 0.05.)

Tree species	Rhizobium strain treatment	Plant height (cm)	Increment/control
Albizia adianthifolia	STM 916	14.05^a	+69.7%
	Uninoculated	8.28^b	–
Albizia altissima	STM 916	11.82^a	+17.5%
	Uninoculated	10.06^b	–
Albizia ferruginea	STM 922	20.50^a	+35.3%
	Uninoculated	15.15^b	–
Albizia zygia	STM 923	14.88^a	+58.8%
	Uninoculated	9.37^b	–
Erythrophleum guineensis	STM 934	22.37^a	+20.3%
	Uninoculated	18.60^b	–
Erythrophleum ivorensis	STM 934	17.23^a	+37.3%
	Uninoculated	12.55^b	–
Millettia rhodantha	STM 931	26.10^a	+63.1%
	Uninoculated	16.00^b	–
Millettia zechiana	STM 851	27.12^a	+113.0%
	STM 926	18.47^b	+44.8%
	Uninoculated	12.73^c	–

Considering the endangered situation of tropical rainforests and the urgent need for research on their symbioses, a closer focus was given to bradyrhizobia nodulating legume trees from African (Republic of Guinea) and South American (French Guyana) rainforests. A nodulation survey was conducted on 156 West African and 52 South-American leguminous tree species, of which a large number had never before been examined for nodulation. The survey revealed the nodulation capability of several new plant taxa, i.e., *Chidlowia sanguinea, Aubrevillea platycarpa, Tetrapleura tetraptera,* and *Amphimas pterocarpoides* in Guinea and *Spirotropis longifolia* in Guyana. Fifty-seven *Bradyrhizobium* strains were directly isolated (following rehydration in sterile distilled water) from collected nodules and compared at the molecular level (ITS) with respect to both their continental and plant taxon origins. Seven clades, all of which were distinct from the genomic *Bradyrhizobium* species described by Willems et al. (2003) were found, thus confirming the wide microbial diversity pool of tropical rainforests. Moreover, for each of the seven new clades, there is no intermixing between the Guyana and Guinea geographical origins. Host plant taxons can be distributed in different clades, as observed with either *Millettia* and *Albizia* in Guinea or *Erythrina* in Guyana, but some clades only contain strains from a specific host plant, such as *Chidlowia sanguinea*. Diversity within the *Bradyrhizobium* genus could thus be much wider than expected, each plant species or genus selecting its own bacterial partner.

The relevance of such studies in the framework of forest-restoration programs was tested in Guinea by setting up a field inoculation experiment using *Albizia adianthifolia, A. altissima, A. ferruginea, A. zygia* (strains STM916 , 922 and 923), *Milletia Zechiana, M. rhodantha* (strains STM926, 931 and 851), *Erythrophleum guineensis* and *E. ivorensis* (strain STM934). The height of the plants was measured after 4 months of growth in the nursery (Diabaté et al., 2005). There were 30 replicates per treatment and the data are presented in Table 1. They show a relative infra-generic specificity in plant-growth response between the strains as exemplified by *Albizia altissima* and *A. adianthifolia* inoculated with strain STM916 and exhibiting growth responses of 17.5% and 69.7% over the uninoculated control, respectively. The impact of inoculation may be extremely beneficial, reaching 113% over the controls with *Milletia zechiana* and strain STM 851.

References

Diabaté M et al. (2005) New Phytol. 166, 231–239.
Doyle JJ and Lückow MA (2003) Plant Physiol. 131, 900–910.
Doyle JJ et al. (1997) Am. J. Bot. 84, 541–554.
Willems et al. (2003) System. Appl. Microbiol. 26, 203–210.
Wojciechowski MF et al. (2004) Am. J. Bot. 91, 1846–1862.

INNOVATIVE MICROBIAL APPROACHES TO THE MANAGEMENT OF *ACACIA SENEGAL* TREES TO IMPROVE AND SUSTAIN GUM-ARABIC PRODUCTION IN SUB-SAHARAN AFRICA

D. Lesueur[1]*, A. Faye[2], S. Sall[3], J.-L. Chotte[4] and A. Sarr[2]**

[1]Forest Department of CIRAD, UPR – Ecosystems of Plantations, Laboratoire Commun de Microbiologie (UCAD/IRD/ISRA), Dakar, Senegal; [2]Laboratoire Commun de Microbiologie (UCAD/IRD/ISRA), Dakar, Senegal; [3]Laboratoire d'Ecologie Microbienne des Sols Tropicaux (IRD/UCAD/ISRA), Dakar, Senegal; [4]UR SeqBio, IRD – Carbon Sequestration and Soil Bio-functioning, Montpellier, France *Present address: Forest Department of CIRAD, UPR- Ecosystems of Plantations, Tropical Soil Biology and Fertility of Soils Institute of CIAT, Nairobi, Kenya; **Present address: UMR Microbiology and Geo-Chemistry of Soil, INRA/University of Bourgogne, Dijon, France

A potential solution to land degradation is to promote the utilization, regeneration, and planting of the native legume tree, *Acacia senegal*, which is the main species in the world producing the internationally traded gum-arabic. This tree species is very important for the livelihoods of many rural populations and has potential for wider use. Sarr et al. (2005a) assessed the rhizobial populations in soils from natural *A. senegal* forests in Mauritania and the Senegal River Valley. They showed that soils from Mauritania were less rich in native rhizobia than the soils from Senegal. In some cases, no native rhizobia were detected. The low numbers of native rhizobia in the soils of *A. senegal* forests may have a negative implication on the growth of the trees in very poor soils. In this case, we need to inoculate with selected rhizobial strains to improve the growth of the trees/seedlings. Sarr et al. (2005b) determined the most efficient methods of inoculation to improve nodulation and growth of *A. senegal* under greenhouse conditions. Their results showed that inoculation, using dissolved alginated beads containing rhizobia, significantly improved the growth of *A. senegal* and was better than the growth of plants in other treatments, such as liquid inoculum. The same authors demonstrated an interaction between tree provenances and the effect of inoculation with rhizobia. Their results suggest that: (1) it may be possible to improve growth and yield of *A. senegal* by careful selection of the symbiotic partner; and (2) nursery-grown seedlings of *A. senegal* should be inoculated, just after sowing, with dissolved alginate beads containing the selected rhizobia. Sarr et al. (2003c) assessed the efficiency of this

F. D. Dakora et al. (eds.), *Biological Nitrogen Fixation: Towards Poverty Alleviation through Sustainable Agriculture.*
© Springer Science + Business Media B.V. 2008

inoculum in the field and recorded a significant positive impact of rhizobial inoculation on total biomass production by *A. senegal* trees under irrigation.

The effect on gum-arabic production is currently being investigated. In natural conditions, two field trials set up in Northern Senegal with inoculated seedlings did not confirm the previous positive impact of rhizobial inoculation on the growth of *A. senegal* seedlings in plantation by comparison with non-inoculated seedlings; this, however, may be the consequence of a severe drought following months after the planting. Rhizobial inoculation of mature trees has also been tested to explore the possible relationship between tree growth and gum-arabic yield. Faye et al. (2006) demonstrated that it may be possible to increase significantly, in natural conditions in Northern Senegal, gum-arabic production through rhizobial inoculation of 10-year-old *A. senegal* trees. In Senegal and Niger, D. Diouf and A. Alzouma (personal communication, 2006) confirm the positive impact of the rhizobial inoculation of mature trees on the yield of gum-arabic. Other studies on the consequences of rhizobial inoculation of 10-year-old *A. senegal* trees on the soil microbial functioning in relation to gum-arabic yields (Marion et al., 2004; D. Lesueur et al., unpublished data, 2006) showed that the soil total microbial biomass under inoculated trees was significantly higher than that measured under non-inoculated trees. This difference was particularly evident at 0–50 cm depth. Furthermore, for the inoculated-tree treatment, our results indicated a positive relationship between gum-arabic production and ammonium produced (in the 0–25 cm soil layer) under laboratory conditions. Also, a positive relationship between nitrification and the production of gum of non-inoculated trees was observed.

In conclusion, our results suggest that the nitrogen activities in the soils under trees may be different depending on if they are inoculated or not with rhizobia, however, these results need to be confirmed by further analyses. If confirmed, inoculation could provide significant improvement in the livelihoods of local populations who can expect to get sustainable benefits through gum-arabic production and should open new ways to mitigate soil erosion and land degradation, which are often observed in the drylands.

References

Faye A et al. (2006) Arid Land Res. Manage. 20, 79–85.
Marion C et al. (2004) Proceedings of 11th Congress de l'Association Africaine sur la Fixation Biologique de l'Azote (AABNF), 22–27 November 2004, Dakar, Senegal.
Sarr A et al. (2005a) Microbial Ecol. 50, 152–162.
Sarr A et al. (2005b) New Forests 29, 75–87.
Sarr A et al. (2005c) Bois et Forêts des Tropiques, 283, 5–17.

ACACIA SPECIES USED FOR REVEGETATION IN SOUTH-EASTERN AUSTRALIA REQUIRE MORE THAN ONE MULTI-STRAIN RHIZOBIAL INOCULANT

A. McInnes[1] and J. Brockwell[2]

[1]Building K29, University of Western Sydney – Hawkesbury Campus, Locked Bag 1797, Penrith South DC, New South Wales 1797, Australia; [2]CSIRO Plant Industry, GPO Box 1600, Canberra, ACT 2601, Australia

Revegetation of degraded land requires either nursery production of tube stock for out-planting or successful establishment of plants after direct seeding, using a range of native species including *Acacia* spp. Inoculation of *Acacia* spp. with effective rhizobia benefits plant establishment and growth in the nursery and the field (Brockwell et al., 1999; Thrall et al., 2005). *Acacia* species exhibit strain specificity for effective nitrogen fixation (Burdon et al., 1999) and the use of a multi-strain inoculant is a practical method for ensuring effective nodulation over a broad host range. In this study, *Acacia* species commonly used for revegetation were inoculated with a multi-strain product in a commercial nursery trial. The trial aimed to determine: (i) whether there was strain selection by the host for nodulation; and (ii) whether the inoculant strains were competitive for nodulation in the presence of a background soil population of *Acacia* rhizobia.

Seeds of 15 *Acacia* species were sown into pasteurised commercial growth medium inoculated with Wattle Grow[TM] (Becker Underwood Pty Ltd, Somersby, New South Wales, Australia), containing four strains of *Bradyrhizobium* (CPBR1, CPBR2, CPBR3 and CPBR6). There were two inoculation treatments for each species. In the first treatment, Wattle Grow was mixed into the growth medium to give 2.33×10^5 inoculant rhizobia g^{-1} medium. In the second treatment, soil taken from beneath a natural stand of *Acacia acinacea*, as well as Wattle Grow, was mixed into the medium to give 2.27×10^4 soil rhizobia g^{-1} plus 2.22×10^5 inoculant rhizobia g^{-1}. Plants were assessed for shoot dry matter, extent of nodulation, and nodule occupancy by inoculant rhizobia and other strains determined by RAPD PCR fingerprinting of nodule isolates (Ballard et al., 2004). Data were analysed by ANOVA and, where F values were significant ($P < 0.05$), LSD was used as the *post hoc* test ($P < 0.05$).

For *A. dealbata*, *A. genistifolia*, *A. implexa*, *A. mearnsii* and *A. rubida*, 76–100% of plants inoculated with Wattle Grow became nodulated, with no change ($P > 0.05$)

F. D. Dakora et al. (eds.), *Biological Nitrogen Fixation: Towards Poverty Alleviation through Sustainable Agriculture*.
© Springer Science + Business Media B.V. 2008

observed when soil rhizobia were added to the growth medium. Nodules of each species were occupied by all four inoculant strains and also by other strains originating from the introduced soil population and from contamination (inevitable under commercial nursery conditions). The only evidence for inoculant strain selection was seen for *A. dealbata*, where strains CPBR2 and CPBR6 dominated nodules in the Wattle Grow treatment (60% and 24% nodule occupancy, respectively) and in the Wattle Grow plus soil treatment (35% and 44% nodule occupancy, respectively). There was no difference in shoot dry matter ($P > 0.05$) between the two treatments for any species. These data indicate that Wattle Grow is an effective multi-strain inoculant for these species.

In contrast, *A. salicina, A. buxifolia, A. decora* and *A. verniciflua* failed to nodulate fully with Wattle Grow despite the large population of rhizobia (2.33×10^5 g^{-1} medium) in the growth medium. This was apparent in increased nodule scores ($P < 0.05$) for all species when the soil rhizobia were added. For *A. salicina, A. buxifolia* and *A. decora*, increases in nodule score with added soil rhizobia corresponded with increased ($P < 0.05$) shoot dry matter. It is not clear whether the increase in nodulation and shoot dry matter is due to the increase in inoculum level with added soil or to minor changes in nodule occupancy by the inoculant strains and other strains. These species may benefit from an alternative more effective multi-strain inoculant.

A. iteaphylla, A. acinacea, A. pycnantha and *A. flexifolia* nodulated preferentially with strains other than the four inoculant strains. In the Wattle Grow treatment, nodule occupancy by other strains (presumably contaminants) ranged from 45% to 90% and, when soil rhizobia were added, nodule occupancy by other strains ranged from 63–90%. Not surprisingly, there was no difference ($P > 0.05$) in nodulation parameters or shoot dry weight between inoculation treatments. These species may benefit from the development of an alternative multi-strain inoculant.

A. doratoxylon failed to nodulate properly when inoculated with Wattle Grow (3% plants nodulated; nodule score 0.01 out of 5). Addition of soil rhizobia improved ($P < 0.05$) nodulation, but it remained suboptimal (52% plants nodulated; nodule score 0.62 out of 5), and there was no increase in shoot dry matter ($P > 0.05$). Nodule occupancy in the Wattle Grow plus soil treatment was 5% for CPBR2, 5% for CPBR6 and 90% for other strains (no isolates were obtained for the Wattle Grow treatment). These data show that *A. doratoxylon* has different strain requirements from the other species and may benefit from identification and incorporation of a specific *A. doratoxylon* strain in a multi-strain inoculant.

References

Ballard RA et al. (2004) Soil Biol. Biochem. 36, 1347–1355.
Brockwell J et al. (1999) Proc. 12th Aust. Nitr. Fix. Conf., pp. 73–75.
Burdon JJ et al. (1999) J. Appl. Ecol. 36, 398–408.
Thrall PH et al. (2005) J. Appl. Ecol. 42, 740–751.

MANAGEMENT OF SYMBIONTS IN CALLIANDRA (*CALLIANDRA CALOTHYRSUS* MEISN., LEGUMINOSAE) BASED AGROFORESTRY SYSTEMS TO IMPROVE GROWTH, PRODUCTIVITY, QUALITY OF FODDER, AND BIOLOGICAL N_2 FIXATION

D. W. Odee[1], D. Lesueur[2], X. Poshiwa[3], D. Walters[4] and J. Wilson[5]

[1]Kenya Forestry Research Institute, P.O. Box 20414-00200, Nairobi, Kenya;
[2]CIRAD-Foret, c/o Tropical Soil Biology and Fertility Institute (CIAT), World Agroforestry Centre, Nairobi, Kenya;
[3]Grasslands Research Station, Marondera, Zimbabwe;
[4]Scottish Agricultural College, Edinburgh, UK;
[5]Centre for Ecology and Hydrology, Penicuik, Midlothian, UK
(Email: dodee@africaonline.co.ke)

In eastern Africa, continuous land use, due to the shortage of land and the lack of resources to use chemical fertilisers, has induced a decline in small-scale farm productivity. Therefore, as part of an INCO-DEV project (Contract No. SAFSYS Project No. ICA4–2000-20037), the long-term goal is to find low-input ways of sustainably improving tree establishment, growth, and fodder production, which are likely to be widely adopted by the resource-poor small-scale farmers in eastern Africa. *Calliandra calothyrsus* (calliandra), an N_2-fixing agroforestry tree, is being used because of its increasing socio-economic and environmental importance for the resource-poor small-scale farmers in the region. In common with many other tree species, mycorrhizal and rhizobial inoculation is beneficial to its growth, but participatory research with tree nurseries, farmers, and other stakeholders is necessary to develop appropriate inoculation methods and improve uptake pathways. We have undertaken a multidisciplinary approach and utilised effective strains of rhizobia and arbuscular mycorrhizal (AM) inoculants to study the persistence of inoculants and evaluate their effects on growth, productivity, and quality of calliandra fodder in field conditions.

Standard microbiological procedures were used to trap, enumerate, isolate, characterize, and screen rhizobia and AM for effectiveness (Ingleby and Mason, 1999; Vincent, 1970). Several laboratory/glasshouse and field experiments were undertaken in various conditions in Kenya, Senegal, and Zimbabwe namely: (i) to evaluate inoculants of *Calliandra calothyrsus* under field conditions: (ii) to determine the persistence of efficient rhizobial and mycorrhizal inoculants for establishment and sustainable fodder

F. D. Dakora et al. (eds.), *Biological Nitrogen Fixation: Towards Poverty Alleviation through Sustainable Agriculture*.
© Springer Science + Business Media B.V. 2008

productivity of *Calliandra calothyrsus* under field conditions; and (iii) to establish adoption and uptake pathways for inoculant technology.

Assessments of indigenous rhizobia in various experimental field sites in Kenyan and Zimbabwean soils indicated low population sizes for calliandra ($\leq 3.1 \times 10^3$ rhizobia g^{-1} soil), whereas for common bean (*Phaseolus vulgaris*), also used as a calliandra intercrop in most farms, a wide variation in population sizes of bean rhizobia (1.0×10^2 to 1.8×10^5 rhizobia g^{-1} soil) was found. Calliandra inoculant strain KWN 35 effectively nodulated both calliandra and the bean crop under glasshouse conditions, hence it was used in the intercrop persistence studies. Similarly, 24 AM fungal species/types were isolated from the same experimental field sites in Zimbabwe and Kenya; *Glomus etunicatum* was the most dominant in all the soils analysed. AM fungi populations ranged from 2 to 145 'live' spores per 50 g of soil. Dual inoculation with both selected rhizobia and AM fungal inoculants improved growth of calliandra seedlings in nursery conditions. Nodule occupancy by inoculant rhizobial strains ranged from 40–78%, and the spread of AMF inoculant from calliandra to associated crops was also demonstrated in a trough experiment under glasshouse conditions. There was high microsymbiont diversity in the field sites from which effective inoculants were developed. In most cases, the inoculants enhanced growth of calliandra seedlings in both nursery and potted field-soil trials in Kenya, Zimbabwe and Senegal.

The results of several field experiments showed variable growth responses depending on the rhizobial and mycorrhizal inoculants, either singly or in combination. In a rock phosphate field trial in Kenya, there were significant inoculation effects on fodder quality in terms of the P, Mn, Fe, C and Na concentrations of twigs and leaves. In addition, a generally higher nodulation score was obtained in either single rhizobia (R) or dual inoculated (R + M) trees in rock phosphate-fertilized treatments compared to non-inoculated trees. The field trial that tested the effect of provenance (Flores or San Ramón) and inoculation (rhizobia, mycorrhiza, dual inoculation, or control) on tree regrowth (stem heights after previous cuts) was significantly influenced by provenance and the time of harvest (9, 12, 18 and 24 months after transplanting) at both the Grasslands and Africa University sites in Zimbabwe. San Ramón was generally taller (3.1–10.8%) than Flores at Grasslands at 9, 12, 18 and 24 months after transplanting in the field. In a rock phosphate field experiment at the Grasslands Research Station, there was significant ($p < 0.05$) effects of rock-phosphate application on tree height and root collar girth, and both nitrogen and phosphorus contents of the leaf and twig fractions. Rock phosphate application at 100 kg P/ha induced a 3% increase in tree height over the control (0 kg/ha).

References

Ingleby K and Mason PA (1999). Training manual: Production of arbuscular mycorrhizal inoculum in the glass house and nursery. Institute of Terrestrial Ecology, Midlotian, UK (21 pp.).
Vincent, JM (1970). A Manual for the Practical Study of Root-Nodule Bacteria. Blackwell, Oxford, UK (164 pp.).

LEGUME TREES IN THE COFFEE AGROECOSYSTEM OF PUERTO RICO

M. A. Arango, M. Santana, L. Cruz and E. C. Schröder

BNF Laboratory, P.O. Box 9030, Agronomy and Soils, University
of Puerto Rico, Mayagüez, PR 00681-9030, USA
(Email: eschroder5596@yahoo.com)

Coffee (*Coffea arabica*) probably originated in Abyssinia (now Ethiopia) and, after travelling around the world, arrived in Puerto Rico in 1755 (Mondoñedo, 1957). Up to the 1960s, it was grown under the shade of trees. In an effort to increase production, the trees were cut and the Department of Agriculture recommended the sun-grown system (Vicente-Chandler et al., 1968).

Coffee in Puerto Rico is produced mainly in small and medium farms with limited resources. The growing area is located in the mountainous region, mostly on acid and highly erodable soils (Conjunto Tecnológico, 1999). The negative effects of the "modern" monocropping system on the environment and coffee quality are well known (Perfecto et al., 1996). A small number of farmers never adopted the sun-grown systems and now environmental organizations and sustainably-minded farmers are supporting shade-grown coffee. Species composition of abandoned coffee plantations have been surveyed, but the species composition in active farms is unknown. To obtain detailed information about the tree species remaining in use for coffee shade, we conducted a survey in farms using shade growth. Additionally, we evaluated legume tree species for their potential to fix N_2.

Field work was done between June 2005 and March 2006. Farms were selected at random in the 22 municipalities with larger areas of shaded coffee, and 1,000 m^2 plots inventoried. The survey showed the presence of 66 tree species with diameter at breast height (DBH) >10 cm.

Soils representative of the coffee region (Humatas and Los Guineos Series) were placed in 20-L plastic pots and seeds of Mimosoideae and Papilionoideae planted. Plants were evaluated for growth at three months and harvested at six months. Nodule number, nodule weight, and acetylene-reduction activity (ARA) were measured. Bacteria were isolated from representative nodules in yeast extract manitol agar with Congo Red. The predominant species found were *Inga vera* (29.36%), *Andira inermis* (12.57%), and

F. D. Dakora et al. (eds.), *Biological Nitrogen Fixation: Towards Poverty Alleviation through Sustainable Agriculture.*
© Springer Science + Business Media B.V. 2008

Citrus sp (10.48%). Other species of *Inga* were important, and some fruit trees were also present. Species abundance varied at different elevations. This agrosystem would be classified as mixed (Moguel and Toledo 1999).

The selected legume tree species responded differently to the different soils. Among the Mimosoideae, *Inga vera* and *Inga spectabilis* showed the highest nodule number, whereas *Clitoria fairchildiana* and *Flemingia macrophyla* were the species with highest nodulation among the Papilionoideae. Several species (*Adenanthera pavonina*, *Anadenanthera peregrina* and *Parkia pedunculata*) lacked nodules in both soils at three and six months old. Nitrogenase activity was highest in *Flemingia macrophyla*, *Clitoria fairchildiana*, and *Erythrina variegata*. Approximately 80 strains of putative *Rhizobium* strains have been isolated so far.

Further studies are needed to select the best shade species for coffee production at different altitudes and soils in Puerto Rico.

References

Conjunto Tecnológico para la Producción de Café (1999) Publication 104 (Revised edition). Agric. Exp. Stn., University of Puerto Rico, Rio Piedras (29 pp.).
Moguel R and Toledo VM (1999) Conserva. Biol. 13, 11–21.
Mondoñedo JR (1957) Revista de Agricultura de Puerto Rico XLIV (2), 3–7.
Perfecto I et al. (1996) Bioscience 46, 598–608.
Vicente-Chandler J et al. (1968) Agric. Exp. Stn., University of Puerto Rico, Bulletin 211.

NODULATION ADAPTED TO HABITAT SUBMERGENCE

M. Holsters[1], W. Capoen[1,2], J. D. Herder[1], D. Vereecke[1], G. Oldroyd[2] and S. Goormachtig[1]

[1]Department of Plant Systems Biology, Flanders Institute for Biotechnology, Ghent University, B-9052 Gent, Belgium; [2]John Innes Centre, Norwich Research Park, Norwich NR4 7UH, England, UK

The semi-tropical legume *Sesbania rostrata* grows in the Sahel region of West Africa, thriving in temporarily flooded habitats, and exhibits versatility in nodulation. In both model and some crop legumes, nodulation occurs in root zone 1 with developing root hairs. Azorhizobia invade the epidermis via root hair curling (RHC) and membrane invagination. These responses are triggered by rhizobial Nod factors (NFs) perceived by plant receptors of the LysM-RLK type (Geurts et al., 2005). Intracellular root-hair invasion (zone-1 nodulation) also occurs in *S. rostrata* but only under well-aerated root growth (Goormachtig et al., 2004). Submergence interferes with RHC nodulation by accumulation of inhibiting ethylene concentrations. In natural habitats, nodules form by default at both lateral root bases (LRBs) of hydroponic roots and at the bases of stem-located adventitious rootlets. Here, azorhizobial invasion occurs via direct intercellular cortical colonization at epidermal cracks. The cortical invasion skips epidermal responses, requires less stringent NF features, and depends on ethylene to induce local cell death and so create space for microbial infection (D'Haeze et al., 2003).

A comparative transcriptome analysis of the different invasion processes was performed using cDNA-AFLP to analyze expression patterns (W. Capoen et al., unpublished data, 2006). Hierarchical clustering of 646 differential tags showed the onset of each invasion strategy by specific gene expression; at later stages, many commonalities corresponded to nodule primordium formation. One early common tag matched the *S. rostrata* ortholog of the *M. truncatula Lyr3* gene (Arrighi et al., 2006), a novel member of the family of LysM-RLKs to which the NF receptors also belong. RNAi knock-down of *SrLyr3* expression in transgenic *S. rostrata* roots interfered with nodule development; the nodules remained small and contained few infected cells. The *MtLyr3* promoter was fused to the *uidA* reporter and gave a novel expression pattern in outer apical cell layers of the indeterminate nodules. The number of RHC-specific tags was three-fold higher than crack entry-specific ones, suggesting that epidermal entry adds complexity to invasion or LRB nodule initiation is predisposed, perhaps due to the hormone landscape

F. D. Dakora et al. (eds.), *Biological Nitrogen Fixation: Towards Poverty Alleviation through Sustainable Agriculture.*
© Springer Science + Business Media B.V. 2008

(Mathesius et al., 2000). RHC-specific PIN2-, ADP-ribosylation factor-, and ROP-GTPase-related tags could be related to polar tip growth. Crack entry-specific hexose transporter and papain-like cysteine-protease tags suggest a role in local cell death.

Epidermal signal perception transduction for nodule initiation is well documented (Kaló et al., 2005). The hydroponic nodulation system of *S. rostrata* allows several aspects of NF signaling in sub-epidermal cells to be investigated. First, the leucine-rich repeat-type RLK *SrSymrk* is the ortholog of *LjSymrk* and *MtDmi2*, which are essential for nodulation and mycorrhization, whereas mutations result in an epidermal block and the receptors are involved in NF-signal transduction in the epidermis. Functional knockdown of *SrSymrk* expression did not prevent formation of LRB nodules, but the cortical infection threads showed aberrant features. The most striking phenotype was the loss of bacterial uptake for symbiosome formation, implying an essential role of *Symrk* in the internalisation process (Capoen et al., 2005). Second, the putative role of NFs downstream of the epidermal responses could be investigated because the cortical infection pockets, which form at LRBs, allow inter-bacterial complementation between mutant strains that are defective in different nodulation functions. Hence, a non-NF-producing strain can invade when complemented by a non-invading mutant with altered surface polysaccharides to give functional nitrogen-fixing nodules that were chaotically organized with loss of coordination between invasion and organ development. Nevertheless, the internalization of the NF⁻ bacteria was not hampered (Den Herder et al., 2007). Third, calcium spiking is an early integral NF response of epidermal root hairs (Oldroyd and Downie, 2006). We have studied calcium spiking in hydroponic root hair initials that occur at LRB positions of *S. rostrata*. These axillary initials respond to NFs by root hair outgrowth but are not invaded. By microinjection of a calcium-sensitive dye, calcium spiking responses were followed after NF application. The spiking pattern was different from patterns observed in zone-1 root hairs of *M. truncatula*: the shape was more symmetric and the period shorter (Capoen et al., in preparation). Application of jasmonate or ethylene inhibitors rendered the spiking pattern more similar to the *M. truncatula* zone-1 root hair pattern. Under these conditions bacterial invasion of axillary root hairs could be observed, demonstrating a strict correlation between spiking signature and invasion type.

References

Arrighi J-F et al. (2006) Plant Physiol. 142, 265–279.
Capoen W et al. (2005) Proc. Natl. Acad. Sci. USA 102, 10369–10374.
D'Haeze W et al. (2003) Proc. Natl. Acad. Sci. USA 100, 11789–11794.
Den Herder J et al. (2007) Mol. Plant-Microbe Interact. 20, 129–137.
Geurts R et al. (2005) Curr. Opin. Plant Biol. 8, 346–352.
Goormachtig S et al. (2004) Proc. Natl. Acad. Sci. USA 101, 6303–6308.
Kaló P et al. (2005) Science 308, 1786–1789.
Mathesius U et al. (2000) Mol. Plant-Microbe Interact. 13, 617–628.
Oldroyd GED and Downie JA (2006) Curr. Opin. Plant Biol. 8, 351–357.

PART 1D

STRESS TOLERANCE
AND BIOREMEDIATION

MECHANISM OF QUICK AND REVERSIBLE INHIBITION OF SOYBEAN NODULE GROWTH AND NITROGEN FIXATION ACTIVITY BY NITRATE AND ITS METABOLITES

T. Ohyama[1], A. Yamazaki[2], N. Yamashita[1], T. Kimura[2], S. Ito[2], N. Ohtake[1] and K. Sueyoshi[1]

[1]Faculty of Agriculture; [2]Graduate School of Science and Technology, Niigata University, Japan

The inhibitory effect of nitrate on both nodulation and nitrogen fixation of leguminous plants has been well recognized, but the precise mechanism has not been fully elucidated. It was suggested that there are multiple effects of nitrate inhibition, such as a decrease in nodule number, nodule mass, and N_2-fixation activity, as well as the acceleration of nodule senescence or disintegration. In addition, the effect of nitrate on nodule growth is influenced by nitrate concentration, placement and treatment period, as well as legume species. Many hypotheses have been proposed for the cause of nitrate inhibition of nodulation and N_2 fixation, e.g., carbohydrate-deprivation in nodules, feed-back inhibition by nitrate metabolites, and the decreased O_2 diffusion into nodules which restricts the respiration of bacteroids. We have found the quick and reversible inhibition of both soybean root-nodule growth and nitrogen-fixation activity by 5 mM nitrate using solution culture (Fujikake et al., 2002, 2003). The increase in the diameter of root nodules was almost completely stopped after 1 day of supplying 5 mM nitrate, however, nodule growth quickly returned to the normal growth rate following withdrawal of nitrate from the solution. On the other hand, the systemic effect of a long-term supply of nitrate from the lower part of the root system on the growth of nodules attached to the upper part of roots depended on the concentration of nitrate (Yashima et al., 2003, 2005). The continuous supply of 5 mM nitrate decreased the nodule mass in the upper part, but a 1 mM supply increased the nodule dry weight through the promotion of plant growth.

In this study, the quick and reversible nitrate inhibition was also observed at the lower nitrate concentration (1 mM) in solution. The depression of nodule growth was observed within several hours after 1 mM nitrate treatment, and recovered completely after withdrawal of nitrate from solution within several hours. On the other hand, apparent nitrogen-fixation activity of intact plants, monitored by H_2 evolution, showed that 1 mM nitrate gave the inhibitory effect after 1 or 2 days of nitrate supply. These

F. D. Dakora et al. (eds.), *Biological Nitrogen Fixation: Towards Poverty Alleviation through Sustainable Agriculture.*
© Springer Science + Business Media B.V. 2008

results indicate that the inhibitory effect of nitrate on nodule growth is more immediate than the effect on nitrogen-fixation activity. The addition of the metabolites (1 mM as N) of nitrate was compared with direct nitrate supply. The addition of ammonium and glutamine to the solution also exhibited quick and reversible inhibition of nodule growth and acetylene-reduction activity, although the inhibitory effects were moderate compared with nitrate. A similar effect was observed on addition of 1 mM as N of either urea or asparagine.

^{13}C and ^{15}N partitioning in nodulated soybean was traced to evaluate the contribution of C and N transport to the inhibitory effect on nodule growth. A culture solution containing a ^{15}N-labeled 1 mM N source [$Na^{15}NO_3$, ($^{15}NH_4)_2SO_4$, both amide^{15}N and amino^{15}N glutamine, or ^{15}N-urea] was supplied for 3 days from 12 DAP (days after planting) until 15 DAP. The whole shoot of the 14-DAP plant was enclosed in a plastic bag (about 2 L), and 2 mL of $^{13}CO_2$ (99.0 atom %) was injected into each plastic bag. Plants were exposed to the $^{13}CO_2$ for 60 min. At 26 h after $^{13}CO_2$ exposure, plants were harvested and the ^{15}N and ^{13}C enrichment in the plant parts was determined using an elementary analyzer-IRMS coupling system.

The total amount of ^{15}N of the plant supplied with glutamine was highest of all the treatments, and the glutamine-derived ^{15}N was highly accumulated in the roots. The total amount of ^{15}N in the nitrate and ammonium treatments was almost the same. The total amount of ^{15}N was lowest in the plants with urea treatment. The amount of ^{15}N ($\mu g/g$ DW) in nodules was higher in glutamine (43.5) and ammonium (40.0) treatments than in the nitrate (9.9) and urea (5.2) treatments. On the other hand, the average amount of ^{13}C ($\mu g/g$ DW) in nodules was highest in the control (4.8), followed by ammonium (1.3), urea (1.2), glutamine (1.0) and nitrate (0.7) treatments. Based on the results obtained, the quick and reversible nitrate (and its metabolites) inhibition of nodule growth and nitrogen-fixation activity may be due to systemic regulation, and primarily associated with the changes in carbohydrate supply to the nodules rather than the feed-back inhibition of nitrate metabolites.

References

Fujikake et al. (2002) Soil Sci. Plant Nutr. 48, 211–217.
Fujikake et al. (2003) J. Exp. Bot. 54, 1379–1388.
Yashima et al. (2003) Soil Sci. Plant Nutr. 49, 825–834.
Yashima et al. (2005) Soil Sci. Plant Nutr. 51, 981–990.

INHIBITION OF SYMBIOTIC NITROGEN FIXATION BY DARK CHILLING IN SOYBEAN

P. D. R. van Heerden[1], U. Schlüter[2], P. W. Mokwala[3], K. Kunert[2] and C. H. Foyer[4]

[1]Section Botany, North-West University, Potchefstroom 2520, South Africa; [2]Forestry and Agricultural Biotechnology Institute (FABI), University of Pretoria, Pretoria 0002, South Africa; [3]Department of Biodiversity, University of Limpopo, Sovenga 0727, South Africa; [4]School of Agriculture, Food and Rural Development, The University of Newcastle upon Tyne, Newcastle upon Tyne, NE1 7RU, UK

Legume crops are an important source of protein, oil and secondary metabolites and they are also used as a natural nitrogen source in agriculture, particularly in Africa, because of presence of nitrogen-fixing bacteria in specialized organs called "nodules". The symbiotic association between the plant and nitrogen-fixing bacteria in nodule formation has been the subject of intensive study but much less information is available on the mechanisms that cause the breakdown of symbiosis particularly during stress. Although the symptoms and progression of nodule senescence has been described, much remains to be discovered regarding the mechanisms that trigger the end of symbiosis and the genes and proteins that underpin nodule senescence (Puppo et al., 2005).

As in other warm-climate crop species, such as cucumber, tomato and maize, exposure to sub-optimal night growth temperatures (dark chilling) in the range from 6°C to 15°C inhibits photosynthesis and limits growth in soybean [*Glycine max* (L.) Merr.]. The sensitivity of soybean to dark chilling causes changes in metabolism with decreased growth, development, and yield (Musser et al., 1983, 1984; Van Heerden et al., 2003). A single night of dark chilling, with minimum temperatures of 8°C, is sufficient to inhibit pod formation in soybean (Hume and Jackson, 1981). We have previously characterised the adverse effects of dark chilling on soybean leaf photosynthesis and metabolism (Van Heerden et al., 2003). However, it has long been recognised that dark chilling causes rapid inhibition of symbiotic nitrogen fixation (SNF) in the nodules (Duke et al., 1979; Walsh and Layzell, 1986). Exposure to even relatively mild temperatures (15°C) can inhibit SNF by up to 45% (Walsh and Layzell, 1986). However, the mechanisms that contribute to the high sensitivity of SNF to chilling remain to be elucidated. One possibility is that chilling modifies nodule O_2 homeostasis (Vessey et al., 1988).

F. D. Dakora et al. (eds.), *Biological Nitrogen Fixation: Towards Poverty Alleviation through Sustainable Agriculture.*
© Springer Science + Business Media B.V. 2008

Chilling stress increases nodule O_2 concentrations (O_i) and the fractional oxygenation of leghemoglobin (Kuzma et al., 1995). High O_i values will inhibit nitrogenase activity directly as the enzyme is extremely sensitive to high O_i.

Soybean genotypes vary greatly in their ability to tolerate low growth temperatures and there is much variation in the capacity to maintain optimal metabolism under such conditions. However, relatively few studies to date have investigated genotypic variations in chilling-induced inhibition of SNF (Lynch and Smith, 1994). Recently, Strauss et al. (2006) evaluated the dark chilling response of 30 South African soybean genotypes on the basis of changes induced in fast fluorescence rise kinetics during exposure to dark chilling. Large genotypic differences in dark chilling response were observed, especially in the two genotypes, PAN809 and Highveld Top, which were subsequently classified as dark chilling "sensitive" and "tolerant", respectively (Strauss et al., 2006). In studying the effects of dark chilling in these genotypes, we have found a marked inhibition of nitrogenase activity, leading to a rapid decline in shoot and nodule ureide contents and loss of photosynthetic competence in the sensitive genotype PAN809, but not in the tolerant genotype Highveld Top. Conversely, dark chilling does not affect nodule carbohydrate contents or metabolism in either genotype. Nodulation rates were unaffected by chilling treatments and new nodules were gradually able to compensate for loss of SNF capacity in the chilled organs, which never recovered. The marked genotypic differences in the response of nitrogenase activity to dark chilling suggest that nitrogenase activity and/or ureide levels might have a crucial role in the regulation of nodule senescence.

References

Duke SH et al. (1979) Plant Physiol. 63, 956–962.
Hume DJ and Jackson AKH (1981) Crop Sci. 21, 689–692.
Kuzma MM et al. (1995) Plant Physiol. 107, 1209–1216.
Lynch DH and Smith DL (1994) Physiol. Plant. 90, 105–113.
Musser RL et al. (1983) Plant Physiol. 73, 778–783.
Musser RL et al. (1984) Plant Physiol. 74, 749–754.
Puppo A et al. (2005) New Phytol. 165, 683–701.
Strauss AJ et al. (2006) Environ. Exp. Bot. 56, 147–157.
Van Heerden PDR et al. (2003) Plant Cell Environ. 26, 323–337.
Vessey KJ et al. (1988) Physiol. Plant. 73, 113–121.
Walsh KB and Layzell DB (1986) Plant Physiol. 80, 249–25.

DO NODULE PHOSPHATASE AND PHYTASE LINK WITH THE PHOSPHORUS USE EFFICIENCY FOR N_2-DEPENDENT GROWTH IN *PHASEOLUS VULGARIS* ?

J. J. Drevon[1], S. Kouas[2], L. Amenc[1], N. Alkama[1], S. Beebe[3], L. Bouhmana[1], A. Lopez[4], C. Plassard[1], P. Rodino[5] and G. Viennois[1]

[1]INRA-Agro M, Montpellier, France; [2]INRST, Hammam Lif, Tunisia; [3]CIAT, Cali, Colombia; [4]UNAM, Cuernavaca, Mexico; [5]MBG-CSIC Pontevedra, Spain

Grain legumes can contribute to cropping systems through their ability to fix atmospheric N_2 in their root-nodules. However, the symbiotic nitrogen fixation process requires additional phosphorus. Thus, the production of grain legumes, particularly common bean, is limited by P-deficiency in many soils, mostly in tropical and Mediterranean areas. Recombinant inbred lines (RILs – F8) from a cross of common bean parents BAT477 with DOR364 were selected for improved phosphorus-use efficiency and N_2-dependent growth in glasshouse hydro-aeroponic culture and tested in farmers' field. The present work with *Phaseolus vulgaris* as a model grain legume shows a significant difference in the overall phosphatase activity in nodules, including a relatively high level of phytase activity, between two recombinant inbred lines that contrast in their efficiency in utilization of P for symbiotic nitrogen fixation.

In order to assess where the phosphatase activities are most expressed in nodules, an *in situ* PCR was performed with degenerate primers designed from the most conserved part of the sequence of various acid phosphatases. Figure 1 shows that the expression of acid phosphatase varies among tissues. It is particularly high in the cells of the inner cortex, where the specific nodule carbonic anhydrase is located (Schumpp et al., 2003). Thus, it may link with the change of nodule permeability that is associated with the adaptation to P deficiency (Vadez et al., 1996) and may be osmoregulated. The acid phosphatase signal was increased by P deficiency, and was higher in cortical cells of the more efficient recombinant inbred lines for P utilization, namely RIL115 rather than in RIL147 (data not shown). Acid phosphatase activities were higher in the nodule extract from RIL115 compared to RIL147. Relatively high phytase activity was also detected in these extracts by measuring the amount of Pi generated from phytate supplied to the extract. This nodule pytase activity was again higher in the most efficient recombinant inbred line, i.e., higher in the extract from RIL 115 compared to RIL 147 (Figure 2).

F. D. Dakora et al. (eds.), *Biological Nitrogen Fixation: Towards Poverty Alleviation through Sustainable Agriculture.*
© Springer Science + Business Media B.V. 2008

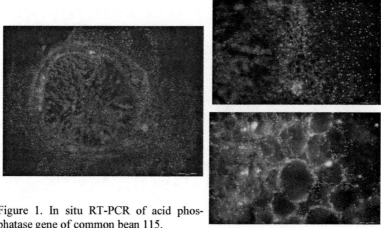

Figure 1. In situ RT-PCR of acid phosphatase gene of common bean 115.

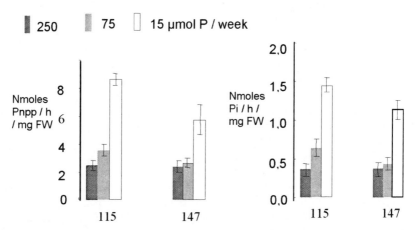

Figure 2. Acid phosphatase and phytase activities in contrasting RILs of the BAT477 × DOR364 cross.

Thus, nodule acid-phosphatase activity may contribute to the efficiency of P use for N₂ fixation in nodules. The nodulated-root phytase activity, described here for the first time, will be explored for additional RILs.

References

Schumpp O et al. (2003) Agronomie, 23, 1–6.
Vadez V et al. (1996) Plant Physiol. Biochem. 34, 871–878.

THE ALTERNATIVE SIGMA FACTOR RPOH2 IS REQUIRED FOR SALT TOLERANCE IN *SINORHIZOBIUM* SP. STRAIN BL3

P. Tittabutr[1], W. Payakapong[1], N. Teaumroong[1], N. Boonkerd[1], P. W. Singleton[2] and D. Borthakur[3]

[1]School of Biotechnology, Suranaree University of Technology, Nakhon Ratchasima 30000, Thailand; [2]Department of Tropical Plant and Soil Sciences, University of Hawaii at Manoa, Honolulu, HI 96822, USA; [3]Department of Molecular Biosciences and Bioengineering, University of Hawaii at Manoa, Honolulu, HI 96822, USA

The objectives of this investigation were to isolate the *rpoH2* gene encoding an alternative sigma factor from *Sinorhizobium* sp. BL3 and to determine its role in exopolysaccharide (EPS) synthesis, salt tolerance, and symbiosis with *Phaseolus lathyroides*. The *rpoH2* gene of *Rhizobium* sp. strain TAL1145 is known to be required for EPS synthesis and effective nodulation of *Leucaena leucocephala*.

Three overlapping cosmid clones containing the *rpoH2* gene of BL3 were isolated by complementing an *rpoH2* mutant of TAL1145 for EPS production. From one of these cosmids, *rpoH2* of BL3 was identified within a 3.0-kb fragment by subcloning and sequencing. The cloned *rpoH2* gene of BL3 restored both EPS production and nodulation defects of the TAL1145 *rpoH2* mutants.

Three *rpoH2* mutants of BL3 were constructed by transposon-insertion mutagenesis. These mutants of BL3 grew normally in complete or minimal medium and were not defective in EPS synthesis, nodulation, and nitrogen fixation, but they failed to grow in salt-stress conditions. The mutants complemented with cloned *rpoH2* from either BL3 or TAL1145 showed higher levels of salt tolerance than BL3. The expression of *rpoH2* in BL3 started increasing during the exponential growth phase and reached the highest level in the mid stationary phase.

These results indicate that RpoH2 is required for salt tolerance in *Sinorhizobium* sp. BL3, and it may have additional roles during the stationary phase.

F. D. Dakora et al. (eds.), *Biological Nitrogen Fixation: Towards Poverty Alleviation through Sustainable Agriculture.*   95
© Springer Science + Business Media B.V. 2008

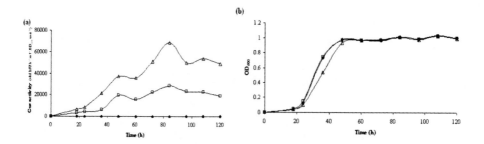

Figure 1. Expression of *rpoH2* at different growth phases of wild-type BL3 (closed diamond); the *rpoH2* mutant RUH180 (open square); and the transconjugant BL3: pUHR345 (open triangle) in YEM medium at 28°C. Expression of *rpoH2* is as (a) β-glucuronidase activity, and (b) cell density (OD at 600 nm).

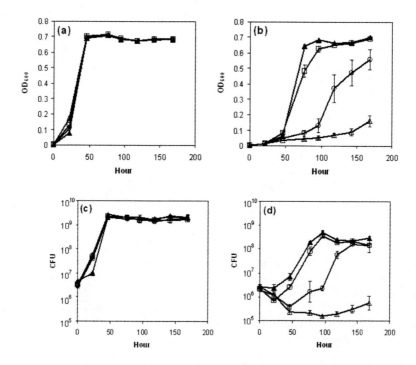

Figure 2. The growth of BL3 (open circle), the *rpoH2* mutant RUH180 (open triangle), the *rpoH2* mutant RUH180 complemented with pUHR37 (close triangle), and pUHR339 (open square) in YEM broth (a and c), and YEM broth containing 0.5 mM NaCl (b and d) with shaking at 28°C. The growth data were recorded as OD at 600 nm (a and b), and as viable cell counts (c and d) from the same experiment. Each data point represents the mean of three replications and the vertical bars indicate standard deviations.

ROOT-NODULE BACTERIA OF ARID-ZONE LEGUMES FOR USE IN REHABILITATION IN THE SHARK BAY WORLD HERITAGE AREA

Y. Hill[1], G. W. O'Hara[1], E. Watkin[2] and K. Dixon[3]

[1]Centre for Rhizobium Studies, Murdoch University, Murdoch, Western Australia; [2]School of Biomedical Sciences, Curtin University of Technology, Perth, Western Australia; [3]Kings Park and Botanic Garden, Perth, Western Australia

The Shark Bay World Heritage Property (SBWHP) is a transition zone between the South West and Eremaean Biogeographic regions and was listed in 1991 due to its great geological, botanical and zoological importance (UNESCO, 2002). Shark Bay Salt is located within the SBWHP. Its lease area includes Useless Inlet and Useless Loop and there has been salt-mining production since 1965 (EPA, 1991). Over this time, borrow pits have been mined in areas surrounding the evaporation ponds, the majority of which were decommissioned over 15 years ago. Many of these pits remain in a highly disturbed state when compared to the surrounding undisturbed flora (Figure 1a, b).

The detrimental effect borrow-pit mining has had on the root-nodule bacteria numbers present was determined by Most Probable Number analysis on soils collected from a

Figure 1a. Undisturbed area adjacent to borrow pit G; June. 2006.

Figure 1b. Borrow pit G south west sector; June, 2006.

F. D. Dakora et al. (eds.), *Biological Nitrogen Fixation: Towards Poverty Alleviation through Sustainable Agriculture.*
© Springer Science + Business Media B.V. 2008

representative borrow pit and from adjacent undisturbed soils. Unlike undisturbed temperate area Australian soils, which can contain root-nodule bacteria densities measuring
thousands per gram of soil (Thrall et al., 2001), it was found that very low numbers
were present in the undisturbed soils at the time the samples were collected and negli-
gible numbers were detected in the disturbed soils. The deficiency of symbiotic microbes
in the borrow pits may be an additional contributing factor to the lack of recruitment of
flora from the surrounding pristine areas.

There is a significant contribution of nitrogen by *Acacia* species to arid and semi-arid
ecosystems that are dominated by these species (Beadle, 1964). *Acacia ligulata* and
A. tetragonophylla were selected for this study as pioneer species for introduction
into borrow pits that are yet to be rehabilitated effectively. Their selection was based on
their presence in the surrounding undisturbed vegetation associations and the availability of sufficient provenance seed. These two species, as well as *A. rostellifera* and
Templetonia retusa, which are also distributed in the area but for which no provenance
seed is available, were grown in glasshouse conditions in undisturbed soils collected
from three sites adjacent to borrow pits. The majority of the trap hosts formed effective
nodules. Of the 368 isolates obtained, 32 have been pre-selected for further study based
on the vigour of the trap plants.

Numerous desiccated nodules were collected from *A. ligulata* growing in the Shark
Bay Salt lease area and 11 isolates obtained. From re-authentication and effectiveness
trials, it is apparent that there are numerous strains capable of producing vigorous
growth. One of the isolates, after inoculation of *A. ligulata*, resulted in a mean shoot dry
weight 2.75-times greater than the nitrogen-starved uninoculated control. With further
trials, it is envisaged that several strains could be developed for a multi-strain inoculum,
which would have the potential to improve the survival and vigor of plants introduced
into the degraded areas of Shark Bay Salt Lease area. This inoculum might also be of
use in the rehabilitation strategies of other future major developments in the Pilbra
region.

References

Beadle NCW (1964) Proc. Linn. Soc. NSW. 89, 273–286.
EPA (1991) EPA WA Bulletin No. 542.
Thrall PH et al. (2001) Ecol. Man. Rest. 2, 233–235.
UNESCO (2002) World Heritage Periodic Reporting Asia-Pacific Region, 2003.

EFFECT OF VEGETATION CLEARING ON THE NITROGEN CYCLE AND WATER RESOURCE QUALITY IN SOUTH AFRICA

F. Rusinga, S. Israel and G. Tredoux

Council for Scientific and Industrial Research, P.O. Box 320, Stellenbosch 7599, South Africa

Clearing nitrogen-rich vegetation is known to result in increased mineralization and consequently to generate nitrate pools in excess of biota requirements. Nitrate from this pool can potentially leach into water resources. Nitrates in water pose a risk to humans and animals that depend on the resource for drinking water. The CSIR and the Universities of the Western Cape (UWC) and Stellenbosch (US) are currently conducting a 2 year (2006–2008) research project to investigate the impacts that the clearing of leguminous invasive alien vegetation, particularly *A. saligna* (Port Jackson willow), has on groundwater and surface water quality. Groundwater resources offer an alternative source, often of better quality than surface sources, such as ponds and streams. But it can become contaminated and nitrates are among the key groundwater contaminants. At high levels (20 mg N/L (as N) is the upper limit in South Africa), ingested nitrate causes methaemoglobinaemia, which may be fatal for infants and livestock. Most nitrate problems are due to pollution but, in semiarid to arid regions, high groundwater nitrate has been reported from areas where anthropogenic influences can be excluded. This indicates that also under natural conditions, nitrogen fixation may be followed by nitrification with subsequent leaching of nitrate into the groundwater. However, anthropogenic activities, such as deforestation, can modify the N-cycle by producing excessive mobile nitrate. Faillat and Rambaud (1991) linked the groundwater nitrogen in the rural area of Côte d'Ivoire, West Africa, to deforestation.

Conrad et al. (1999) studied the application of sewage sludge to land in an experimental plot located near Atlantis in South Africa. Following the first rain, they observed an unexpectedly higher nitrate level at a control site, which was located in uncultivated area with degrading *A. Saligna*, than at a site located within the cultivated area, where sludge and inorganic fertilizer was applied. The degrading *A. saligna* biomass was identified among the suspected factors accounting for the observations at the control site.

F. D. Dakora et al. (eds.), *Biological Nitrogen Fixation: Towards Poverty Alleviation through Sustainable Agriculture.*
© Springer Science + Business Media B.V. 2008

The joint CSIR, UWC, and US research is intended to follow up on the findings of Conrad et al. (1999) to more specifically investigate the impacts of clearing the alien vegetation on groundwater quality in light of the nationwide Working for Water (WfW) programme, which is aimed at restoring the ecosystem integrity. The objectives of this research are to assess systematically the nitrogen stocks in soils under alien vegetation and nitrogen movement in soils, sub-soils, groundwater, and surface water before and after clearing the vegetation and the impact on the quality of water resources.

The hypothesis is that clearing the invasive alien vegetation produces a large episodic input of fresh litter rich in N, it eliminates a large N sink, and it changes the local micro-climate as well as soil chemistry and physics existing under the alien vegetation, thereby modifying the nitrogen cycling within the plant-microorganism-soil system. The effect of clearing vegetation on the nitrogen cycle has been studied (Vitousek, 1981). Figure 1 illustrates the expected net effect on the internal nitrogen cycle of clearing nitrogen-rich vegetation. Excessive mineralization may generate nitrate pools in excess of the requirements of the biota and this pool can potentially leach to below-ground water resources.

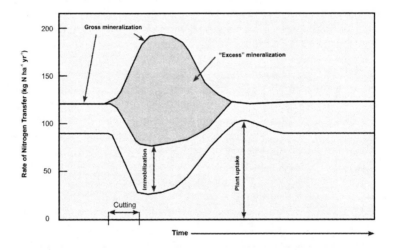

Figure 1. The hypothetical effect of clear-cutting to the internal N of deciduous forest (Vitousek, 1981).

References

Conrad et al. (1999) Water Research Commission (WRC) Report No. 641/1/99, 6-6.97.
Faillat JP and Rambaud A (1991) Environ. Geol. Water Sci. 17 (2), 133–140.
Vitousek PM (1981) Terrestrial Nitrogen Cycles, Ecol. Bull. (Stockholm) 33, 631–642.

BIORHIZOREMEDIATION OF HEAVY METALS TOXICITY USING *RHIZOBIUM*-LEGUME SYMBIOSES

E. Pajuelo[1], M. Dary[1], A. J. Palomares[1], I. D. Rodriguez-Llorente[1], J. A. Carrasco[2] and M. A. Chamber[2]

[1]Microbiology Department, Faculty of Pharmacy, University of Seville, Professor Garcia Gonzalez 2, 41012 Seville, Spain; [2]CIFA Las Torres, IFAPA, Junta de Andalucia, Alcala del Rio, 41200 Seville, Spain. (Email: mangel.chamber@juntadeandalucia.es or palomares@us.es)

Massive amounts of acidic waters and mud ($pH \sim 3$) containing toxic metals, such as zinc, lead, arsenic, copper, cadmium, mercury and selenium, were released as a consequence of the mine tailings spill accident in Aznalcollar close to Seville in Spain along a 62-km stretch of the Guadiamar river (Grimalt et al., 1999). Approximately 2,000 ha of croplands were contaminated. In spite of work done to clean the area, contamination by heavy metals (HM) in the soil in the area is still high. Although mechanical and chemical processes can be adopted to deal with toxic spills, most traditional remediation methods do not provide acceptable solutions for the removal of metal from soils. Contamination by As and heavy metals, such as Cd, Cu, Pb and Zn, is persistent (Martín et al., 2002).

More recently, phytoremediation has emerged as a promising ecoremediation technology, which uses plants and their associated rhizospheric microorganisms to remove pollutants from contaminated environments (Kuiper et al., 2004; Click, 2004). In this context, we have recently proposed the use of the symbiotic *Rhizobium*-legume interaction as a bioremediation system to clean-up metals and metalloids from contaminated soils (Carrasco et al., 2005). An experimental highly polluted area has been established by Andalusia government for research and several *Rhizobium*-legume symbiosis interactions have been set up here in an *in situ* phytoremediation experimental plot at the Guadiamar riverbed near the Aznalcollar mine spills.

The levels of contaminating HM have been determined in the contaminated area and found to be well over the limits established by Andalusian environmental law (Table 1). Also, a wide collection of *Rhizobium* and some other bacteria (*Pseudomonas* sp. and *Ochrobactrum* sp.) have been isolated from the rhizosphere of legumes growing on the

F. D. Dakora et al. (eds.), *Biological Nitrogen Fixation: Towards Poverty Alleviation through Sustainable Agriculture.*
© Springer Science + Business Media B.V. 2008

polluted area (Table 2). Many of these *Rhizobium* strains showed very high resistance to as and high nitrogen-fixation effectiveness (Carrasco et al., 2005). Further, the number of nodules decreased in the presence of HM, but not the nitrogen-fixation activity.

Table 1. Determination of the concentration of toxic elements in soils (ppm).

Element	Non-contaminated soil	Contaminated soil		
		Zone 0	Zone 1	Zone 2
As	6.6	320	100	504
Pb	25.8	295	122	1,229
Cu	23.8	176	65	1,059
Zn	36.6	932	281	1,317
Cd	1.1	6.0	2.5	11.0
Ni	17.7	25.1	15.7	34.0
Cr	28.4	28.9	39.1	44.4

Table 2. Resistance of *Ochrobactrum* sp. and other rhizospheric strains to heavy metals.

Isolate	As (mM)	Cd (mM)	Co (mM)	Cu (mM)	Hg (μM)	Ni (mM)	Pb (mM)	Zn (mM)
Rhizobium sp.	1–10	0.1–1	0.1–4	0.5–2	5–20	0.25–1	3–6	0.5–2
Pseudomonas sp.Az13	3	<0.25	1	3.5	25	n.d.	2	2
Ochrobactrum sp.	2–3	1–1.5	1	2.5–3	50–75	0.5	6–7	10
Acinetobacter rhizospherae	2	0.25		4	75	n.d.	n.d.	10
Chryseobacterium sp.	4	2	0.5	4	150	n.d.	n.d.	12.5

An "*in situ*" pilot rhizoremediation experiment has been performed at the contaminated area. Several legume plants (native and commercial legumes) belonging to seven genera were grown at the contaminated area and inoculated with compatible *Rhizobium* and other rhizospheric bacteria that are resistant to As and several HM in order to establish the phytoremediation capacity of these legumes as well as the effect of the inoculated bacteria. Among the legume species studied, *Vicia sativa* showed the highest capability to accumulate and translocate HM, although lupines produced the largest biomass (Table 3).

Coinoculation of *Lupinus* plants with *Bradyrhizobium* sp. and other rhizospheric bacteria isolated from the contaminated area improved yields (Table 4), but decreased the accumulation of HM by plants (Table 5). *Ochrobactrum* sp. strains isolated from legumes roots at the polluted area showed high resistance to HM and were able to nodulate alfalfa (Table 6).

Table 3. Comparison of total phytoremediation capacity of native and commercial legumes (n.d., not detected).

Legume	Inoculant	Total biomass	Total phytoremediation (mg/m^2)			
			Cu	Pb	Cd	Zn
Lupinus albus	*Bradyrhizobium sp. lupinus 750*	3,800	38	n.d.	n.d.	304
Lupimus luteus	*Bradyrhizobium sp. lupinus 750*	3,750	52.5	n.d.	3.7	562
Vicia sativa	*Vicz1.1*	1,300	32.5	19.5	5.2	1,040
Trifolium subterraneum	*RTL15*	800	20	11.2	2.6	200

Table 4. Effects of inoculation with *Rhizobium* and other rhizospheric bacteria on yields of lupines cropped on the polluted plot.

Inoculant	Nitrogen Content (%)	Plant biomass (g)	Total biomass (g)
Non inoculated	1.73	11.4	307
Bradyrhizobium sp. *Lupinus*	2.40 (139%)	17.1 (150%)	377
Bradyrhizobium sp.*Lupinus +* *Ochrobactrum* sp. *OmedZ 2.1*	2.50 (144%)	17.8 (156%)	515
Bradyrhizobium sp *lupinus +* *Ochrobactrum* sp. OmedZ 2.1 + *Pseudomonas* sp. Az13	2.40 (139%)	23.8 (209%)	691

Table 5. Effect of co-inoculation with *Rhizobium* and other rhizospheric bacteria on metal accumulation by plants of lupines.

Inoculant		Accumulation (ppm)			
		Cd	Cu	Pb	Zn
Non inoculated	Shoots	2.1	27.5	8.4	447
	Roots	4.0	70.0	50.0	642
Bradyrhizobium lupini	Shoots	1.6	21.5	n.d.	472
	Roots	4.0	65.0	27.0	755
Bradyrhizobium lupini +Ochrobactrum sp. Shoots		1.6	23.2	5,6	421
	Roots	3.0	72.0	34.0	462
Bradyrhizobium lupini + Ochrobactrum sp. *+Pseudomonas sp.* Shoots		0.6	17.8	n.d.	185
	Roots	3.0	49.0	21.0	320

Table 6. Most important features of nodulation of *Medicago sativa* by *S. meliloti* and *Ochrobactrum sp.* strains resistant to HM.

Inoculant	Number of nodules	1st nodule (cm)	2nd nodule (cm)	3rd nodule (cm)	Primordia (dpi)	Nodules (dpi)
Sinorhizobium meliloti MA11	2–3	1.5–3.5	3–4.5	4–5.5	5–6	9–10
Ochrobactrum sp. Azn6-2	1–2	7–9	8.5–13	–	20–21	24–26
Ochrobactrum sp. ESC1	2–3	10–11.5	10.5–12	11–12	22–24	27–29

Conclusions

1. *S. meliloti* strains isolated from polluted soils show high resistant to As and were symbiotically effective, however, the number of nodules decreased in HM-polluted soils.
2. *Ochrobactrum* sp. Strains are highly resistant to HM and possess *nifH* and □*od* rhizobial genes, being able to form late and ineffective nodules in alfalfa.
3. *Vicia sativa* was the legume with the higher accumulation and translocation capability, although lupines produced the largest biomass.
4. Co-inoculation of *Lupinus* plants with *Bradyrhizobium* sp. and other rhizospheric bacteria isolated from the contaminated area improved yields, but decreased accumulation of HM.

References

Carrasco et al. (2005) Soil Biol. Biochem. 37, 1131–1140.
Click (2004) Nature Biotechnol. 22, 526–527.
Grimalt et al. (1999) Sci. Total Environ. 242, 3–11.
Kuiper et al. (2004) Mol. Plant-Microbe Interact. 17, 6–15.
Martín et al. (2002) Nucl. Instrum. Method. Phys. Res. 188, 102–105.

PART 1E

SUMMARY PRESENTATIONS

THE COMMERCIALISATION OF SMALL HOLDER AGRICULTURE THROUGH BIOLOGICAL NITROGEN FIXATION (BNF) IN SOYBEAN

E. Musimwa and O. Mukoko

Agricultural Research Council Zimbabwe, #79 Harare Drive, Marlborough, Harare, Zimbabwe
(Email: ideaazim@ecoweb.co.zw or mukoko@africaonline.co.zw)

Small-scale farming in Zimbabwe has largely been practiced for subsistence purposes because growers cannot afford the high costs of inputs and transport. Poverty reduction among resource-poor soybean growers depends on the adoption and utilization of less expensive alternatives that do not compromise yields. BNF, a system whereby atmospheric nitrogen is converted into nitrogen that can be used by the plant, is an economically attractive option that is 80% cheaper than using equivalent inorganic fertilizers. In this paper we reviewed research on the existing contribution and potential of BNF as a tool for enhancing the commercialisation of small holder agriculture whilst using soybean as the high value commodity and vehicle for poverty reduction.

Research carried out in three different sites of Njiri, Somnene, and Bumba, using the same inoculated variety indicated that resultant yields for the inoculated soybean in-creased by 600kg/ha due to direct utilisation of nitrogen contributed through BNF. Yields for the succeeding rotational crops (maize and millet) increased by an average 600 and 300kg/ha respectively due to the residual nitrogen stored in the soil. Though BNF is dependant on the moisture of the soil, timely application which seeks to balance the rest of the biological factors and the nutritional conditions will increase the actual and residual soil nitrogen pool, resultant crop yields and will maintain soil fertility. Improved harvests resulted in increased household processing, utilisation and marketing of agro-products thereby increasing income generated at household level.

In our research, we learned that small scale farmers can adopt soybean (as a high value commodity) and use of BNF to commercialise agriculture. The factors affecting BNF in the three different sites are based on the physical environment, climatic constraints (dry land) and application of technology by users. These have been the basis of our suggestions on how to commercialise small holder agriculture.

F. D. Dakora et al. (eds.), *Biological Nitrogen Fixation: Towards Poverty Alleviation through Sustainable Agriculture*.
© Springer Science + Business Media B.V. 2008

ESTIMATES OF N_2 FIXATION IN COWPEA GROWN IN FARMERS' FIELDS IN THE UPPER WEST REGION OF GHANA

J. B. Naab[1], S. B. M. Chimphango[2] and F. D. Dakora[3]

[1]Savanna Agricultural Research Institute, P.O. Box 494, Wa, Ghana; [2]Botany Department, University of Cape Town, Rondebosch 7701, South Africa; [3]Cape Peninsula University of Technology, Cape Town 8000, South Africa

Few studies have assessed the levels of symbiotic N nutrition in legumes from farmers' fields, especially in Africa. The objective of this study was to estimate the N_2 fixed by cowpea crops grown in farmers' fields at various locations within the five districts of the Upper West Region of Ghana.

On-farm surveys were conducted in 2003 in the Upper West Region of Ghana. In each farmer's field, shoot material was sampled for dry-matter determination from three 1-m^2 quadrants around the time of peak biomass. Weeds were removed and retained as non-N_2-fixing reference material. All plant samples were processed and N_2 fixation assessed using the ^{15}N natural abundance technique.

On average, the $\delta^{15}N$ values of cowpea were -0.498 ± 0.40 for Jirapa, -0.233 ± 0.11 for Nadowli, 0.154 ± 0.19 for Lawra, 0.231 ± 0.17 for Sissala and 0.960 ± 0.36 for Wa district. The proportional reliance by cowpea on symbiotic fixation for its N nutrition ranged from 12.1–99.7% across locations, with the majority of cowpea crops meeting more than 60% of their N requirements from atmospheric N_2 fixation. Estimates of the legume's N derived from fixation were 66.3% for Wa, 89.9% for Nadowli, 79.4% for Lawra, 78.9% for Sissala and 80.9% for Jirapa district. Total N in the shoot ranged from 10.2–59.0 kg ha^{-1}. The amount of fixed-N in shoots ranged from as low as 1.2 to as much as 54.4 kg N ha^{-1} at some sites. Averaged across farm localities, fixed-N in shoots was 19.8 ± 1.33 for Wa, 16.5 ± 0.58 for Nadowli, 21.1 ± 0.90 for Lawra, 23.0 ± 0.76 for Sissala and 17.6 ± 0.10 kg N.ha^{-1} for Jirapa districts, respectively.

These variations in levels of N_2 fixation in the different cowpea varieties from farmers' fields clearly indicate the need for screening of germplasm for superior symbiotic performance under farmer conditions in Ghana.

We thank the McKnight Foundation for funding this study.

INTEGRATING N$_2$-FIXING LEGUMES, SOILS, AND LIVELIHOODS IN SOUTHERN AFRICA: A CONSORTIUM APPROACH

P. Mapfumo

Soil Fertility Consortium for Southern Africa (SOFECSA),
CIMMYT-Zimbabwe, P.O. Box MP 163, Mt. Pleasant, Harare, Zimbabwe
(Email: p.mapfumo@cgiar.org)

Nitrogen-fixing legumes can contribute up to 300 kg N ha^{-1} year^{-1}, but current contributions rarely exceed 5 kg N ha^{-1} year^{-1} on the majority of smallholder farms. The same old constraints of inherently low soil nutrients, underdeveloped legume seed systems and farm labour, a general lack of information and knowledge, poor markets, and an unfavourable policy environment, continue to frustrate current research efforts. For instance, N$_2$-fixing legumes have often been erroneously taken as substitutes for mineral fertilizers, even in constraining soil environments. Mono-disciplinary approaches in the development of legume-based soil-fertility technologies have also led to a general exclusion of important elements of human nutrition and markets that can potentially drive adoption.

The Soil Fertility Consortium for Southern Africa (SOFECSA) is promoting Integrated Soil Fertility Management (ISFM) innovations in Malawi, Mozambique, Zambia, and Zimbabwe through: (i) identification and prioritization of regionally relevant agricultural research and development issues; (ii) establishment of coordinated inter-disciplinary county teams to facilitate farmer participatory research; (iii) setting up learning centres and adaptive trials that integrate national priorities; and (iv) development of innovation platforms to address complex natural-resource management issues underpinning sustainable agriculture and rural livelihoods, such as fertilizer market development, linking farmers to markets, knowledge management, and technology transfer.

Key ISFM technologies promoted under SOFECSA initiatives include efficient use mineral fertilizers (including new formulations), animal manures, green manure legumes, grain legumes, agroforestry-based technologies, and selection of nutrient use-efficient crop varieties. The SOFECSA work hinges on farmer participatory research and development approaches coordinated at both national and regional levels.

F. D. Dakora et al. (eds.), *Biological Nitrogen Fixation: Towards Poverty Alleviation through Sustainable Agriculture.*
© Springer Science + Business Media B.V. 2008

DISTRIBUTION OF PROMISCUOUS SOYABEAN RHIZOBIA IN SOME ZIMBABWEAN SOILS

J. S. Mukutiri, S. Mpepereki and F. Makonese

Department of Soil Science and Agricultural Engineering, Faculty of Agriculture, University of Zimbabwe, P.O. Box MP167, Mt. Pleasant, Harare, Zimbabwe

Zimbabwean farmers use commercial rhizobia inoculants in the production of soyabeans. Although the demand for commercial rhizobia inoculants is high, farmers in Zimbabwe cannot access these due to distribution bottlenecks and the cost of travel to the factory. Indigenous promiscuous rhizobia offer an alternative option through exploiting their ability to nodulate compatible improved soyabean varieties, assuming that effective compatible rhizobia are present in sufficient numbers in local soils. This study evaluated the potential for using promiscuous rhizobia on commercial soyabean varieties to obviate the need for inoculation. Virgin soils were sampled from 33 sites in three agro-ecological zones and nine promiscuous soyabean varieties were checked for nodulation in the sampled soils 65 days after planting. Positive and negative controls (with and without *Bradyrhizodium japonicum* strain MI 491, respectively) were used to check for contamination and growth conditions.

Magoye, a promiscuous soyabean variety, nodulated in all soils, whereas seven improved varieties nodulated in >75% of the soils. SC Edamame nodulated in the least number of soils (50%). Thus, local soils harbour indigenous soyabean rhizobia suitable for all improved varieties and Magoye has the potential to be grown in agro-ecological zones IIA, III, and IV without rhizobial inoculation.

If promiscuity has a role to play in soyabean production in Zimbabwe, there is no need to determine the agro-ecological limits of soyabean and rhizobia in local soils. Further research to assess both levels of N_2 fixed and symbiotic effectiveness will be relevant before breeding for promiscuity.

F. D. Dakora et al. (eds.), *Biological Nitrogen Fixation: Towards Poverty Alleviation through Sustainable Agriculture*.
© Springer Science + Business Media B.V. 2008

ECOPHYSIOLOGICAL STUDIES OF LEGUMES IN BOTSWANA: TYPES OF BACTERIAL STRAINS FROM ROOT NODULES OF COWPEA (*VIGNA UNGUICULATA* L. WALP.) AND AN ASSESSMENT OF THE SYMBIOTIC STATUS OF GRAIN LEGUMES USING ^{15}N NATURAL ABUNDANCE

F. Pule-Meulenberg[1], T. Krasova-Wade[2] and F. D. Dakora[1]

[1]Cape Peninsula University of Technology, Cape Town 8000, South Africa; [2]IRD, BP 1386, Dakar, Senegal

Six genera (*Rhizobium*, *Bradyrhizobium*, *Sinorhizobium*, *Mesorhizobium*, *Azorhizobium* and *Allorhizobium*) within the Rhizobiaceae have the ability to infect roots and form N_2-fixing nodules on members of the Leguminosae and in the non-legume *Parasponia* (Ulmaceae). The major aim of this study was to determine whether the different agro-ecological zones of Botswana have an effect on the natural distribution of rhizobia and the local Leguminosae.

Leaves of various grain legumes were sampled from the five agro-ecological zones to assess their symbiotic status using ^{15}N natural abundance. The test legume species included Bambara groundnut (*Vigna subterranean* L. Verdc.), peanut (*Arachis hypogea* L), and cowpea (*Vigna unguiculata* L. Walp.). Root nodules were also sampled and the rhizobia isolated and characterized. PCR/RFLP of the 16S–23S intergenic rDNA in crushed nodules was performed (Krasova-Wade et al., 2003). Significant differences occurred between the $\delta^{15}N$ values of the legume species from the five agro-ecological zones of Botswana as did the %Ndfa (percent N derived from the atmosphere). A total of 616 bacterial isolates with differing colony morphologies were obtained from the collected nodules. Of these, 90% were fast growers, 10% slow, 86.3% were dome-shaped, 9.1% flat, 2.3% conically-shaped, 55.7% opaque, 33.8% translucent, 4.2% yellow, 1.4% orange, and 0.9% pink in colour. Additionally, 72.6% were buttery in texture and 17.4% elastic, 28.3% produced gum and 61.7% did not. Colony characteristics and PCR-RFLP data of the 16S-23S IGS region of rDNA showed considerable diversity among these nodule bacteria.

F. D. Dakora et al. (eds.), *Biological Nitrogen Fixation: Towards Poverty Alleviation through Sustainable Agriculture.*
© Springer Science + Business Media B.V. 2008

We thank the McKnight Foundation, USA, for a grant and the Botswana College of Agriculture for a Fellowship (to FPM).

References

Krasova-Wade T et al. (2003) Afr. J. Biotech. 2, 13–22.

SYMBIOTIC N NUTRITION AND YIELD OF COWPEA GENOTYPES UNDER DIFFERENT PLANT DENSITIES AND CROPPING SYSTEMS

J. H. J. R. Makoi[1], S. B. M. Chimphango[2] and F. D. Dakora[1]

[1]Cape Peninsula University of Technology, Cape Town 8000, South Africa; [2]Botany Department, University of Cape Town, Rondebosch 7701, South Africa

Symbiotic legumes are an important component of the cropping systems in tropical environments because of their ability to transfer fixed-N to non-legume crops. The total amount of N-fixed per unit area in intercropping systems can be low due to sub-optimal legume plant density. The choice of legume cultivar can also influence the potential contribution of fixed-N to the cropping system. This study assessed the effect of plant density and different cowpea genotypes on N_2 fixation in a sorghum-based cropping system.

A three-factor field trial involving five cowpea genotypes (Bensogla, ITH98-46, Sanzie, TVU1509, and Omandaw), two legume densities (83,333 plants ha^{-1} and 166,666 plants ha^{-1}), and two cropping systems (mono- and inter-cropping) was conducted at Nietvoorbij (33°54' S and 18°14' E) in Stellenbosch, South Africa. Plants were sampled at 67 days after planting for determination of growth and symbiotic performance.

Increasing cowpea plant density significantly ($p \leq 0.05$) decreased plant growth, grain yield, %N, and $\delta^{15}N$ values, but increased %Ndfa (percent N derived from the atmosphere). Similarly, inter-cropping significantly ($p \leq 0.05$) decreased plant growth, grain yield, %N, and $\delta^{15}N$ values, but again increased %Ndfa. Cowpea genotypes showed differences in plant growth, grain yield %N, $\delta^{15}N$, and %Ndfa.

Our data suggest that optimizing legume density in cropping systems can increase N_2 fixation and contribute to the N economy of agricultural soils.

This study was supported by a grant from the McKnight Foundation, Minnesota, USA.

F. D. Dakora et al. (eds.), *Biological Nitrogen Fixation: Towards Poverty Alleviation through Sustainable Agriculture.*
© Springer Science + Business Media B.V. 2008

THE DIVERSITY OF ROOT NODULE BACTERIA ASSOCIATED WITH *LEBECKIA* SPECIES IN SOUTH AFRICA

F. L. Phalane[1], E. T. Steenkamp[2], I. J. Law[1] and W. F. Botha[1]

[1]Plant Protection Research Institute, Private Bag X134, Queenswood 0121, South Africa; [2]Department of Microbiology and Plant Pathology, University of Pretoria, Pretoria, South Africa

Lebeckia is a shrubby herbaceous legume of the tribe Crotalarieae, indigenous to the Southern and Western Cape regions of South Africa (Germishuizen and Meyer, 2003). The genus contains 33 species and is undergoing taxonomic revision (B.-E. van Wyk, personal communication, 2006). Species are divided into trifoliate-leaf species in the sections Calobota and Stiza (Bentham, 1844) and needle-leaf species in the section *Lebeckia* (Le Roux and van Wyk, 2007). In the veld, species, like *L. spinescens* and *L. multiflora*, provide nutritious grazing, whereas species like *L. cytisoides* are toxic. Our aim was to determine the diversity of root-nodule bacteria associated with these different *Lebeckia*.

Seventy-six *Lebeckia* rhizobia were isolated from ten *Lebeckia* species. Nodulation and nitrogen-fixation assays showed that all isolates fixed nitrogen on their *Lebeckia* host, whereas only about 65% of the isolates were effective on cowpea and siratro. All nodule isolates had similar colony morphology to the original rhizobial cultures used as inoculants. DNA of the 76 isolates was compared by random amplified PCR using primers RPO1 and CRL7 and a combined UPGMA dendogram drawn using bionumerics. Cluster analysis showed that isolates from the same species of *Lebeckia* generally grouped together in individual clusters with a few outliers. Isolates from each *Lebeckia* species were selected from clusters within the UPGMA tree and subjected to 16S rRNA gene sequencing. The resulting sequences were compared to those in public domain nucleotide databases using *BLASTn* and subjected to phylogenetic analysis. The respective 16S sequences were most closely related to members of the α-proteobacterial genera *Bradyrhizobium*, *Sinorhizobium* and *Mesorhizobium* as well as to the β-proteobacterial genus *Burkholderia*, indicating that the indigenous *Lebeckia* is nodulated by diverse rhizobia from both the α- and β-proteobacteria.

F. D. Dakora et al. (eds.), *Biological Nitrogen Fixation: Towards Poverty Alleviation through Sustainable Agriculture.*
© Springer Science + Business Media B.V. 2008

References

Bentham G (1844) in Hook, Lond. J. Bot. 3, 338–365.
Germishuizen G and Meyer NL (2003) Strelitzia 14, 523–524.
Le Roux M and van Wyk BE (2007) S. Afr. J. Bot. 73, 118–130.

RHIZOBIAL DIVERSITY ASSOCIATED WITH SOUTH AFRICAN LEGUMES

M. A. Pérez-Fernández[1], A. J. Valentine[2] and S. Muñoz García-Mauriño[1]

[1]Área de Ecologia, Dpto. de Sistemas Físicos, Químicos y Naturales, Universidade Pablo de Olavide, Carretera a Utrera, Km 1, 41013 Sevilla, Spain; [2]Plant Physiology Group, South African Herbal Science Institute, University of the Western Cape, Belleville 7535, South Africa

N_2-fixing legumes play a major role in the natural ecosystems in areas of Mediterranean-type climates, accounting for 80% of biologically-fixed N (McInroy et al., 1999). However, information is needed particularly for shrubby legumes and for those that, like many acacias, succeed in new environments and out-compete the native flora. Although mutualisms, involving N_2-fixing bacteria, could constrain legume invasion success, they have not been investigated in this context (Yates et al., 2004). This work aims to identify rhizobial strains associated with six South African shrubby legumes and to investigate the effect of *Bradyrhizobium* on the growth of *Acacia karoo* Hayne and *Acacia cyclops* A.Cunn. ex G.Don. *Bradyrhizobium* strains were isolated from root nodules of all the selected species and were characterized morphologically and by PCR and 16S rDNA. Seeds from the two acacias were inoculated with 2 mL of the appropriate bacterial inocula, which consisted of bacterial strains from the six South African and six Spanish species. Ten plants per treatment were grown in a greenhouse for 20 weeks and watered with a sterile N-free nutrient solution. Morphological and molecular characterization show that all extracted strains are different bradyrhizobia. Both acacias are promiscuous hosts and form efficient nodules with *Bradyrhizobium* strains from the six South African and six Spanish shrubby legumes. The number of nodules and resulting biomass production was higher in *Acacia cyclops* than in *Acacia karoo* when inoculated with any of the strains. The better ability of *A. cyclops* to produce more effective nodules probably accounts for the better ability of *A. cyclops* to colonize new environments. Once *Acia cyclops* is established, it dominates other shrub vegetation and alters grassland and woodland under-storey habitats by modifying light and soil moisture regimes and nitrogen dynamics, thus preventing other plants establishing. The lack of specificity of *A. cyclops* suggests that isolates that establish early in the plant's growth are more likely to contribute inoculum for other nodules as the roots growth.

F. D. Dakora et al. (eds.), *Biological Nitrogen Fixation: Towards Poverty Alleviation through Sustainable Agriculture.*
© Springer Science + Business Media B.V. 2008

References

McInroy S (1999) FEMS Microb. Lett. 170, 111–117.
Yates RJ et al. (2004) Soil Biol. Biochem. 36, 1319–1329.

DIVERSITY AND COMPETITIVENESS OF RHIZOBIA
OF THE LEGUME TREE *ACACIA NILOTICA* IN SENEGAL

R. T. Samba[1], M. Neyra[1], D. Francis[1], S. Sylla[1,2] and I. Ndoye[1,2]

[1]Laboratoire Commun de Microbiologie, IRD/ISRA/UCAD, Centre de Recherche de Bel-Air, B.P. 1386 – CP 18524 Dakar, Sénégal; [2]Département de Biologie Végétale, Université Cheikh Anta Diop, BP. 5005 Dakar, Sénégal

Acacia nilotica (var. *adansonii* and var. *tomentosa*) is a tree species belonging to the family Leguminosae, subfamily Fabaceae (Allen and Allen, 1981). They are endemic to the sahelian arid and Sudanean semi-arid tropical regions, respectively. They have high socio-economic and ecological importance as sources of forage and wood for local populations and might improve the fixed-N status of soil due to their nitrogen-fixing symbiosis with rhizobia (Lal and Khanna, 1993). They are salt tolerant and adapted to temporarily waterlogged sites (Nabil and Coudret, 1995).

A collection of 34 strains, isolated from root nodules of *A. nilotica tomentosa* and *A. nilotica adansonii*, was characterized by PCR/RFLP. Four distinct groups were formed and two separate strains. One strain representing each group was selected and its 16S rDNA sequenced to determine the taxonomical position of these groups.

Competition among *Acacia nilotica* strains in single and multi-strain inoculants were studied in the greenhouse. Plants were cultivated in potted soil for two months and the ability of six fast-growing strains of rhizobium to compete for nodule sites against native strains was measured using PCR/RFLP of the intergenic 16S-23S rDNA on crushed nodules. Analysis of RFLP profiles revealed that inoculated strains are more competitive than indigenous strains in all treatments and had higher nodule occupancy. Biomass analysis used to measure for efficiency of the inoculated strains indicated that the more efficient were plants that were inoculated with either a mixture of two strains, ORS3180 + ORS3168, or with a mixed inocula of six strains.

References

Allen ON and Allen EK (1981) The Leguminoseae, MacMillan, London.
Lal B and Khanna S (1993) Can. J. Microbiol. 39, 87–91.
Nabil M and Coudret A (1995) Physiol. Plantarum 93, 217–224.

F. D. Dakora et al. (eds.), *Biological Nitrogen Fixation: Towards Poverty Alleviation through Sustainable Agriculture.*
© Springer Science + Business Media B.V. 2008

EFFECTS OF ROCK PHOSPHATE ON INDIGENOUS RHIZOBIA ASSOCIATED WITH *SESBANIA SESBAN*

O. Sacko[1], R. T. Samba[2], I. Yattara[3], N. Faye[4], F. do Rego[5], T. Diop[6], M. Neyra[7] and M. Gueye[8]

[1,3]Biology Department, Faculty of Sciences and Technics, Bamako University, BP: E 3206, Senegal; [2,4,5]LCM, IRD, Dakar, Senegal; [6]LCB, UCAD, Dakar, Senegal; [7]IRD UR 40, Senegal; [8]MIRCE/LCM, IRD, Dakar, Senegal. (Email: ousmane_sacko@yahoo.fr)

Low soil fertility, mainly low levels of P and N, seems to be a major constraint to the establishment and growth of leguminous trees used in agroforestry systems (Sanginga, 1992; Bâ et al., 1996). *Sesbania sesban* is an important multipurpose legume, which is often used as cover tree (Desaeger and Rao, 2001) or green manure (Carsky et al., 2001), but its effectiveness may be hindered by a P deficiency (Carsky et al., 2001). P may be in soluble or insoluble forms. Although the former can be immediately taken up by plants, the latter is used mostly because of lower cost. In Africa, the latter form exists in great quantity with Tilemsi rock phosphate (PNT) from Mali as one of the best in West Africa (Truong et al., 1978). In this work at LCM, Dakar, we studied over 105 days, the effects of 150 kg $P_2O_5.ha^{-1}$ of PNT on the diversity of indigenous rhizobia associated with *Sesbania sesban* on soil from Nioro (Senegal), which contains 23.4 ppm available P. Nodules were harvested every 21 days, crushed, and PCR-RFLP used to extract DNA from the 16S-23S IGS region. The PCR products were digested and analysed. Sixteen IGS types were detected, but they varied depending on whether insoluble-P was present or not. It seems that PNT has an effect on rhizobia diversity in contrast to soluble phosphate (TSP) (O. Sacko et al., unpublished data, 2005). Insoluble-P, despite its low cost and its usefulness in agroforestry (Bâ et al., 2001), disturbed the indigenous rhizobia associated with *S. sesban*. This result may be due to competition for soluble-P, which is insufficiently in most soils of West Africa. However, this idea must be confirmed.

This study was supported by INCO-ENRICH (ICA4-CT-2000-30022) and AUF.

References

Bâ A et al. (2001) Fruits 56, 261–269.
Bâ AM et al. (1996) Cahiers Scientifiques N° 2, Cirad-Forêt, pp. 237–244.
Carsky RJ et al. (2001) Nutr. Cycl. Agroecosys. 59, 151–159.
Desaeger J and Rao MR (2001) Crop Prot. 20, 31–44.
Sanginga N (1992) Fert. Res. 31, 165–173.
Truong B et al. (1978) Agron. Trop. 33, 1–136.

EFFECT OF ADDED FERTILIZERS ON N_2 FIXATION AND PLANT PERFORMANCE OF SHRUBBY LEGUMES

M. A. Pérez-Fernández[1], A. J. Valentine[2], E. Calvo[1], S. Muñoz[1] and C. Vicente[1]

[1]Area de Ecología, Facultad de Ciencias Experimentales, University Pablo de Olavide, Cartera de Utrera, Km 1, 41013, Seville, Spain; [2]Western Cape University, Cape Town, South Africa

Most shrubby legumes under a Mediterranean-type climate have a remarkable capacity to withstand drought and the associated Mediterranean summer stresses, which make them good invaders and good candidates to be used in rehabilitation projects by reducing costs of plant care and plant reposition (Herrera et al., 1993; Lafay and Burdon, 2006). However, studies on how seedling performance is affected by nutrient availability or by the presence of nodules in these species are lacking. Also an understanding of the responses of these legumes to soil nutrient availability would give insight into the mechanisms involved in their ability to colonise new environments.

We have investigated the effect of fertilization and biological nitrogen fixation by rhizobial nodules on seedling development in six Australian and six Spanish shrubby legumes. A factorial experiment was designed with: (i) two plant origins; (ii) five fertilisation regimes; and (iii) one inoculant plus a control (no inoculation); with five nutrient availability treatments (Osmocote Mini 2–3 months) and eight replicates per treatment. Legumes were grown for 28 weeks in Forest Pot containers filled with sterile sand (pH ca. 6.7). The higher nodule production in the Spanish plants accounts for their higher biomass production. Low and intermediate fertilization enhanced biomass production and N accumulation, whereas high fertilization exerts a toxic effect on all plants, but more so, on the Australian species. Australian species develop in nutrient-poor soils and any increase in soil fertilization kills them. Spanish species are fertiliser-tolerant, which would help if introduced into new environments, as for *C. scoparius* (Lafay and Burdon, 2006). Thus, either high fertilisation or inoculation must be applied to enhance seedling growth and quality, but both methods are mutually exclusive and fertilisers have negative environmental also (Dileep et al., 2001).

F. D. Dakora et al. (eds.), *Biological Nitrogen Fixation: Towards Poverty Alleviation through Sustainable Agriculture.*
© Springer Science + Business Media B.V. 2008

M. A. Pérez-Fernández et al.

References

Dileep K et al. (2001) Plant Soil 229, 25–34.
Herrera MA et al. (1993) Appl. Environ. Microbiol. 59, 129–133.
Lafay B and Burdon JJ (2006) J. Appl. Microbiol. 100, 1228–1238.

GENETIC DIVERSITY OF *ACACIA SEYAL* DEL. RHIZOBIAL POPULATIONS INDIGENOUS TO SENEGALESE SOILS IN RELATION TO SALINITY AND pH OF THE SAMPLING SITES

D. Diouf [1,2], R. Samba-Mbaye[2], D. Lesueur[3], A. Bâ[1], B. Dreyfus[4], P. de Lajudie[4] and M. Neyra[4]

[1]Département de Biologie Végétale, UCAD, BP 5005, Dakar, Senegal; [2]LCM IRD/ISRA/UCAD, BP 1386, Dakar, Senegal; [3]CIRAD-Foret/UPR 80 "ETP" TSBF/CIAT, P.O. Box 30677, Nairobi, Kenya; [4]IRD/UMR 113, LSTM, Baillarguet, TA 10/J, 34398 Montpellier, France. (Email: diegane.diouf@ird.sn or ddiegane@ucad.sn)

An understanding of the diversity of symbiotic rhizobia should facilitate selection of efficient strains for improving establishment of multi-purpose tree species commonly advocated to overcome salt-stress problems (Giri et al., 2003). This research deals with reliable methods of rehabilitating degraded soils in the Groundnut Basin of Senegal. The genetic and phenotypic diversity of rhizobial populations associated with *Acacia seyal*, a multi-use, potentially salt-tolerant species, in relation to pH and salinity of the sampling soils was examined. The diversity of *A. seyal* rhizobial populations from 42 soils samples, collected throughout the Groundnut Basin of Senegal, was characterized by PCR-RFLP of IGS rDNA of DNA extracted from nodules. The 16S and 16S-23S rDNA sequences of rhizobial strains were analyzed. The effect of inoculation with *Mesorhizobium* strains on the growth of *A. seyal* plants was evaluated after 4 months on non-sterile soil. Fifteen IGS-RFLP profiles were identified among 138 nodules analyzed. Clustering of rhizobial populations did not reflect their geographical origin. Analysis of 16S and 16S-23S rRNA gene sequences of 15 strains showed that, in these soils, *A. seyal* is nodulated by *Mesorhizobium* (64%) and *Sinorhizobium* (29%) species. *Agrobacterium* and *Burkholderia* strains were also found with rhizobia within the nodules, which resulted probably in more opportunities to exchange their genomic information (Bala and Giller, 2001). The *Mesorhizobium* strains showed high diversity and inoculation with them significantly improved nodulation. Our results showed a widespread occurrence of compatible rhizobia associated to *A. seyal*. Hence, inoculation with these rhizobia might improve establishment and salt tolerance of *A. seyal* seedlings in local saline environments.

F. D. Dakora et al. (eds.), *Biological Nitrogen Fixation: Towards Poverty Alleviation through Sustainable Agriculture*.
© Springer Science + Business Media B.V. 2008

References

Giri B et al. (2003) Biol. Fertil. Soils 38, 170–175.
Bala A and Giller KE (2001) New Phytol. 149, 495–507.

POTENTIAL EFFECTS OF ELEVATED UV-B RADIATION ON N₂-FIXING LEGUMES IN AGRICULTURAL SYSTEMS

S. B. M. Chimphango[1], C. F. Musil[2] and F. D. Dakora[3]

[1]Botany Department, University of Cape Town, Rondebosch 7701, South Africa; [2]South Africa National Botanical Institute, P/B X7, Claremont 7735, South Africa; [3]Cape Peninsula University of Technology, Cape Town 8000, South Africa

The overwhelming evidence seems to paint a negative picture of UV-B effects on terrestrial plants, including agricultural species. This study assesses UV-B effects on eight symbiotic legumes.

Three experiments were conducted and compared to a control with two levels of elevated UV-B radiation that simulated 15% and 25% depletions in the total column of ozone above Cape Town. In two of the three experiments, a N treatment was introduced where some plants received 2 or 1 mM nitrate, whereas others relied entirely on symbiotic N_2 fixation for their N nutrition.

Elevated UV-B radiation did not alter whole-plant biomass, nodule dry matter, %Ndfa, and seed yield. However, both the leaf and total dry matter of nitrate-fed *L. luteus* plants increased ($P \leq 0.05$) with exposure to UV-B$_{166}$ relative to a UV-B$_{100}$ control. With *C. maculata*, however, both the stem and total dry matter of nitrate-fed plants were decreased by both UV-B$_{134}$ and UV-B$_{166}$. This result was in contrast to plants relying purely on symbiotically fixed N, which were decreased ($P \leq 0.05$) by only UV-B$_{166}$. Tissue concentration of N in the stem and leaf, and Ca^{2+} in the latter increased in *V. unguiculata* plants exposed to UV-B$_{166}$ relative to ambient and UV-B$_{134}$ treatment. However, plants exposed to UV-B$_{134}$ showed decreased concentration of P in stems relative to both ambient and UV-B$_{166}$ treatment.

Nitrate application to plants exposed to elevated UV-B radiation increased the sensitivity of *C. maculata* to UV-B. UV-B-induced changes in tissue concentration of nutrients could potentially alter the rate of litter decomposition, thereby affecting nutrient cycling at the ecosystem level.

The authors thank AAU and DAAD for the fellowship awarded to SMBC.

ARBUSCULAR MYCORRHIZAL COLONIZATION AND NODULATION IMPROVE FLOODING TOLERANCE IN *PTEROCARPUS OFFICINALIS* JACQ. SEEDLINGS

L. Fougnies[1], S. Renciot[1], F. Muller[1,2,3], C. Plenchette[4], Y. Prin[2], S. M. de Faria[5], J. M. Bouvet[3], S. Sylla[6], B. Dreyfus[2], I. Ndoye[6] and A. Bâ[1,2]

[1]LBPV, FSEN, UAG, Pte-à-Pitre, Guadeloupe; [2]LSTM, UMR 113 IRD/INRA/AGRO-M/CIRAD/UM2, Montpellier, France; [3]LGF, CIRAD- Forêt, Montpellier, France; [4]INRA, UMR BGA, Dijon, France; [5]CNPAD/EMBRAPA, Seropédica Itaguai, Seropédica, RJ 23890-000, Brazil; [6]LCM ISRA/IRD/UCAD, Dakar, Senegal

P. officinalis is the dominant wetland tree species of the seasonally flooded swamp forests in Guadeloupe, the Lesser Antilles (Muller et al., 2006). The establishment and population maintenance of *P. officinalis* are affected by the variations in salinity and hydrology as well as differences in soil microtopography in swamp forests (Eusse and Aide, 1999). This tree species forms bradyrhizobial nodules and arbuscular mycorrhizas (AMs) on lateral roots of buttresses both above and below the water table (Bâ et al., 2004; Saint-Etienne et al., 2006). We hypothesized that nodulation and AMs could improve the performance of *P. officinalis* seedlings under flooding. Two questions were addressed. First, are *P. officinalis* seedlings adapted to flooding? Second, do AMs and N_2-fixing nodules increase the performance of *P. officinalis* seedlings under flooding?

P. officinalis seedlings in pots were inoculated with *Bradyrhizobium* sp. (UAG 11A) and/or one AM fungus, *Glomus intraradices*. After 13 weeks of flooding, several changes in plant seedlings were observed including formation of hypertrophied lenticels, aerenchyma tissue and adventitious roots on submerged portions of the stem. Flooding induced nodules both on adventitious roots and stems and resulted also in an increase in the total biomass of seedlings, regardless of inoculation. Flooding did not affect root-nodule formation and N_2 fixation. However, there was no additive effect of AMs and nodulation on plant growth and nutrition under either flooding or non-flooding. These results suggest a competitive interaction between endophytes and that *Pterocarpus officinalis* seedlings develop some adaptive mechanisms to flooding.

F. D. Dakora et al. (eds.), *Biological Nitrogen Fixation: Towards Poverty Alleviation through Sustainable Agriculture.*
© Springer Science + Business Media B.V. 2008

References

Bâ et al. (2004) Rev. Ecol. 59, 163–170.
Eusse AM and Aide TM (1999) Plant Ecol. 145, 307–315.
Muller et al. (2006) Mol. Ecol. Notes 6, 462–464.
Ruiz-Lozano JM et al. (2001) New Phytol. 151, 493–502.
Saint-Etienne et al. (2006) Forest Ecol. Manag. 232, 86–89.

MYCORRHIZAL COLONIZATION IMPROVES GROWTH AND NUTRITION OF AN INVASIVE LEGUME IN A MEDITERRANEAN ECOSYSTEM

A. J. Valentine[1], M. R. Le Roux[1] and M. A. Pérez-Fernández[2]

[1]Plant Physiology Group, South African Herbal Science Institute, University of the Western Cape, Private Bag X17, Belleville 7535, South Africa; [2]Area de Ecología, Facultad de Ciencias Experimentales, University Pablo de Olavide, Cartera a Utrera, Km 1, 41013, Seville, Spain
(Email: alexvalentine@mac.com)

The Fynbos biome is a unique flora. It is the only kingdom contained in its entirety within a single country and is characterised by a high richness of some 8,700 species and a high endemicity with 68% are confined to the Cape Floristic Kingdom (CFK). Many of the aliens present in the Western Cape originate from Australia. One of the main invaders is a woody legume, *Acacia cyclops* A.Cunn. ex G.Don. Since low soil fertility, due to nitrogen (N) and phosphorus (P) deficiencies and low soil organic matter, is the major limiting factor in Mediterranean-type ecosystems, symbiotic association with both *Rhizobium* and arbuscular mycorrhizae (AM) may be a major advantage to these invading legumes.

It is well known that AM can improve the P nutrition required by developing nodules, but this double symbiosis may impose a high C drain on the host. This work studies the effects of AM colonisation on the nutritional physiology of *A. cyclops* in the Fynbos.

The interaction between mycorrhizal inoculum, *Glomus mossae*, and a *Rhizobium* strain cultured from *A. cyclops* nodules, and their effects on *A. cyclops* was studied in P- and N-deficient sand culture. Plants were grown for five months on a 10% Long Ashton solution under glasshouse conditions. The double-symbiosis caused a greater decline in colonization by both symbionts compared with either symbiont alone. In spite of the potential C costs of both symbionts, the benefits of AM colonisation was an improved biomass production in the nodulated plants due to improved N and P nutrition by the AM. In particular, the AM symbiont improved N uptake and incorporation by the plant as evidenced by increased xylem export of inorganic NH_4^+ from nodules.

F. D. Dakora et al. (eds.), *Biological Nitrogen Fixation: Towards Poverty Alleviation through Sustainable Agriculture.*
© Springer Science + Business Media B.V. 2008

However, the xylem export of assimilated organic N compounds, such as ureides and amino acids, from nodules was not significantly increased by AM symbionts, which might be due to the C costs imposed by the double symbiosis that may restrict the energetically expensive biological production of NH_4. These findings suggest that the AM symbiosis is an important partner of the invasive woody legume, *Acacia cyclops* in P-poor soils.

NUTRITIONAL AND PHOTOSYNTHETIC PERFORMANCE OF INVASIVE AND INDIGENOUS LEGUMES IN A MEDITERRANEAN ECOSYSTEM

A. J. Valentine[1], M. A. Pérez-Fernández[2], G. Shanks[1] and P. E. Mortimer[1]

[1]Plant Physiology Group, South African Herbal Science Institute, University of the Western Cape, Private Bag X17, Belleville 7535, South Africa; [2]Area de Ecología, Facultad de Ciencias Experimentales, University Pablo de Olavide, Cartera a Utrera, Km 1, 41013, Seville, Spain
(Email: alexvalentine@mac.com)

Exotic plant invasions have large impacts on native species and ecosystems (Richardson et al., 2000). Invasions often reduce native species diversity or alter species composition (Richardson, 2001), but the mechanisms underlying these impacts are rarely elucidated. Better understanding of mechanisms underlying these impacts is critical to restoring native ecosystems and in understanding why some invaders have larger impacts than others. In the nutrient-poor soil of the Cape Floristic Region (CFR), the success of the invasive legume *Acacia cyclops* may depend on improved N and P nutrition. This work aims: (i) to identify the traits that make *Acacia cyclops* a better coloniser than the native *Virgilia oroboides*; and (ii) to relate invasive ability to the photosynthetic and nutritional physiology of *A. cyclops* as compared to that of the indigenous *V. oroboides*.

Photosynthetic traits, plant nutritional status and plant performance were evaluated in seedlings of *A. cyclops* and *V. oroboides*. The experiment consisted of two factors: (1) species (*Acacia* and *Virgilia*); and (2) harvest time (3 and 7 months). Plants were grown in a glasshouse on sterile sand and supplied with a 20% Long Ashton nutrient solution (modified to contain zero P and N) once per week. During the harvests at 3 and 7 months, gas exchange and biomass parameters were determined. Although *Virgilia* produced more root nodules than *Acacia*, they were less effective as evidenced by their low δN. The lower biomass accumulated by *Acacia* is related to a lower photosynthetic water-use efficiency (PWUE) and a constant tissue construction cost over time. Compared to *Virgilia*, the higher photosynthetic rate of *Acacia* does not contribute to its biomass because it relies more on energetically expensive N_2 fixation in its nodules. This can account for the C-drain from growth and the sink-stimulation of photosynthesis. In a field situation, the greater amount of N fixed by *Acacia* modifies the natural N cycle, being more beneficial for *Acacia* and toxic for native species.

F. D. Dakora et al. (eds.), *Biological Nitrogen Fixation: Towards Poverty Alleviation through Sustainable Agriculture.*
© Springer Science + Business Media B.V. 2008

References

Richardson D (2001) Encyclopedia of Biodiversity, Vol. 4, pp. 677–688.
Richardson et al. (2000) Invasive Species in a Changing World, pp. 303–349.

ECOLOGY AND PHYTOCHEMICAL ANALYSIS OF THE MEDICINAL LEGUME, *SUTHERLANDIA FRUTESCENS* (L.)R. Br., AT TWO LOCATIONS

Q. Johnson[1], A. J. Valentine[1] and M. A. Pérez-Fernández[2]

[1]Plant Physiology Group, South African Herbal Science Institute, University of the Western Cape, Private Bag X17, Belleville 7535, South Africa; [2]Area de Ecología, Facultad de Ciencias Experimentales, University Pablo de Olavide, Cartera a Utrera, Km 1, 41013, Seville, Spain
(Email: alexvalentine@mac.com)

The Fynbos biome is a unique flora with great richness in plant species, some of which include legumes of medicinal importance, e.g., *Sutherlandia frutescences* (L.)R. Br. or "cancer bush". Sutherlandia is high in the non-protein amino acid, canavanine, an active medicinal compound. An applied ethno-botanical study was conducted with experimental cultivation trials to develop a community-based model for management of medicinal plant cultivation programmes. We studied the growth and the amino acid profile of *S. frutescens* at two sites in South Africa, Bitterfontein and Nuwerust.

Plants from Bitterfontein site were cleaned of soil and weighed. Large variations were noted in plant mass (2–83 g), in their lengths (17–54 cm), and in volume. Of the randomly analyzed specimens, 64% showed well-developed side branches, the other 36% had a single stem. All plants showed good new leaf formation and development. No flowers or potential flowering were observed in any plant at this stage and no pests. In plants from Nuwerus site, less variation in plant mass, compared to Bitterfontein, was noted for individual plants (1–20 g). Plant length ranged from 14 to 37 cm. There appears to be no correlation between root development and shoot length. Of the collected specimens, 32% showed well-developed side branches, whereas the remaining 68% were single stemmed plants. Most of the plants showed good new leaf formation and new leaf development. No flowers or potential flowering were observed in any of the plants. No pests or abnormal growth were detected on any of the plants.

The spray reagent, ninhydrin, used in the TLC method, detected many amino acids. *Sutherlandia* contains high concentrations of amino acids, of which canavanine and its analogues are the most prominent. TLC plates of *Sutherlandia* from both locations show an almost identical spectrum of amino acids with slightly different concentrations. The biochemical profiles of the medicinal plant species are critical. Raw or processed products need to be of high quality in terms of phytobiochemical profiles to be competitive in national and international markets.

F. D. Dakora et al. (eds.), *Biological Nitrogen Fixation: Towards Poverty Alleviation through Sustainable Agriculture*.
© Springer Science + Business Media B.V. 2008

IMPROVING CHICKPEA INOCULATION IN THE NORTHERN CROPPING REGION OF AUSTRALIA

A. McInnes[1], C. Douglas[2], J. Thies[3], J. Slattery[4], P. Thompson[5] and K. McCosker[6]

[1]Building K29, University of Western Sydney – Hawkesbury Campus, Locked Bag 1797, Penrith South DC, NSW 1797 Australia; [2]Queensland Department of Primary Industries and Fisheries, Locked Bag 1, Biloela, Qld 4715, Australia; [3]Department of Crop and Soil Sciences, 719 Bradfield Hall, Cornell University, Ithaca, NY 14853, USA; [4]Department of Primary Industries, Rutherglen Centre, RMB1145, Chiltern, Valley Road, Rutherglen, Victoria 3581, Australia; [5]Queensland Department of Primary Industries and Fisheries, Hermitage Research Station, 604 Yangan Road, Warwick, Qld 4370, Australia; [6]Queensland Department of Primary Industries and Fisheries, P.O. Box 6014, Rockhampton, Qld 4702, Australia

Chickpeas are an important rotation legume in the northern Australian grain-cropping region, with an average of 156,000 ha sown per year in New South Wales (NSW) and Queensland (Qld) over the last 5 years. Chickpeas are routinely inoculated at sowing with commercial preparations of *Mesorhizobium ciceri* strain CC1192 that we found to be effective on a wide range of temperate and sub-tropical cultivars in glasshouse tests. Effectiveness was persistent over time and field reisolates of CC1192 being as effective as a laboratory culture on chickpea in glasshouse tests.

However, in northern NSW and southern Qld (but not central Qld), strain CC1192 is often a minor nodule occupant in field-grown or soil-grown bait plants 0–3 years after sowing inoculated field trials and crops. The more dominant strains from these sites were either equally effective or less effective than strain CC1192 on chickpea, with specificity evident in different cultivar × strain combinations. Nodule dry weight and nodule occupancy by strain CC1192 were improved using clay granular inoculants or water injection of inoculants directly into soil rather than seed inoculation with peat. Nodule occupancy by CC1192 may also be improved by a break of at least 3 years between chickpea crops (to reduce soil population numbers of competing strains) and by ensuring that all subsequent crops are reinoculated.

F. D. Dakora et al. (eds.), *Biological Nitrogen Fixation: Towards Poverty Alleviation through Sustainable Agriculture.*
© Springer Science + Business Media B.V. 2008

OPTIMIZING CHICKPEA NODULATION FOR NITROGEN FIXATION AND YIELD IN NORTH-WESTERN NEW SOUTH WALES, AUSTRALIA

N. Elias[1,2], A. McInnes[1] and D. Herridge[2]

[1]University of Western Sydney, Penrith South DC, NSW 1797, Australia;
[2]Tamworth Agricultural Institute, 4 Marsden Park Rd, Calala, NSW 2340, Australia
(Email: natalie.elias@dpi.nsw.gov.au)

Nodulation of chickpea in the northern grains belt of New South Wales is sub-optimal. This is limiting the total N_2 fixed and reducing the value of chickpea as a rotation crop in the largely cereal-based farming systems. This project is taking a novel approach to the problem by combining an understanding of the dynamics of rhizobial populations with evaluation of farmer practice and inoculant technologies.

Once introduced to the soil the fate of commercial rhizobial strain, CC1192, is largely unknown. Thirteen soils with varied histories of CC1192 inoculation were sampled. Using the Whole Soil Inoculation Technique, effectiveness of the *Mesorhizobium ciceri* populations present were evaluated. Variation in the N_2-fixing capacity of the soils was evident, with 80% of this variation attributed to variation in *M. ciceri* numbers (enumerated using the MPN plant-infection technique). The remaining 20% may be due to variations in effectiveness of resident populations. Despite soils being originally inoculated CC1192, genetic change could have occurred over time.

RAPD-PCR is being used to identify isolates of *M. ciceri* to determine the extent of strain diversity. Of particular importance is whether *M. ciceri* strains less effective than CC1192 are present. If so, competition for nodulation may be reducing yield. The impact of farmer practice on nodulation and yield is being determined by benchmarking of commercial chickpea fields. Chickpea yield, nodulation and N_2 fixation will be correlated with agronomic variables such as sowing date, plant density, plant height, disease, weed density and stubble cover. Determination of soil nitrate levels in 12 chickpea fields showed all were above the 50 kg ha^{-1} (0–120 cm) threshold for optimising N_2 fixation. This suggests poor paddock selection by farmers may be a significant factor in inhibiting N_2 fixation.

Issues still to be examined include the impact of inoculant technologies on: (i) distribution of nodules; and (ii) nodule occupancy in competitive environments.

F. D. Dakora et al. (eds.), *Biological Nitrogen Fixation: Towards Poverty Alleviation through Sustainable Agriculture.*
© Springer Science + Business Media B.V. 2008

INVESTIGATING NITROGEN FIXATION IN THE *MEDICAGO-SINORHIZOBIUM* SYMBIOSIS

J. J. Terpolilli, R. P. Tiwari, M. J. Dilworth, G. W. O'Hara and J. G. Howieson

Centre for *Rhizobium* Studies, Murdoch University, Murdoch 6150, Western Australia

The *Medicago* genus is of global importance to agriculture, with the perennial *M. sativa* being the most widely cultivated and studied member. After many years of studying this plant along with its microsymbiont *Sinorhizobium meliloti*, it became clear that another host was required to allow simultaneous study of the genetic determinants of both symbiotic partners. *M. sativa* was unsuited to this role as it is autotetraploid, allogamous and shows strong in-breeding depression, making the analysis of recessive mutations no easy task. Researchers identified the annual medic *M. truncatula* as a viable alternative as this host is diploid, autogamous and possess a rapid generation time, among other traits. Consequently, this organism was chosen for sequencing.

In routine strain-host effectiveness trials, we found that the symbiosis between *M. truncatula* and the sequenced *S. meliloti* 1021 (Sm1021) microsymbiont, may not be optimally matched for N fixation. Comparisons of *M. truncatula* inoculated with Sm1021, WSM1022 (an effective *S. meliloti* from Naxos, Greece) and the sequenced *S. medicae* strain WSM419, showed Sm1021 produced lower plant shoot dry weights and shoot N concentration. Nodule number was almost double and nodules were small and pale to green in colour, as opposed to the large, pink nodules observed on *M. truncatula* inoculated with WSM1022 or WSM419. The timing of nodulation was delayed with Sm1021 on *M. truncatula*. We also tested the effectiveness of the galactoglucan-producing derivative of Sm1021 (Rm8530 $expR^+$), to see if restoration of a mucoid colony phenotype would increase N fixation. We found no difference with Rm8530 on *M. sativa*, but substantially reduced effectiveness on *M. truncatula*, when compared to Sm1021. This reflects a poorly effective symbiosis between Sm1021 and *M. truncatula*, which cannot be overcome by expression of the *expR* gene. Sm1021 and *M. truncatula* are not well matched for N fixation. If investigation of an efficient symbiosis is required, researchers could utilise the *S. medicae* WSM419 genome (soon to be released) or a more effective *S. meliloti* strain could be sequenced (e.g., WSM1022).

This work was supported by a Grains Industry Research Scholarship from GRDC to JT.

F. D. Dakora et al. (eds.), *Biological Nitrogen Fixation: Towards Poverty Alleviation through Sustainable Agriculture.*

145

© Springer Science + Business Media B.V. 2008

NON-SYMBIOTIC NITROGEN FIXATION IN SOIL, LITTER, AND PHYLLOSPHERE IN DRY CHACO FOREST OF ARGENTINA

A. Abril and L. Noe

Microbiologia Agricola, FCA – Universidad Nacional de Cordoba, Argentina
(Email: aabril@agro.uncor.edu)

It is assumed, on the basis of measurements of BNF in soils, that N input through non-symbiotic biological fixation is of low significance in terrestrial ecosystems. However, there is evidence that BNF occurs in other habitats, like in decomposing litter and on the leaf surface (phyllosphere), but their magnitude is scarcely known.

We evaluated the abundance of N fixers and the activity of nitrogenase (ARA method) in leaves (trees, shrubs, and grasses), litter (trees and grasses), and soil (under trees, under shrubs, and bare soil) in a dry forest of Argentina. Biological N fixation was always lower in soils than on leaves and litter. The highest value was detected in leaves of grasses and the lowest in the soil under trees. The abundance of nitrogen-fixing bacteria was similar at all sites, suggesting that populations could fix N_2 only under conditions suitable for nitrogenase activity (Figure 1). Our results show that BNF in arid environments may have been underestimated because non-symbiotic fixers in litter and phyllosphere were not considered.

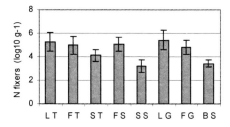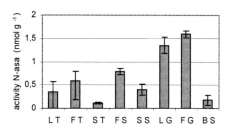

Figure 1. Abundance of N fixers (left) and nitrogenase (N-ase; right) activity in the analyzed situations. (LT, tree litter; FT, tree leaves; ST, soil under trees; FS, shrub leaves; SS, soil under shrubs; LG, grass litter; FG, grass leaves; and BS, bare soil.)

F. D. Dakora et al. (eds.), *Biological Nitrogen Fixation: Towards Poverty Alleviation through Sustainable Agriculture.*
© Springer Science + Business Media B.V. 2008

147

INOCULATION AND COMPETITIVENESS OF A *BRADYRHIZOBIUM* SP. STRAIN IN SOILS CONTAINING INDIGENOUS RHIZOBIA

C. Bonfiglio[1], P. Bogino[2] and W. Giordano[2]

[1]Síntesis Química SAIC, Buenos Aires, Argentina. [2]Dpto de Biología Molecular, Universidad Nacional de Río Cuarto, Río Cuarto, Córdoba, Argentina
(Email: wgiordano@exa.unrc.edu.ar)

The successful nodulation of legumes by a particular rhizobial strain is determined by the competitive ability of that strain against a mixture of other native and inoculant rhizobia. Selected *Bradyrhizobium* sp strains inoculated on peanut seeds often fail to occupy a significant proportion of nodules when a competitor rhizobial population is already established in soil. Such failure may result either from a genetic or physiological advantage of the adapted soil population over the introduced inoculant, or from a positional advantage, i.e., the soil population occupies the soil profile where the roots will penetrate, whereas the inoculant remains concentrated around the seeds.

We studied competition among native rhizobia and a *Bradyrhizobium* sp. strain inoculant. The seed-inoculated strain formed fewer nodules (9%) on peanuts than did bradyrhizobia inoculated in-furrow (78%). The rest of the nodules were formed by indigenous strains or both strains (inoculated and indigenous), indicating the positional advantage of soil populations or bradyrhizobia inoculated in-furrow for nodulation. We subsequently assessed the contribution of this positional effect using a laboratory model in which a rhizobial population is stabilized in sterile vermiculite. Our results demonstrate the importance of bradyrhizobial distribution into soil.

International interest in environmentally sustainable development by using renewable resources has drawn attention to the ecologically "benign" nature of biological nitrogen fixation and its potential role in supplying nitrogen for agriculture. Full potential of the symbiotic system may be realized by increasing the number of effective rhizobia through improved inoculation technology. In furrow inoculation offers upper benefits that seed inoculation; for this reason is a technologic tool recommended to inoculate peanut and legumes crops when naturalized rhizobia are present in the soil.

F. D. Dakora et al. (eds.), *Biological Nitrogen Fixation: Towards Poverty Alleviation through Sustainable Agriculture.*
© Springer Science + Business Media B.V. 2008

DIVERSE ROLE OF DIAZOTROPHS IN THE RHIZOSPHERE

F. Y. Hafeez, S. Hameed, M. S. Mirza, F. Mubeen, S. Yasmin, A. Aslam, M. Gull and K. A. Malik

National Institute for Biotechnology & Genetic Engineering (NIBGE), P.O. Box 577, Jhang road, Faisalabad, 38000-Pakistan
(Email: fauzia@nibge.org)

Over the last several years, many diazotrophs have been isolated in our laboratories from nodules of legumes, plant roots, and rhizospheric soil (Hafeez et al., 2005; 2006). In addition to fixing N_2 (0.02–153 μmol mg^{-1} protein h^{-1}), many can solubilize phosphate (189–204 μg/mL). Some strains can produce indole acetic acid at concentrations as high as 977 μg/mL. Such strains, when used as bio-inoculants for cereals, enhance root growth and significantly increase the N and P contents in grain and straw. Some diazotrophs degrade ACC deaminase and result in lower ethylene synthesis. Other selected diazotrophs, such as K1, ER20, JCM 1270 and Ca18, successfully mobilize Zn, e.g., *Enterobacter* sp. 3.1.1.C solubilized both $Zn_3(PO_4)_2$ and ZnO in plate assays. Some diazotrophs produce antifungal metabolites, like antibiotics, siderophores, and HCN and have biocontrol activity against fungal pathogens. Greenhouse evaluations suggest that *Pseudomonas putida* StS3, *Pseudomonas* sp. SPR7, *P. fluorescens* Mst 8.2, and *Enterobacter* sp. 3.1.1.C have potential as biocontrol agents with decreased disease incidence of potato black scurf, rice sheath blight, and wheat root rot. Other diazotrophs produce the broad spectrum antifungal 2,4-diacetylphloroglucinol (to 100 ppm). Some (4.1.1.B, 6.1.1.B, S4, S10, 17A1, 18 A1, etc.) produce chitinase, indicating mycolytic activity. Others inhibit growth of less competitive bacteria by producing bacteriocins, the genes for which have been identified. One consortium of diazotrophs has great potential for promoting rice plant growth and vigor with consequent suppression of weed competition. The most effective of these diazotrophs are used as the biofertilizer 'Biopower' in Pakistan. The use of 'Biopower', plus half the recommended N-fertilizer, resulted in considerable yield increase. These results increase our understanding of the diversity and role of diazotrophs in the rhizosphere and will help improve the quality of beneficial microorganism as inoculants for sustainable agriculture.

References

Hafeez FY et al. (2005) Environ. Exp. Bot. 54, 142–147.
Hafeez FY et al. (2006) Agron. Sus. Develop. 26, 143–150.

F. D. Dakora et al. (eds.), *Biological Nitrogen Fixation: Towards Poverty Alleviation through Sustainable Agriculture.*
© Springer Science + Business Media B.V. 2008

ROLE OF NATURE'S FERTILITY PARTNERS WITH CROP PROTECTANTS FOR SUSTAINABLE AGRICULTURE

F. Mubeen[1], A. Aslam[1], V. Radl[2], M. Schloter[2], K. A. Malik[1] and F. Y. Hafeez[1]

[1]National Institute for Biotechnology and Genetic Engineering (NIBGE), P.O. Box 577, Jhang Road, Faisalabad, Pakistan; [2]GSF – National Center for Environment and Health, Institute of Soil Ecology, Munich, Germany

Plant growth-promoting rhizobacteria (PGPR) with their beneficial properties act as Nature's fertility partners and are used as biofertilizer (Hafeez et al., 2002). Fungicides are crop protectants but unfortunately, these chemicals affect the beneficial bacteria especially when used with the biofertilizer. This study is the first report concerning the interaction of PGPR and fungicides.

Fungicide-resistance patterns, studied through the disc-diffusion method showed that Darosal and Mancozeb inhibited the growth of phosphate-solubilizing *Agrobacterium* strain Ca-18 and diazotrophic strains of *Pseudomonas* and *Azospirillum*, respectively, whereas all other strains were resistant to these fungicides. Estimation of most probable number and plate count of viable bacterial cells in fungicide-supplemented media showed similar results. Their effect on bacterial activities, like nitrogen fixation and growth-hormone production, showed that Alert Plus, Mancozeb and Darosal were overall negative, whereas Benlate had a positive effect on physiological efficiency, but this effect varied among strains. A genotoxicity study through the comet assay (Mubeen et al., 2006) indicated that no fungicide caused significant DNA damage. Fungicides affect rhizospheric bacterial communities differently. Using Denaturing Gradient Gel Electrophoresis, the biofertilizer-treated bacterial community was most similar to the community in the presence of Benlate, but different from that with Darosal. Pesticides/ fungicides at the recommended doses did not significantly reduce indigenous microbial activities and sometimes, as with Benlate, they are stimulatory. The recommended dose usually decreases bacterial growth *in vitro* but under field conditions, does not significantly affect crop yield. Thus, the interaction between pesticides and soil micro-flora may differ under field conditions and in the laboratory.

F. D. Dakora et al. (eds.), *Biological Nitrogen Fixation: Towards Poverty Alleviation through Sustainable Agriculture*.
© Springer Science + Business Media B.V. 2008

153

References

Hafeez FY et al. (2002) Tech. for Sust. Agri. 67–73.
Mubeen F et al. (2006) Pak. J. Bot. 38, 1261–1269.

CHARACTERIZATION OF BRADYRHIZOBIA ISOLATED FROM ROOT NODULES OF *CYTISUS TRIFLORUS* IN THE RIF OCCIDENTAL OF MOROCCO

R. Chahboune[1], M. R. El Akhal[1], A. Arakrak[1], M. Bakkali[1], A. Laglaoui[1], J. J. Pueyo[2] and S. Barrijal[1]

[1]Équipe de Recherche Valorisation Biotechnologique des Microorganismes, Département des Sciences de la Vie, FST Tangier, Morocco; [2]Laboratorio de Ciencias Medioambiantales, CSIC, Madrid, Spain

Over the last 3 decades, soil in the north of Morocco has been heavily degraded as a result of human activities. In the north of Morocco, particularly the Rif occidental, there is strong anthropozoogenic pressure, through the practice of clearing the land to profit from the extension of Cannabis culture and an increase in the frequency of forest fires. These factors threaten the survival of the shrubby legume in its ecological niche.

To help decrease this pressure on the Moroccan forest, which annually provides a fodder reserve estimated at 1.5–2 billion fodder units and is likely to put the national livestock industry at risk, an evaluation of our sylvo-pastoral natural resources is necessary. Many legumes, such as *Cytisus triflorus*, establish symbiosis with soil rhizobia forming nodules capable of fixing atmospheric N_2. *C. triflorus* L'Hérit is a shrubby legume in regions with a Mediterranean-type climate and is often associated with *Quercus canariensis* or *Q. Pyrenaica*. In Mediterranean areas, shrubby legumes are key components of natural ecosystems and are good candidates for revegetation projects, including the Rif occidental because it is autochthonous and highly valued by goats. However, the population of symbiotic micro-organisms needs to be assessed to select suitable symbiotic partners and for introduction into new environments. In Morocco, no study has been carried out on the *C. triflorus-Rhizobium* symbiosis and the symbiotic effectiveness of indigenous *C. triflorus* rhizobia needs evaluation. As a first step, phenotypic and genotypic characteristics of *C. triflorus* rhizobial populations need to be determined.

Fifty-two strains isolated from root nodules of *C. triflorus* were genetically characterized by 16S rDNA. PCR amplification products were sequenced. All strains clustered into a group consisting of slow-growing bradyrhizobia. The results show that the Moroccan shrubby legume *C. triflorus* L'Hérit is nodulated by bradyrhizobia strains, which show 99% homology with *Bradyrhizobium genosp*. AD. This is the first study to indicate that *C. triflorus* is nodulated by bradyrhizobia in Morocco.

F. D. Dakora et al. (eds.), *Biological Nitrogen Fixation: Towards Poverty Alleviation through Sustainable Agriculture.*
© Springer Science + Business Media B.V. 2008

DIVERSITY OF RHIZOBIA ISOLATED FROM NODULES OF PEANUT (*ARACHIS HYPOGAEA* L.) IN MOROCCO

M. R. El Akhal[1,2], A. Rincón[2], F. Arenal[2], A. Laglaoui[1], N. El Mourabit[3], J. J. Pueyo[2] and S. Barrijal[1]

[1]Equipe de Recherche Valorisation Biotechnologique des Microorganismes, Département des Sciences de la Vie, FST Tanger, Morocco; [2]Departamento de Fisiología y Ecología Vegetal, Instituto de Recursos Naturales, Centro de Ciencias Medioambientales, CSIC, Madrid, Spain; [3]Institut National de Recherche Agronomique (INRA), Tanger, Morocco
(Email: barrijal@yahoo.fr)

Peanut (*Arachis hypogaea*) was introduced into Morocco 100 years ago and it is mostly cultivated in the north-western region. Despite the local importance of this crop, the diversity of rhizobial isolates that nodulate peanut in Moroccan soils has not been investigated so far. The diversity and phylogeny of 64 isolates obtained from root nodules in different locations of Morocco were studied using different fingerprinting techniques, such as restriction fragment length polymorphism (RFLP), analysis of the 16S rRNA gene, IGS-PCR, tDNA-PCR, as well as sequence analysis of the 16S rRNA gene.

The results revealed *Bradyrhizobium* as the predominant genus that nodulates *Arachis hypogaea*, with a low degree of diversity at the species level. Restriction patterns of the 16S RNA gene obtained with four endonucleases allowed clustering of the isolates into two major groups: Group I consisted of slow-growing bradyrhizobia; and Group II included fast-growing rhizobia. Results obtained from both IGS-PCR and tDNA-PCR, were jointly analyzed to obtain a combined dendrogram, which grouped the isolates according to their geographic origin, and indicated a high degree of similarity within isolates of Group I. Analysis of 16S DNA sequences of isolates from Group I revealed that they were phylogenetically more related to *B. japonicum* than to other *Bradyrhizobium* species.

F. D. Dakora et al. (eds.), *Biological Nitrogen Fixation: Towards Poverty Alleviation through Sustainable Agriculture.*
© Springer Science + Business Media B.V. 2008

DIVERSITY OF RHIZOBIA THAT NODULATE *COLUTEA ARBORESCENS* ISOLATED FROM DIFFERENT SOILS OF MOROCCO

O. Mohamed[1], H. Benata[1], H. Abdelmoumen[1], M. Amar[2] and M. M. El Idrissi[1]

[1]Laboratorie de Biologie des Plantes et des Microorganismes, Faculte des Sciences, Universite Mohamed Premier, Oujda, Morocco; [2]Laboratoire de Microbiologie et Biologie Moleculaire, Center National de Recherche Scientifique et Technique, Rabat, Morocco
(Email: mssbah49@yahoo.fr)

Eighteen strains of rhizobia were isolated from root nodules of *Colutea arborescens* grown in the different soils of the eastern region of Morocco. These strains are able to grow under different stress conditions. Many strains tolerate high salt concentrations and all of them grow at 35°C and in media with pH values ranging from 6 to 9. They also show resistance to several antibiotics at the concentrations used. The tolerance level to heavy metals is diverse and depends on the strains and the concentrations used.

Among 17 sugars tested as sole carbon sources, *Colutea arborescens* rhizobia metabolize monosaccharides and disaccharides more easily than gluconic and citric acids and polysaccharides. All the strains are catalase- and oxidase-positive and none of them grew on 8% KNO_3. The numerical analysis of the data showed a correlation between phenotypic and genetic approaches. Although, the rep PCR method, using BOX AIR primer, permitted the identification of five strains as sinorhizobia, the other strains require other analytical methods, such as the sequencing of 16S rDNA, to determine their taxonomic position.

F. D. Dakora et al. (eds.), *Biological Nitrogen Fixation: Towards Poverty Alleviation through Sustainable Agriculture.*
© Springer Science + Business Media B.V. 2008

DIVERSITY OF RHIZOBIA THAT NODULATE TWO MEDITERRANEAN LEGUMES, *ADENOCARPUS DECORTICANS* AND *RETAMA MENOSPERMA*

H. Abdelmoumen[1], M. Neyra[2], R. T. Samba[3] and M. M. El Idrissi[1]

[1]Laboratoire de Biologie des Plantes et des Microorganismes, Faculte des Sciences, Universite Mohamed Premier, Oujda, Morocco; [2]Laboratoire des Symbiosses Tropicales et Mediterraneennes, UMR113, IRD/INRA/CIRAD/UM2/Argo-M, Campus International de Baillarguet, Montpellier F34398, France; [3]Laboratoire Commun de Microbiologie, IRD/ISRA/UCAD, Dakar, Senegal

We analyzed the diversity of rhizobia nodulating two Mediterranean legumes, *Adenocarpus decoritcans* and *Retama monosperma*. These genera are native to the Mediterranean area and tropical Africa. There is no information concerning their N_2-fixing symbiotic bacterial partners. The nodulating species are *A. complicatus* and *A. foliolosus* in South Africa and *A. decorticans* in Great Britain. In Morocco, best known are *A. decorticans* in the Beni Znassen Mountains (1,400 m altitude), *A. anageryfolius* in the High Atlas, and *A. bacqueii* in the region of Figuig, an arid to Sahelian area. *Retama monosperma* (known as *Genista monosperma* or bridal broom) is a tall ornamental woody shrub with white papilionaceous flowers. The plant is native to Morocco and other countries of North Africa. It grows on sandy soils at the coast and border marshes. It crowds out other vegetation and can dominate grasslands and disturbed habitats.

Thirty strains of rhizobia were isolated from *A. decorticans* root nodules and 30 strains from bridal broom root nodules. We analyzed 72 phenotypic (physiological, biochemical and cultural characteristics) parameters of these and some reference strains. The resulting dendrogram showed high phenotypic diversity of the strains, which were all different from the reference strains. The genetic diversity of the strains was assessed by RFLP-PCR of the 16S and 23S rDNA genes. The purified PCR products were individually restricted with RsaI, MspI, HaeIII and CFO1. All endonucleases produced polymorphic restriction patterns. Combination of the restriction patterns obtained with each endonuclease produced only two dominant patterns with *Adenocarpus* rhizobia and two patterns with *Retama* rhizobia. The sequencing of the 16S rDNA gene of a strain representing the most dominant pattern in *Adenocarpus* rhizobia showed that it is much closer to *Ralstonia* and *Pseudomonas*-like rhizobia, whereas the sequencing of a strain of *Retama* rhizobia showed that it is affiliated to the unclassified rhizobia. More studies need to be undertaken to indicate the taxonomic position of these strains.

F. D. Dakora et al. (eds.), *Biological Nitrogen Fixation: Towards Poverty Alleviation through Sustainable Agriculture.*   161
© Springer Science + Business Media B.V. 2008

RESPONSE OF COMMON BEAN AND FABABEAN TO EFFECTIVE NODULATION IN SALT- AND DROUGHT-STRESSED AREAS AROUND THE NILE DELTA

Y. Yanni, M. Zidan, R. Rizk, A. Mehesen, F. A. El-Fattah and H. Hamisa

Sakha Agricultural Research Station, Kafr El-Sheikh 33717, Egypt
(Email: yanni244@hotmail.com)

Symbiotic performances of some common and fababean rhizobial strains, collected from slightly to highly saline and dry soils around the Nile delta, varied with different strain/variety combinations in 64 locations of salt- and drought-affected areas in the east and west of the delta, where the soil textures were calcareous, sandy-loam, sandy, and clay-loamy maintained at 50–60% of their field capacities.

Positive responses to inoculation were observed as increased plant growth, seed yield, straw, harvest index (percentage of seed yield/seed + straw), and fertilizer N-use efficiency (kg seed yield/kg fertilizer-N). The observed superiority of researchers' recommendations over the conventional farmers practices was attributed to integrated crop management practices involving inoculation with high performing strains, proper stand densities, optimized NPK fertilization, integrated pest management with minimum amounts of pesticides, and new responsive varieties. An additional positive effect was the extent and fluency of knowledge dissemination to junior researchers, who were directly supervising field experimentation sites, to the extension specialists and to the cooperating farmers.

F. D. Dakora et al. (eds.), *Biological Nitrogen Fixation: Towards Poverty Alleviation through Sustainable Agriculture.*   163
© Springer Science + Business Media B.V. 2008

SECTION 2

NODULE ORGANOGENESIS
AND PLANT GENOMICS

PART 2A

HOST RESPONSE TO INVASION

NOD-FACTOR PERCEPTION IN *MEDICAGO TRUNCATULA*

J. Cullimore[1], B. Lefebvre[1], J. F. Arrighi[1], C. Gough[1], A. Barre[2], J. J. Bono[2], P. Rougé[2], E. Samain[3], H. Driguez[3], A. Imberty[3], A. Untergasser[4], R. Geurts[4], T. W. J. Gadella, Jr.[5], J. Cañada[6] and J. Jimenez-Barbero[6]

[1]Laboratoire des Interactions Plantes Microorganismes, INRA-CNRS; [2]Surfaces Cellulaires et Signalisation chez les Végétaux, CNRS-UPS, 31326 Castanet-Tolosan, France; [3]Centre de Recherches sur les Macromolécules Végétales, CNRS (affiliated with Université Joseph Fourier), 38041 Grenoble, France; [4]Laboratory of Molecular Biology, Department of Plant Science, Wageningen University, 6703HA, The Netherlands; [5]Section of Molecular Cytology and Centre for Advanced Microscopy, SILS, University of Amsterdam, 1098 SM Amsterdam, The Netherlands; [6]CSIC, Centro de Investigaciones Biológicas, 28040 Madrid, Spain

Nod factors of *Rhizobium* have been described as the key to the legume door (Relic et al., 1994) and consequently, Nod factor receptors can be considered as the locks which allow *Rhizobium* access. This concept lies at the heart of the mechanisms of partner recognition leading to the establishment of the legume-*Rhizobium* symbiosis. Since the discovery of *Rhizobium* Nod factors in 1990 (see Dénarié et al., 1996), much attention has been paid to how these intriguing molecules interact with plants and this has led to the recent identification of legume symbiotic receptor genes. These studies were aided by focussing on model systems in which genetic and genomic approaches could be used. In this short review, we present our current knowledge of Nod factor perception in *Medicago truncatula*, a model legume that is nodulated by *Sinorhizobium meliloti*.

Sinorhizobium meliloti and its Nod Factors

The major Nod factor produced by *S. meliloti* is an *O*-sulphated, *O*-acetylated lipochito-tetra-oligosaccharide (LCO) termed NodSm-IV(Ac,S,C16:2). Studies on *S. meliloti* host-specificity mutants have shown that the *nodH* gene, encoding a sulpho-transferase, is the major specificity determinant of nodulation on *Medicago* species, whereas the *nodL* and the *nodFE* genes, determining *O*-acetylation and the production

F. D. Dakora et al. (eds.), *Biological Nitrogen Fixation: Towards Poverty Alleviation through Sustainable Agriculture.*
© Springer Science + Business Media B.V. 2008

of the specific C16:2 fatty acid, respectively, are required for infection. Studies on these mutants and with the corresponding LCOs led to the suggestion that *M. truncatula* may have two types of receptors for Nod factors: (i) the signalling receptor, which is required for initial responses; and (ii) the entry receptor, which is required for infection (Ardourel et al., 1994). The entry receptor is predicted to have more stringent requirements for the structure of the Nod factor (*O*-acetylation and the specific C16:2 acyl chain), whereas the signalling receptor requires the *O*-sulphate, but is less specific for these other structures. Because Nod-factor responses occur at concentrations as low as 10^{-13}M, their receptors are expected to have high affinities for Nod factors and should have high specificities in order to discriminate between closely related molecules.

Studies with fluorescent Nod factors have shown that they bind to cell walls of *M. truncatula* roots but this binding is relatively non-specific because it is independent of the *O*-sulphate (Goedhart et al., 2003). Biophysical studies have shown that these fluorescent LCOs rapidly incorporate into plant and artificial membranes (Goedhart et al., 1999), but it is not clear if this attribute is related to the mechanism of Nod-factor perception. In solution, Nod factors are monomers at sub-micromolar concentrations and a combination of NMR spectroscopy and molecular mechanics and dynamics calculations have shown that, in aqueous solution, they preferentially form structures in which the acyl chain and the oligosaccharide backbone are in a quasi-parallel orienttation (Gonzalez et al., 1999). However, differences in the saturation of the acyl chain and in the polarity of the solvent may lead to changes to this orientation. The importance of the acyl-chain structure for determining the shape of the Nod factors suggests that the lipid moiety may play an active role in receptor binding and ligand specificity rather than a more passive role such as membrane-anchoring (Groves et al., 2005).

Nod-Factor Perception Involves LysM-Receptor-Like Kinases

A combination of forward genetics and analysis of allelic variation has led to the identification of genes involved in Nod-factor perception and signal transduction (see Geurts et al., 2005). In *M. truncatula*, a single gene has been identified which is required for all Nod factor responses. This gene is called *NOD FACTOR PERCEPTION (NFP)* and has recently been shown to encode a lysin-motif receptor-like kinase (LysM-RLK) (Arrighi et al., 2006). In pea, a gene, *SYM2*, has been identified in which different allelic variants control the interaction with Nod factors differing in the presence of an *O*-acetate on the reducing sugar. The corresponding genic region has been cloned from *M. truncatula* through genetic synteny and this region has been shown to contain a cluster of seven LysM-RLK genes, which have been termed the *LYSM-RECEPTOR KINASE (LYK)* genes. The gene encoding LYK3 is required for infection of *M. truncatula* by a rhizobial *nodFE* strain, but not for initial responses (Limpens et al., 2003). These results are consistent with NFP and LYK3 constituting part of the signalling and entry receptors, respectively.

Both NFP and LYK3 are predicted to have similar overall structures with three LysM domains in the extracellular region, separated from an intracellular kinase-like domain by a single transmembrane spanning helix. Studies on NFP have shown that it is a membrane-located highly glycosylated protein (Mulder et al., 2006). LysM domains are small domains of about 45 residues and, because they occur in proteins and enzymes that interact with GlcNAc oligomers including a plant chitin receptor (Kaku et al., 2006), it has been postulated that the symbiotic LysM-RLKs directly bind Nod factors. Using molecular modelling, structures of the three LysM domains of NFP have been proposed and surface analysis and docking calculations have been used to predict the most favoured binding modes for chitooligosaccharides and Nod factors (Mulder et al., 2006; Arrighi et al., 2006). The relative size of a LysM domain and a Nod factor suggests that a maximum of one Nod factor could bind per domain. Studies are in progress to determine whether this protein and/or LYK3 bind Nod factors.

Studies on the extensive sequence data bases of *M. truncatula* (expressed sequence tags and genomic sequences) have shown that this model legume contains at least 17 LysM-RLK genes, most of which can be divided between two clades, based on kinase domains; these are the *LYK* genes and the *LYK-related (LYR)* genes, which includes *NFP*. Unlike the *LYK* genes, the *LYR* genes seem to have aberrant kinase domain structures. Indeed, studies on the intracellular domains of NFP and LYK3 have shown that only LYK3 has autophosphorylation activity, thus raising the question of how NFP transmits a signal to downstream components if it is enzymatically inactive (Arrighi et al., 2006).

Studies on Nod Factor-Binding Sites Suggest a Role for DMI2/DMI1 in LCO Perception

Forward genetic studies have identified three genes that are required for infection by *Rhizobium* and for Nod factor-induced gene expression, but not for early Nod factor responses, such as root-hair deformation or calcium uptake. These genes are also required for infection by arbuscular mycorrhizal fungi (leading to the beneficial AM symbiosis) and have been termed *DOES NOT MAKE INFECTIONS (DMI)* genes. *DMI2* encodes a leucine-rich-repeat receptor-like kinase, whereas *DMI1* encodes a channel-like protein. Both these genes are required for a calcium spiking response, whereas *DMI3*, which encodes a calcium- and calmodulin-dependent protein kinase, probably interprets this response (see Geurts et al., 2005). Using a radiolabelled Nod factor ligand, Bono and colleagues have characterised three different binding sites for Nod factors in roots and cell cultures of *Medicago*. Remarkably, a high affinity Nod-factor binding site in roots of *M. truncatula*, NFBS3, is absent in roots of *dmi1* and *dmi2* mutants, thus suggesting that DMI1/DMI2 are involved in LCO binding (Hogg et al., 2006). Further work is required to test whether the binding is related to the symbiosis with *Rhizobium* and whether the binding occurs directly to DMI1/DMI2 or involves another protein, such as a LysM-RLK.

A Model for Nod-Factor Perception in *M. Truncatula*

Current work favours a two-step model of Nod-factor perception in root epidermal cells of *M. truncatula* (Figure 1). The first step requires NFP and leads to calcium uptake and root-hair deformation (Had). Due to the aberrant kinase domain of NFP, another LysM-RLK may also act at this step in order to transmit the signal to downstream effectors. *DMI1/DMI2* apparently act downstream of NFP and are required for the induction of calcium spiking and nodulin-gene expression, which are non-stringent early responses. Initiation of infection threads by *S. meliloti* is a stringent response and requires wild-type bacteria that produce Nod factors with the C16:2 acyl chain. This second step requires *LYK3* and *DMI2* and recent studies on RNAi lines of *NFP* suggest that this gene also intervenes during this step. Whether these proteins form multimeric comp-lexes is an interesting possibility. Note DMI1 is not included in the model because it has recently been reported to be located in the nuclear membrane (Riely et al., 2006).

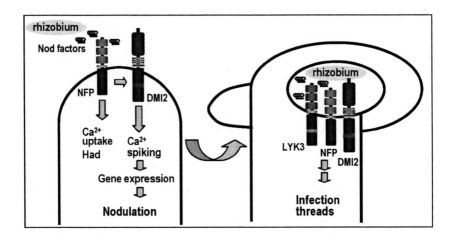

Figure 1. The two-step model for nod-factor perception in *M. trunculata*.

Acknowledgements

The authors gratefully acknowledge funding from the European Union Marie Curie Research Training Network programme (contract NODPERCEPTION) and from the French Agence Nationale de Recherche (contract NodBindsLysM).

References

Ardourel M et al. (1994) Plant Cell 6, 1357–1374.
Arrighi JF et al. (2006) Plant Physiol. 142, 265–279.
Dénarié J et al. (1996) Annu. Rev. Biochem. 65, 503–535.
Geurts R et al. (2005) Curr. Opin. Plant Biol. 8, 346–352.

Goedhart J et al. (1999) Biochemistry 38, 10898–10907.
Goedhart J et al. (2003) Mol. Plant-Microbe Interact. 16, 884–892.
Gonzalez L et al. (1999) Carb. Res. 318, 10–19.
Groves P et al. (2005) Org. Biomol. Chem. 3, 1381–1386.
Hogg B et al. (2006) Plant Physiol. 140, 365–373.
Kaku H et al. (2006) Proc. Natl. Acad. Sci. USA 103, 11086–11091.
Limpens E et al. (2003) Science 302, 630–633.
Mulder L et al. (2006) Glycobiology 16, 801–809.
Relic B et al. (1994) Mol. Microbiol. 13, 171–178.
Riely B et al. (2006) Plant J. 49, 208–216.

NODULATION CONTROL IN LEGUMES

P. M. Gresshoff[1], A. Indrasumunar[1], S. Nontachaiyapoom[1], M. Kinkema[1], Y.-H. Lin[1], Q. Jiang[1], D. X. Li[1], A. Miyahara[1], C. Nguyen[1], D. Buzas[1], B. Biswas[1], P. K. Chan[1], P. Scott[1], T. Hirani[1], M. Miyagi[1], M. Djordjevic[1,4], B. J. Carroll[1,2], A. Men[3] and A. Kereszt[1]

[1]ARC Centre of Excellence for Integrative Legume Research; [2]School of Molecular and Microbial Sciences and LAFS, The University of Queensland, St. Lucia, Brisbane QLD 4072; [3]Australian Genome Research Facility, Brisbane, Australia; [4]Genome Interaction Group, RSBS, Australian National University, Canberra, ACT, Australia

Nodulation and concomitant symbiotic nitrogen fixation are critical for the productivity of the legume, yielding food, feed and fuel. The nodule number in legumes is regulated by numerous factors including the number and efficiency of the interacting *Rhizobium* bacteria and abiotic stresses as well as endogenous processes involving phytohormones, nodulation reception systems and autoregulation of nodulation (AON; Kinkema et al., 2006). The original discovery of the AON-controlling LRR receptor kinases, GmNARK/ LjHAR1/MtSUNN, which is active in leaf tissue of several legu-mes, now has led to an analysis of the mechanism underlying the signal transduction.

Autoregulation Mutants of Legumes

Cloning of chemically-induced, loss-of-function mutant alleles, which led to increased nodulation in soybean, *Lotus* and *Medicago*, implicated the AON-receptor kinases GmNARK, LjHAR1 and MtSUNN in systemic nodulation control (Krusell et al., 2002; Nishimura et al., 2002; Schnabel et al., 2005; Searle et al., 2003). By homology, the pea hypernodulation locus *PsSYM29* was also shown to be mutated in an LRR receptor kinase (Krusell et al., 2002) similar in structure to the other AON-kinases and broadly related to the CLAVATA1 protein of *Arabidopsis*. The *GmNARK* and *LjHAR1* genes were expressed in all plant parts except the apical meristem area and the root-elongation zone (S. Nontachaiyapoom, unpublished data, 2006). Gene fusions of the *GmNARK* as well as *LjHAR1* promoters with reporter genes encoding ß-glucuronidase (GUS) and green fluorescent protein (GFP) indicate vasculature-specific expression in all plant tissues. The mRNA expression levels relative to the actin or ribosomal RNA genes suggest relatively high transcript levels, but this may not be reflected in high protein levels. So

F. D. Dakora et al. (eds.), *Biological Nitrogen Fixation: Towards Poverty Alleviation through Sustainable Agriculture.*
© Springer Science + Business Media B.V. 2008

173

GmNARK protein has not been detected despite enrichment for veins and membrane fractions (T. Hirani, unpublished data, 2006), presumably because receptor kinases possess a high degree of protein turn-over to facilitate rapid responsiveness. The *GmNARK, LjHAR1* and *MtSUNN* promoters share many structural features, including a possible vasculature specifying domain. Deletion analysis places this domain 900-1,700 bp from the transcriptional start. Promoter analysis in soybean suggests that *GmNARK* possesses two transcriptional start sites (S. Nontachaiyapoom, unpublished data, 2006). Promoter deletion analysis failed to define a negative regulatory region that controls the absence of GmNARK expression in the SAM and the root-extension zone. Possibly, no such regulatory region exists and expression is controlled by development itself. Both regions lacking *GmNARK* expression also lack vasculature!

The signals inducing the AON circuit remain uncharacterized, although research into their nature is progressing. Using well-defined symbiotic mutants of pea, coupled with approach-grafts and time-delayed inoculations we are attempting to define the onto-genic stage of nodule initiation that leads to the 'activated state' which triggers AON (Dongxue Li, unpublished data, 2006). Are cell-divisions or even nitrogen fixation required to send the signal? What is the signal? Is it received directly by NARK/HAR1/ SUNN or is there a second message? We have initiated an analysis of xylem-derived proteins using bleeding sap of inoculated and uninoculated wild-type mutant root systems (M. Djordjevic, unpublished data, 2006). Over 20 proteins were identified by proteomic analysis, including a member of the trypsin inhibitor family and a lipid transfer protein (LTP). The latter are implicated in long distance transport of smaller molecules, including jasmonic acid. Interference of mRNA levels (RNAi) of a soybean LTP in transgenic roots resulted in the desired RNA decrease but did not correlate with nodulation control (C. Nguyen, unpublished data, 2006). Other members of the gene family need to be tested to allow generalizations to be made.

The analysis of the down-stream events of the GmNARK receptor kinase are more advanced. Affymetrix gene chips (containing over 37,000 soybean genes) were used to quantify gene expression in wild type and *Gmnark* mutants (M. Kinkema, unpublished data, 2006). *Bradyrhizobium japonicum* inoculation of roots results in suppression of leaf (but not root) gene expression for a number of genes involved in the biosynthesis of jasmonic acid (JA) within a 2–5 day period. This suppression is not observed in *Gmnark* mutants, indicating that the inoculation signal is transmitted through GmNARK to regulate JA levels. In non-inoculated roots of supernodulation mutant *nts1007* (mutated in Q106*) JA was higher than in wild type (Meixner et al., 2005), whereas IAA levels remained identical. Total IAA levels in soybean leaves of inoculated and uninoculated Bragg wild type and inoculated *nts1007* were identical, although proper interpretation will require precise tissue-specific analysis. With *Medicago, Rhizobium* inoculation lowers the amount of auxin translocated from the shoot to the root but this process is absent in the *sunn* mutant (U. Mathesius, personal communication, 2006). We are still uncertain about the causal connection of JA levels and IAA translocation. Possibly, intermediates, such as brassinosteroids, may be involved.

The direct analysis of leaf-derived inhibitors of nodulation progress was facilitated through the development of a liquid feeding method involving the use of the *nts1007* mutant, aqueous extracts, petiole feeding and determination of nodule suppression within a short (8 day) assay period (Y-H. Lin, unpublished data, 2006). An extract from wild-type leaves, harvested from *Bradyrhizobium*-inoculated soybean plants, significantly suppressed supernodulation of *nts1007*, whereas extracts from mutants or uninoculated wild type did not. The extract appeared to be heat stable and of low molecular weight.

Supernodulation and hypernodulation caused by mutations in the AON receptor kinases of several legumes are frequently associated with reduced main root growth (Buzas and Gresshoff, 2007). This level of pleiotropy varies among legumes but is most pronounced in *Lotus japonicus*. We determined that the inhibition of main root growth in the uninoculated state is controlled by both the genotype of the root or shoot, whereas regulation of nodule number is predominantly regulated by the shoot (or leaf more precisely). This may explain the general expression pattern of the AON receptor kinase.

Ethylene and abscisic acid (ABA) are known to alter nodulation responses. We have created both transgenic (using *Arabidopsis ETR1-1* and *abi1*) and induced (EMS mutants) variants of *Lotus japonicus* to evaluate the role of the phytohormones (P-K. Chan and B. Biswas, unpublished data, 2006). Using a verified homozygous double mutant of *Ljhar1-1* (hypernodulating) and *LjETR1-1* (ethylene insensitive; Jiang and Gresshoff, 2005), we separated the hypernodulation and inhibited root-formation traits. The double mutant developed abundant nodules as well as a normal length root system. Reciprocal grafts of the double mutant and the *Ljhar1-1* mutant demonstrated that the suppression of root inhibition caused by the absence of HAR1 function is caused by ethylene insensitivity within the shoot (Q. Jiang, unpublished data, 2006). Additionally, the double as well as *Ljhar1-1* mutants exhibited partial cytokinin insensitivity in shoot growth, further linking ethylene and cytokinin signaling in legumes. Whether the common molecular structure of the ethylene (ETR1) and cytokinin (CRE1=AHK4=LHK4) receptors and related down-stream signaling via AHPs is the reason for signal crosstalk remains to be revealed. We presume that AON receptor kinases have twin functions, one controlling nodule numbers, the other regulating root development; both may involve signaling through the cytokinin receptor, cytokinin regulation of ethylene sensitivity, and ethylene-based control of root meristem growth.

Non-nodulation Mutants of Legumes

Mutants lacking the ability to form nodules in legumes have existed for several decades. The last 5 years have seen the isolation of many causative genes and has led to the development of a model for a signaling cascade starting with the Nod-factor receptor components NFR1 and NFR5 (Radutoiu et al., 2003) and terminating a range of nodulation-essential transcription factors (NSP1, NIN1, etc.). In soybean (Mathews et al., 1989), induced mutant *Gmnod49* (allelic with the naturally occurring *rj1* mutant) and *Gmnod139* (allelic with *Gmnn5* produced by J. Harper) behave a loss-of-function mutants (Kinkema et al., 2006). Mutants *nod139* and *nn5* are mutated in *GmNFR5*, a member of the LysM-like receptor kinase gene family. Interestingly soybean harbors two *GmNFR5* genes, both functional in some genotypes, but inactivated in many commercial varieties by a transposon-like insertion. The mutations were complemented by wild-type *GmNFR5* driven by either its native or CaMV 35S promoter in transgenic roots of soybean (A. Indrasumunar and A. Kereszt, unpublished data, 2006).

The *Gmnod49* and *rj1* mutants have frame-shift mutations in another LysM-like receptor kinase, *GmNFR1α* (A. Indrasumunar and A. Kereszt, unpublished data, 2006). This gene is duplicated in the soybean genome as *GmNFR5β* and expressed at a lower RNA level. *GmNFR5β* also exhibits alternative splicing involving either double codon duplications and/or loss of significant exons. Whether such splice variants produce 'viable' proteins and affect phenotype is unknown. *GmNFR5β* was also mutated naturally in a reference soybean cultivar PI437654, but its nodulation competence was not compromised, suggesting that *GmNFR1α* sufficed for normal nodulation. The question arose as to whether *GmNFR5β* is non-functional owing to its low expression level and/or alternative splicing. We do not think so because the *Gmnod49* mutant still exhibits a phenotype, namely subepidermal cell division (Mathews et al., 1989), which do not progress, but are not seen in uninoculated wild-type plants. Genetic complementation with *GmNFR1α*, but not *GmNFR5β*, resulted in the regaining of nodulation competence. Over-expression with *GmNFR1α* additionally provided the complemented plant with increased nodule number, nitrogen gain, and the ability to nodulate at ultra-low *Bradyrhizobium* inoculant levels. Because abiotic stresses on nodulation and nitrogen fixation appear to be alleviated by application of large amounts of exogenous Nod factor (an agriculturally difficult proposition), the increased sensitivity of soybean plants that over-express GmNFR1α may provide an avenue toward decreased abiotic stress effects on soybean root symbioses.

We thank the Australian Research Council for the Centre for Integrative Legume Research (CEO348212), AUSAID, the University of Queensland, the Queensland Government Smart State Fund, and the Thai Government.

References

Buzas DM and Gresshoff PM (2007) J. Plant Physiol. 164, 452–459.
Jiang Q and Gresshoff PM (2005) Symbiosis 40, 49–53.
Kinkema M et al. (2006) Funct. Plant Biol. 33, 770–785.
Krusell L et al. (2002) Nature 420, 422–426.
Mathews A et al. (1989) Protoplasma 150, 40–47.
Meixner C et al. (2005) Planta 222, 709–715.
Nishimura R et al. (2002) Nature 420, 426–429.
Radutoiu S et al. (2003) Nature 425, 585–592.
Schnabel E et al. (2005) Plant Mol. Biol. 58, 809–822.
Searle IR et al. (2003) Science 299,108–112.

BACTEROID DIFFERENTIATION IS LINKED TO EVOLUTION OF HIGH DIVERSITY OF SECRETED PEPTIDES IN THE HOST PLANT

E. Kondorosi, B. Alunni, Z. Kevei, N. Maunoury, W. van de Velde, M. R. Nieto, P. Mergaert and A. Kondorosi

Institut des Sciences du Végétal, CNRS UPR 2355, Avenue de la Terrasse, 91198 Gif-sur-Yvette, France
(Email: Eva.Kondorosi@isv.cnrs-gif.fr)

Bacteroid Differentiation is Dissimilar in *Medicago truncatula* and *Lotus japonicus* Nodules and is Controlled by Plant Factors

Recent developments in nodulation research have led to the elucidation of more and more steps of nitrogen-fixing nodule development (Barnett and Fischer, 2006), but bacteroid differentiation within the plant cells remained elusive. Therefore, we started to characterize the morphology and physiology of bacteroids in nodules of different legumes and compared them to the corresponding free-living bacteria (Mergaert et al., 2006). In *M. truncatula* and closely related species (belonging to the IRLC clade, also known as galegoid clade), bacteroids undergo a remarkable differentiation process involving cell enlargement (Figure 1A), increased membrane permeability, inability for cell division, all the while keeping an active metabolism. These elongated bacteria are polynucleoid and endoreduplicated as we demonstrated by flow cytometry and comparative genomic hybridization (CGH; see Figure 1B, C). Differentiation of bacteroids is amazingly similar to the host cells that also undergo endoreduplication-driven cell enlargement and exhibit loss of cell-division capacity (Vinardell et al., 2003). Thus, the plant cells and bacteria dance a "pas de deux", both following a common irreversible differentiation mechanism.

Interestingly, bacteroids in *L. japonicus* nodules are very similar to free-living bacteria (Mergaert et al., 2006). To determine whether the differences in bacteroid differentiation are due to the bacterial or the host-plant genetic background, we studied the differentiation of two bacterial strains that were able to interact with both determinate and indeterminate nodule-forming legumes. Comparing the morphology and physiology of the same bacterium in the two nodule types demonstrated that *M. truncatula* and related legumes have evolved a strategy for controlling the fate of the bacteria inside the nodules by using plant factors to force the bacteroids toward a terminal irreversible state of differentiation (Mergaert et al., 2006).

F. D. Dakora et al. (eds.), *Biological Nitrogen Fixation: Towards Poverty Alleviation through Sustainable Agriculture.*
© Springer Science + Business Media B.V. 2008

Bacteroid Differentiation is Independent of Nodule Morphology

In the above experiments, bacteroid differentiation was linked to the formation of indeterminate nodules on galegoid legumes (*Medicago, Vicia* or *Pisum*). To verify whether bacteroid differentiation depends on the nodule morphology, we studied the fate of bacteria in *Sesbania rostrata*, which is closely related to *L. japonicus* and, depending on the growth conditions, can form either determinate or indeterminate nodules. Independently of the nodule type, the bacteroids were similar to free-living bacteria, demonstrating that the phylogenetic origin of the plant is the crucial determinant for bacteroid differentiation rather than nodule architecture (W. Van de Velde et al., unpublished data, 2006).

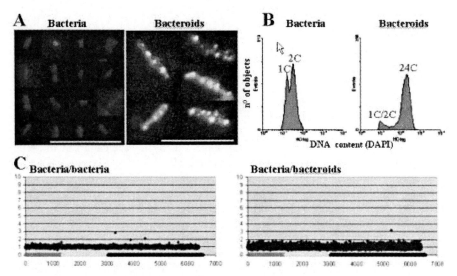

Figure 1. Characterization of *Sinorhizobium meliloti* free-living bacteria and bacteroids isolated from *Medicago truncatula* nodules. A, Microscopic observation of free-living bacteria and bacteroids stained with DAPI (DNA stain). B, Flow cytometry quantification of the DNA content of free-living bacteria (bacteria) and bacteroid cells (the number of C corresponding to the number of genome copies per cell). C, Hybridization ratios of DNA from bacteria/bacteria and bacteria/bacteroids in CGH experiments. The tripartite genome of *S. meliloti* is plotted as follow: dark gray for pSymA; light gray for pSymB; and black for the chromosome (Adapted from Mergaert et al., 2006).

Transcriptomic Identification of Large Gene Families Activated During Bacteroid Differentiation in *M. truncatula* Nodules that are Absent in *L. japonicas*

EST sequencing projects as well as transcriptomic dissections of the nodulation pathway allowed the identification of numerous genes and gene families (Györgyey et al., 2000; Fedorova et al., 2002; Mergaert et al., 2003; El Yahyaoui et al., 2004; Barnett et al., 2004; Lohar et al., 2006). The study of nodulation kinetics and combinations of plant and bacterial mutants, which are halted at different stages of symbiosis, have also been powerful tools in the search for genes that were specifically co-regulated with bacteroid differentiation (N. Maunoury et al., unpublished data, 2006). The majority of these genes, which encode a small secreted peptide, form multigenic families, the largest one of which is the *Nodule-specific Cysteine-Rich* (*NCR*) gene family with more than 300 members (Mergaert et al., 2003). The nodule specific *Glycine-Rich Protein* (*GRP*) gene family is represented by 23 members (Kevei et al., 2002; this work). The NCR and GRP transcripts are present in the infected cells, but members of these gene families display a wide range of spatio-temporal expression (Kevei et al., 2002; Mergaert et al., 2003; Maunoury et al., unpublished data). Thereby, stepwise differentiation events, which convert *Rhizobium* bacteria into functional nitrogen-fixing bacteroids, are accompanied by expression of different sets of nodule-specific NCRs and GRPs. Importantly, these genes have not been found in EST databases of non IRLC-legumes (namely, *L. japonicus* and *Glycine max*), suggesting a role for these genes in specific programs of nodulation that take place only in IRLC-legumes, like bacteroid terminal differentiation.

Nodule-Specific *NCR* and *GRP* Families Appeared and Evolved Within IRLC Legumes

We analyzed the genomic organization of the *NCR* and *GRP* genes as well as their evolutionary dynamics in *M. truncatula* (B. Alunni et al., unpublished data, 2006). One hundred and eight *NCRs* and 23 *GRPs* have been mapped. These included 29 new *NCRs* and 17 new *GRPs*.

Protein alignments and phylogenetic analyses revealed traces of several duplication events in the history of *GRPs* and *NCRs*. Similarly, defensin and disease-resistance genes in plants and animals (Silverstein et al., 2005; Patil et al., 2004; Meyers et al., 2003) were also often clustered and exhibited traces of numerous local duplication events. Another remarkable feature of these genes is that they evolve according to two distinct trajectories; the signal peptide showing purifying selection and the mature domain following diversifying selection (B. Alunni et al., unpublished data, 2006). Moreover, microsyntenic evidence between *M. truncatula* and *L. japonicus* confirm at the gene level that *GRPs* and *NCRs* are absent in *L. japonicus* (Figure 2). PCR analyses confirmed that all genes were nodule specific.

Potential Functions of Nodule-Specific Peptides

NCRs and *GRPs* might be a specific type of antimicrobial peptides (AMP) acting in symbiosis. Like AMPs, NCRs and GRPs could increase the membrane permeability and act as selective bacteriostatic agents controlling the microsymbionts inside the nodule.

As peptides are often involved in cell-to-cell communication, NCRs and GRPs could also monitor the differentiation of bacteroids in accordance with the host cells. Functional studies on these peptides are of particularly high interest as they might provide a large set of biologically active molecules with potential diverse applications in human, animal and plant healthcare.

MtnodGRP2-contig (mth2-18j19) - chr2

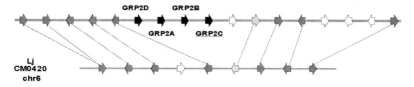

NCR062/074/076/128/145/312/313-contig (mth2-65c4,mth2-94j16) - chr 4

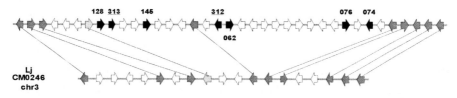

Figure 2. Microsyntenic comparison between *M. truncatula* and *L. japonicas* for one *GRP* and one *NCR* contig.The genes and their orientation are indicated by arrows; intergenic regions are not drawn to scale. Black arrows represent *GRP* or *NCR* genes. Dark grey linked arrows show syntenic genes with the same orientation, light grey ones with inversion. Other non-syntenic genes are shown as white arrows (Alunni et al., 2007).

References

Barnett MJ and Fisher RF (2006) Symbiosis 42, 1–24.
Barnett MJ et al. (2004) Proc. Natl. Acad. Sci. USA 101, 16636–16641.
El Yahyaoui F et al. (2004) Plant Physiol. 136, 3159–3176.
Fedorova M et al. (2002) Plant Physiol. 130, 519–537.
Fernández-López et al. (1998) Proc. Natl. Acad. Sci. USA 95, 12724–12728.
Györgyey J et al. (2000) Mol. Plant-Microbe Interact. 13, 62–71.
Kevei Z et al. (2002) Mol. Plant-Microbe Interact. 15, 922–931.
Lohar DP et al. (2006) Plant Physiol. 140, 221–234.
Mergaert P et al. (2003) Plant Physiol. 132, 161–173.
Mergaert P et al. (2006) Proc. Natl. Acad. Sci. USA 103, 5230–5235.
Meyers BC et al. (2003) Plant Cell 15, 809–834.
Patil A et al. (2004) Physiol. Genomics 20, 1–11.
Silverstein KA et al. (2005) Plant Physiol. 138, 600–610.
Vinardell JM et al. (2003) Plant Cell 15, 2093–2105.

FUNCTIONAL GENOMICS OF SOYBEAN ROOT-HAIR INFECTION

G. Stacey[1,2], L. Brechenmacher[1], M. Libault[1] and S. Sachdev[1]

[1]National Center for Soybean Biotechnology, Division of Plant Science;
[2]Division of Biochemistry, Department of Molecular Microbiology
and Immunology, University of Missouri, Columbia, MO 65211, USA

Soybean root hairs are single tubular-shaped cells formed from the differentiation of epidermal cells, called trichoblasts, on primary and secondary roots. Root hairs improve the capacity of the root to absorb water and nutrients from the soil and function in anchoring the plant into the soil (Gilroy and Jones, 2000). Root hairs are also the site for rhizobia to attach and subsequently initiate a complex developmental cascade resulting in a functional nodule in which the symbiotic rhizobia fix atmospheric nitrogen to aid plant growth (Stacey et al., 2006).

A better understanding of this symbiosis is fundamental for exploiting the benefits of biological nitrogen fixation in sustainable plant-production systems. We are interested in unraveling the mechanisms of soybean root-hair infection by the compatible rhizobium, *Bradyrhizobium japonicum*. Soybean was selected for this study due its agronomic importance and the larger root size, enabling the isolation of gram quantities of root hairs. Different functional, genomic, proteomic, and microscopic (Figure 1) tools, including gene-expression analysis, RNAi silencing, *in situ* hybridization, and confocal microscopy, are being used to determine how the bacteria enter into the single root hair cell via the infection thread, a unique structure formed in response to bacterial invasion.

Soybean Affymetrix chips harboring more than 37,000 *Glycine max* probes were used to identify responsive genes in root hairs 0–72 h after *B. japonicum* inoculation. We are now systematically silencing the key genes using *A. rhizogenes* mediated RNAi silencing (for methods, see Collier et al., 2005). Because the expression level of a gene does not necessarily reflect the protein level, proteomic approaches are also being used to identify gene products responding to *B. japonicum* inoculation.

We are establishing a root hair proteome reference map for soybean (*Glycine max*), using 2-D gel electrophoresis with MS-MS protein identification or MudPIT (Multidimensional Protein Identification Technology) (for methods, see Wan et al., 2005; Lee and Cooper, 2006). By combining the two approaches, we have identified several hundred proteins expressed in uninoculated soybean root hair cells.

F. D. Dakora et al. (eds.), *Biological Nitrogen Fixation: Towards Poverty Alleviation through Sustainable Agriculture*.
© Springer Science + Business Media B.V. 2008

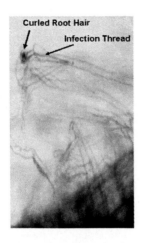

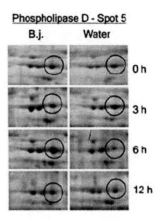

Figure 1. Optical microscopy of a root hair infected by *B. japonicum* trans-formed with *lacZ* gene.

Figure 2. Phospholipase D protein accumulation after *B. japonicum* treatment (From Wan et al., 2005).

A further goal is to identify soybean root-hair proteins responsive to *B. japonicum* inoculation (Figure 2). In our preliminary study, Wan et al. (2005) identified 26 proteins regulated in soybean root hairs from 0–12 h after inoculation. We are currently expanding this original study to analyze protein expression in soybean root hairs from 0–96 h after bacterial inoculation. The proteome of the symbiotic interaction will be analyzed by 2D-DIGE (Differential Gel Electrophoresis) and iTRAQ approaches in order to have a global quantitative view of differential protein accumulation in colonized root hairs.

The above functional genomic tools have already identified several hundred genes/ proteins that are likely involved in the early events of soybean nodulation. The key question remains – what role do they play? To help answer this question, we are developing tools to co-localize nodulation-related proteins to different plant-cell organelles. Translational fusions with the green fluorescent protein, plus specific dyes for organelles and compartments will allow us to localize each protein. Now we have perfected a method to isolate highly enriched preparations of soybean root hairs, the RNA and protein from these root hairs is being analyzed to explore the mechanism of the rhizobial infection process.

Supported (DBI-0421620) by the NSF, Plant Genome Program.

References

Collier R et al. (2005) Plant J. 43, 449–457.
Gilroy S and Jones DL (2000) Trends Plant Sci. 5, 56–60.
Lee J and Cooper B (2006) Mol. Biosystems 2, 621–626.
Stacey G et al. (2006) Curr. Opin. Plant Biol. 9, 110–121.
Wan et al. (2005) Mol. Plant-Microbe Interact. 18, 458–467.

REGULATION OF NODULE ORGANOGENESIS BY AUXIN TRANSPORT AND FLAVONOIDS

G. E. van Noorden[1,2], A. P. Wasson[2], F. I. Pellerone[2], J. Prayitno[1], B. G. Rolfe[1] and U. Mathesius[2]

[1]Research School of Biological Sciences, Australian National University, Canberra, ACT 2601, Australia; [2]School of Biochemistry and Molecular Biology, Australian National University, Canberra, ACT 0200, Australia

Auxin gradients are known to specify meristematic activity throughout the plant. We investigated the role of auxin in nodule initiation and differentiation, as well as in the control of nodule numbers by local and long distance regulation in the indeterminate legume, *Medicago truncatula*. Since the discovery that auxin transport inhibitors can induce spontaneous formation of nodule-like structures in legumes (Hirsch et al., 1989), it has been suggested that auxin transport regulation might be part of the mechanism for nodule organogenesis. Nodule initiation is preceded by local inhibition of auxin transport in indeterminate legumes (Boot et al., 1999; Mathesius et al. 1998), but this might not be required for nodule initiation in determinate legumes (Pacios-Bras et al., 2003; Subramanian et al., 2006). This inhibition is accompanied by changes in PIN gene expression (Huo et al., 2006) and is followed by auxin accumulation at the site where nodules are initiated (Mathesius et al., 1998; Pacios-Bras et al., 2003). An ethylene-insensitive hypernodulating mutant, *sickle*, shows higher auxin accumulation and PIN-gene expression at the site of nodule initiation than the wild type in *M. truncatula* (Prayitno et al., 2006). The role of the accumulation of auxin is likely to be in the regulation of cell division (Kondorosi et al., 2005). Proteome analysis supported a role for auxin at an early stage of nodulation as most of the proteins regulated by rhizobia are also regulated by auxin (G.E. van Noorden et al., unpublished data, 2006).

We genetically separated the local regulation of auxin transport with long distance loading of auxin from the shoot to the root in the supernodulating *SUNN* mutant (van Noorden et al., 2006). Auxin transport experiments suggest that long distance inhibition of auxin transport is part of the autoregulation mechanism and that auxin may be required for sustained nodule initiation, at least in indeterminate legumes.

Flavonoids are known as endogenous auxin-transport regulators in non-legumes (see, e.g., Peer et al., 2004). We showed that silencing of the flavonoid pathway by

F. D. Dakora et al. (eds.), *Biological Nitrogen Fixation: Towards Poverty Alleviation through Sustainable Agriculture.*
© Springer Science + Business Media B.V. 2008

RNA interference in *M. truncatula* prevents nodule initiation and also inhibits the local inhibition of auxin transport by rhizobia (Wasson et al., 2006), suggesting that flavor-noids could be endogenous regulators of local auxin transport during nodulation in indeterminate legumes. However, flavonoids might not be required for auxin transport regulation in determinate legumes (Subramanian et al., 2006). Figure 1 summarizes our model.

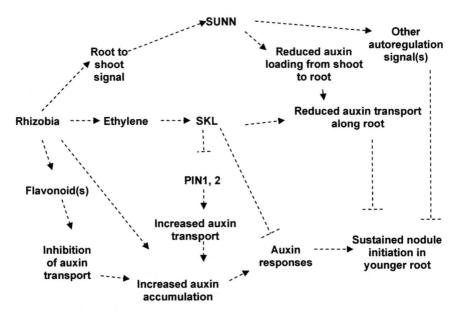

Figure 1. Model for the regulation of local and long-distance auxin transport during nodule initiation in *M. truncatula* by flavonoids, ethylene, and the autoregulation gene, *sunn* (Adapted from Prayitno et al., 2006).

References

Boot KJM et al. (1999) Molec. Plant-Microbe Interact. 12, 839–844.
Hirsch AM et al. (1989) Proc. Natl. Acad. Sci. USA 86, 1244–1248.
Huo X et al. (2006) J. Plant Growth Regul. 25, 156–165.
Kondorosi E et al. (2005) Plant Physiol. 137, 1197–1204.
Mathesius U et al. (1998) Plant J. 14, 23–34.
Pacios-Bras C et al. (2003) Plant Molec. Biol. 52, 1169–1180.
Peer WA et al. (2004) Plant Cell 16, 1898–1911.
Prayitno J et al. (2006) Plant Physiol. 142, 168–180.
Subramanian S et al. (2006) Plant J. 48, 261–273.
van Noorden GE et al. (2006) Plant Physiol. 140, 1494–1506.
Wasson AP et al. (2006) Plant Cell 18, 1617–1629.

LEGUME AGP-EXTENSINS IN *RHIZOBIUM* INFECTION

N. Brewin[1], A. Khodorenko[2], V. E. Tsyganov[2], A. Y. Borisov[2],
I. A. Tikhonovich[2] and E. Rathbun[1]

[1]John Innes Centre, Colney Lane, Norwich NR4 7UH, England, UK;
[2]All-Russia Research Institute of Agricultural Microbiology,
St. Petersburg-Pushkin, Russia
(Email: nick.brewin@bbsrc.ac.uk)

Plant cell-wall remodelling is a fundamental aspect of the *Rhizobium*-invasion process (Brewin, 2004). Some symbiotically defective legume mutants show abnormalities in composition of the infection thread wall and others show modifications of the extracellular matrix. For example, the pea mutant RisFixV (*sym42*) has enlarged infection thread walls that contain callose, whereas SGEFix⁻-2 (*sym33*) develops infection threads with thick walls but without callose. Peroxide-driven cross-linking is apparently another aspect of the infection process. For example, in developing pea nodules, the inside (luminal face) of the infection thread wall is bounded by a girdle of peroxide that co-localises with diamine oxidase, the probable source of this peroxide. As the infection thread matures, the peroxide-associated zone progressively expands through the wall and into the lumen of the infection thread. In mutant lines where the infection process aborts prematurely, there is often an increased abundance of peroxide in infection threads and symbiosomes. Interestingly, however, in mutant SGEFix⁻-2 (*sym33*), which does not release bacteroids from infection droplets, there is no peroxide detectable. These examples point to the need for a systematic analysis of the structure and development of infection threads using molecular probes that recognise components of the plant extracellular matrix.

We have previously described a class of plant glycoprotein that appears to be unique to legumes and may be an essential component of the *Rhizobium* infection process (Gucciardo et al., 2005; Rathbun et al., 2002). Legume AGP-extensins (AGPEs), also termed root nodule extensins (RNEs), are highly glycosylated glycoproteins that are recognised by a suite of monoclonal antibodies, e.g., MAC265. The corresponding antigens are located predominantly in the lumen of *Rhizobium*-induced infection threads where individual encapsulated bacterial cells are embedded in a matrix containing the plant glycoprotein. The same (or similar) glycoproteins can also be detected in inter-cellular spaces in the uninfected tissues of the nodule, particularly in the inner

F. D. Dakora et al. (eds.), *Biological Nitrogen Fixation: Towards Poverty Alleviation through Sustainable Agriculture.*

185

parenchyma, where occlusion of air spaces could serve a physiological role in regulating diffusion of oxygen (and/or) water into the central tissues of the nodule. The glycoprotein is also released from root cap border cells and this proved to be the most convenient source of material for immunopurification. The amino-acid sequence of a chymotryptic peptide fragment was used to design primers for RT-PCR in order to isolate transcript sequences from pea nodules. The cDNA sequences from pea matched closely with previously described nodule-enhanced (early nodulin) genes from Medicago, such as *MtN12*. The cDNA clones derived from a single host legume (pea) appeared to be encoded by a number of related genes of similar sequence but of variable length and with a variable number of internal reiterations. This redundancy implies that it may not be easy to obtain knock-out mutations with a symbiotically defective phenotype. Furthermore, the C-terminal sequence (AYSPPPPAYYYKSPPPPYHH) appeared to be strongly conserved and similar sequences could be identified in cDNA libraries from a number of legumes, but not from any non-legumes. Even the 3'-untranslated sequence of AGPE mRNA was strongly conserved between species, suggesting a possible role in transcript localisation prior to translation and protein secretion. Protein recognition of the 3'-UTR would be similar to the "locasome" model in budding yeast (Schmid et al., 2006) where ASH1-mRNA binds to ribonucleoprotein particles that coordinate secretion a cell-polarised through the endoplasmic reticulum.

The polypeptide sequences predicted from the cDNA clones reveal a new class of (hydroxy)-proline rich glycoprotein that combines arabinosylated extensin motifs (Ser-Pro$_x$) with motifs of alternating proline residues -XPXPXP- characteristic of arabino-galactan proteins. In this respect, they appear to share the biochemical and biophysical properties of soluble gums and emulsifying hydrocolloids (characteristic of AGPs) with the more structural and defensive property of extensins serving to harden cell walls in response to pathogen attack or mechanical stress. Similar structural features have recently been described for the polypeptide component of Gum Arabic, derived from *Acacia senegal*. A computational model for the polypeptide backbone of AGP-extensins is consistent with an extended rod-like structure. These molecular rods are likely to be of variable length because they are encoded by genes with different numbers of reite-rations of the glycopeptide motifs. In view of their strong net positive charge, it is likely that AGP-extensins could serve as emulsifying agents within the lumen of the infection thread. AGPE is able to attach to the negatively charged surface of bacterial cells, although the molecular target for this interaction has not yet been identified. The possibility of physical interactions between AGPE and other plant components of the extracellular matrix has not yet been investigated.

AGP-extensins have many tyrosine residues (but no cysteines). It has been demonstrated *in vitro* that AGPE molecules become cross-linked and insolubilised under oxidative conditions and this property may be relevant to the control of nodule development and the Rhizobium infection process. Vicinal Tyr residues are probably linked by strain-free bonds to form isodityrosine, and these molecules could be involved in a further round of intermolecular cross-linking to form di-isodityrosine bridges (Held et al., 2004). Preliminary experiments involving transgenic tobacco indicate that blocking the

C-terminus of AGPE molecules apparently prevented intermolecular cross-linking and their covalent association with other cell wall components.

Another interesting feature of the AGPE molecule is the abundance of histidine residues. In particular, a very characteristic repeating motif of AGPE is VHTYHOHOVYHSO (where O denotes hydroxyproline, the target for addition of the large arabinogalactan blocks). This oligopeptide was chemically synthesized in order to examine its properties *in vitro*. A fluorescein-tagged derivative was retained on metal-affinity columns, confirming that the His-rich oligopeptide was capable of binding diva-lent cations, such as Ni^{2+} and Cu^{2+}. Furthermore, we obtained spectroscopic evidence that Ni^{2+} and, to a lesser extent, Cu^{2+} ions were able to catalyse the non-enzymic con-version of Tyr residues to dityrosine in the presence of 0.5 mM hydrogen peroxide. Thus, there appear to be two possible mechanisms for peroxide-dependent cross-linking of AGPE molecules, an enzymic method based on extensin peroxidase and a non-enzymic method based on Cu^{2+} or Ni^{2+} catalysis.

In summary, the legume-specific class of AGP-extensin glycoproteins are important components of the extracellular matrix at the plant-microbial interface. These flexible rod-shaped macromolecules of variable length are apparently targeted directly into the infection thread lumen by a cytoplasmic process that requires further study. In the extra-cellular matrix, AGPE can serve either to promote or to prevent colonisation by rhizobium, depen-ding on the circumstances. When held as an emulsion within a fluid matrix, the colonizing rhizobial cells are able to grow and divide. End-to-end linkage of AGPE molecular rods might serve to enhance the lubricant properties at the plant-bacterial cell interface. Further-more, as a result of the observed girdle of hydrogen peroxide in infection threads, it is likely that there is a controlled level of intramolecular tyrosine-cross-linking for AGPE molecules adjacent to the infection thread wall. This could serve to prime cell wall lignification or suberisation. A non-enzymic Ni-catalysed peroxidation of Tyr residues could facilitate this interaction of AGPE with other cell wall components. At a later stage in the maturation of infection threads, there is also the potential for a peroxide-driven gum-to-resin transition that would serve to solidify the matrix in mature infection threads and infection droplets. This cross-linking would prevent further growth of the bacterial cells. Furthermore, on the basis of our observations with SGEFix⁻-2 (*sym33*), it is possible that peroxide-dependent solidification of the matrix in infection droplets is a precondition for successful uptake of bacteria by endocytosis through the host cell plasma membrane.

References

Brewin NJ (2004) Crit. Rev. Plant. Sci. 23, 293–316.
Gucciardo S et al. (2005) Mol. Plant-Microbe Interact. 18, 24–32.
Held MA et al. (2004) J. Biol. Chem. 279, 55474–55482.
Rathbun EA et al. (2002) Mol. Plant-Microbe Interact. 15, 350–359.
Schmid M et al. (2006) Curr. Biol. 16, 1538–1543.

GLYCOSIDE HYDROLASES FROM LEGUMES AND RHIZOBIA AT THE SYMBIOTIC INTERFACE

C. Staehelin, Y.-J. Dong, Y. Tian, J.-S. Xiong and Z.-P. Xie

State Key Laboratory of Biocontrol, School of Life Sciences, Sun Yat-Sen (Zhongshan) University, Guangzhou 510275, China
(Email: cst@mail.sysu.edu.cn)

Extracellular glycoside hydrolases are involved in both the synthesis and degradation of bacterial determinants required for nodulation. Both symbiotic partners contribute to the enzyme cocktail at the symbiotic interface. Glycoside hydrolases from host legumes play an important, but poorly understood, role in inactivation of rhizobial Nod factors. Rhizobial exo-oligosaccharides are symbiotic determinants in certain *Sinorhizobium*-legume interactions. In our laboratory, research work on glycoside hydrolases is focused on Nod factor hydrolases from legumes induced by Nod factor signaling, a symbiosis-related chitinase from *Medicago truncatula*, and rhizobial glycanases required for synthesis of exo-oligosaccharides.

Nod Factor Hydrolases Activated by Nod Factor Signaling

Studies with alfalfa and pea plants showed that active Nod factors induce on host plants their own hydrolytic degradation. After a pre-treatment with nanomolar concentrations of Nod factors, roots displayed an increased Nod factor-cleaving activity in the rhizosphere. The Nod factor hydrolase from alfalfa was found to be a glycoprotein that binds to concanavalin A. The Nod factor hydrolase does not cleave chitin, but rapidly hydrolyzed all Nod factors from *S. meliloti*. Recent work with non-nodulating mutants demonstrated that the Nod factor hydrolase in pea roots is regulated by the Nod factor signaling pathway. In contrast to wild-type peas, mutants defective in the symbiotic genes *sym10*, *sym8*, *sym19*, and *sym9/sym30* did not exhibit any stimulation of the Nod factor hydrolase. Interestingly, the Nod factor hydrolase activity in these mutants was even lower than the constitutive activity of wild-type peas, which were not pretreated with Nod factors (Ovtsyna et al., 2005). We propose that the Nod factor hydrolase contributes to homeostasis of Nod factor levels during the infection process.

F. D. Dakora et al. (eds.), *Biological Nitrogen Fixation: Towards Poverty Alleviation through Sustainable Agriculture.*
© Springer Science + Business Media B.V. 2008

Host Chitinases Involved in Nodulation

Certain chitinases might be involved in Nod factor degradation or play another symbiosis-specific role. In *M. truncatula* for example, a class V chitinase gene (*Mtchit 5*) was specifically induced in the interaction with *S. meliloti*, whereas other chitinases were stimulated in response to pathogen attack (Salzer et al., 2004). The *Mtchit 5* gene is strongly induced in nodules and can be considered as a nodulin gene. Class V chitinases (glycoside hydrolase family 18) in plants have been poorly investigated. The enzyme in tobacco is induced in response to pathogens and cleaves chitin oligosaccharides and certain Nod factors from *S. meliloti*. We are currently characterizing Mtchit 5, which is different from the Nod factor hydrolase mentioned above. Expression analyses of the *chit5* gene suggest that the encoded enzyme plays a role during later symbiotic stages. Chitinases involved in nodulation may have diverse functions. They could exhibit lysozyme activity against rhizobia or play a role in plant developmental processes by cleaving *N*-acetylglucosamine-containing arabinogalactan proteins.

Glycanases Required for Synthesis of Rhizobial Exo-oligosaccharides

Rhizobial glycanases release symbiotically active exo-oligosaccharides by enzymatic degradation of exo-polysaccharide. In contrast to *S. meliloti*, synthesis of exo-oligosaccharides in *(Sino)rhizobium* sp. strain NGR234 only depended on the glycanase ExoK (glycoside hydrolase family 16). Direct synthesis of exo-oligosaccharides by limited polymerization was not observed in the *exoK* mutant of NGR234. Exo-oligosaccharides purified from this strain are repeating subunits of exo-polysaccharide, i.e., octa- and nonasaccharides. Yields of exo-oligosaccharides from culture supernatants were lower than levels of Nod factors (50 µg exo-oligosaccharides L^{-1}). Interestingly, the *exoK* mutant lost its ability to induce nitrogen-fixing nodules on various hosts (e.g., *Albizia lebbeck* and *Leucaena leucocephala*). Hence, exo-oligosaccharides from NGR234 are symbiotic determinants, whereas exo-polysaccharide alone is not sufficient for establishment of symbiosis with various legumes (Staehelin et al., 2006). Future work is required to analyze whether exo-oligosaccharides are signals perceived by specific plant receptors.

Acknowledgments

Research work on glycoside hydrolases in the laboratory of C.S. is supported by the National Natural Science Foundation of China (grant 30671117) and by the Department of Science and Technology of Guangdong Province, China (grant 2006B50104004).

References

Ovtsyna AO et al. (2005) Plant Physiol. 139, 1051–1064.
Salzer P et al. (2004) Planta 219, 626–638.
Staehelin C et al. (2006) J. Bacteriol. 188, 6168–6178.

TRANSCRIPT PROFILING OF NODULE SENESCENCE IN *MEDICAGO TRUNCATULA* DELINEATES DISTINCT PHASES AND MECHANISMS

S. Goormachtig, J. C. P. Guerra, K. D'haeseleer, W. Van de Velde, A. De Keyser and M. Holsters

Department of Plant Systems Biology, Flanders Institute for Biotechnology, Ghent University, B-9052 Gent, Belgium

Leguminous plants can grow under nitrogen-limiting conditions because of their ability to establish endosymbiosis with rhizobia. To fix atmospheric nitrogen, the bacteria internalize into cells of the central tissue, get surrounded by a plant-derived membrane, and differentiate into bacteroids. As such, they exist as intracellular organelles, called symbiosomes, and exchange fixed nitrogen for carbon sources. The symbiotic relationship is lost after some time and nodule senescence is visible as a color change in the N_2-fixing zone from pink (associated with functional leghemoglobin) to green (associated with heme degradation; Roponen, 1970). Changes indicating degradation of bacterial and plant cells have also been observed. A typical hallmark for senescence is the triggering of a wide range of proteolytic activities that cause large-scale protein degradation (Pladys and Vance, 1993) and finally, death of bacteroids and nodule cells. The signal-transduction cascades and regulatory functions that control nodule senescence are unknown. To unravel the molecular processes that govern nodule senescence, developmental nodule senescence in the model legume *Medicago truncatula* has been described (Van De Velde et al., 2006). In-depth light and electron microscopy analyses show two stages; an early stage in which symbiosomes degrade but the plant cells stay rigid, followed by a later stage wherein the plant cells disintegrate. As described for leaf senescence, nodule senescence is slow and consists of different developmental stages. In addition to this slow disintegration, some cells also collapse directly and die, and their number increases as the senescence process proceeds.

A transcript-profiling analysis was also performed via a modified cDNA-amplified fragment length polymorphism (AFLP) protocol (Breyne et al., 2003). Using a very specific sampling method, 508 cDNA-AFLP tags were identified whose transcript profiles were significantly modulated by nodule senescence. Hierarchical average linkage clustering analysis of the differential expressed genes revealed three clusters that were upregulated and one that was downregulated. The three upregulated gene clusters corresponded to three developmental stages of the senescence process. Cluster 1 grouped

F. D. Dakora et al. (eds.), *Biological Nitrogen Fixation: Towards Poverty Alleviation through Sustainable Agriculture.*
© Springer Science + Business Media B.V. 2008

gene tags the expression of which was already upregulated in tissues of senescing nodules outside the senescing zone, such as the meristem, infection zone, and fixation zone, and might correspond to genes involved in triggering the process. In view of delaying senescence, tags with homology to genes coding for regulatory proteins, such as transcription factors belonging to this cluster, are important for future research. Cluster 2 contained genes that were expressed at the early and later stages, whereas cluster 3 had of genes that were only expressed at the later stage. These results indicate that cluster 2 genes might be involved in bacterial degradation, whereas cluster 3 genes might function at the very end of the life of the nodule cells. Within clusters 2 and 3, tags involved in protein degradation as well as other catabolic functions were over-represented. Genes encoding representatives of five protease families were retrieved. Also, ubiquitin-mediated protein degradation was very important, because five tags were found corresponding to genes coding for F-box proteins which determine the specificity of the Skp1-cullin-F-box- specific ubiquitin/26S proteasome. Genes coding for SINA E3 ligases, also involved in ubiquitin-mediated protein degradation, were evenly expressed at the onset of nodule senescence (G. Den Herder, personal communication, 2006). Thus, a large battery of protein-degradation pathways is activated.

Besides the plethora of catabolic genes, genes coding for proteins involved in transport of a wide variety of molecules, such as ATP-binding cassette proteins and specific transporters of phosphate, amino acids, and metal ions, are also activated, which indicates that nodule senescence, like leaf senescence, involves nutrient mining and recycling and marks the transition of the nodule from carbon sink to general nutrient source. The analysis showed that ethylene might be important during the senescence process. The positive role of ethylene is illustrated by the upregulation of ERF transcription factors and ethylene biosynthetic genes, such as S-adenosyl-methionine synthetase and 1-aminocyclopropane-1-carboxylate oxidase. Together, this study revealed an abun-dance of genes that might be good markers for the senescence process. The challenge is to find the main molecular players that control the process. Surely, tags that belong to cluster 1 are good candidates and will be subjected to functional analysis by RNAi and ectopic expression. Moreover, nodules are also subjected to stress-induced senescence due to environmental factors, such as drought, nitrate, and darkness (Gogorcena et al., 1997; González et al., 1998; Matamoros et al., 1999). It will be interesting to analyze whether these tags are also involved in this type of senescence.

References

Breyne P et al. (2003) Mol. Genet. Genomics 269, 173–179.
Gogorcena Y et al. (1997) Plant Physiol. 113, 1193–1201.
González EM et al. (1998) J. Exp. Bot. 49, 1705–1714.
Matamoros MA et al. (1999) Plant Physiol. 121, 97–111.
Pladys D and Vance PC (1993) Plant Physiol. 103, 379–384.
Roponen I (1970) Physiol. Plant. 23, 452–460.
Van de Velde W et al. (2006) Plant Physiol. 141, 711–720.

COMPARISON OF LEAF GROWTH OF HYPERNODULATION SOYBEAN MUTANTS, NOD1-3, NOD2-4 AND NOD3-7, WITH THEIR PARENT CV. WILLIAMS

S. Ito[1], N. Ohtake[2], K. Sueyoshi[2] and T. Ohyama[2]

[1]Graduate School of Science and Technology; [2]Faculty of Agriculture, Niigata University, Japan

Autoregulation of nodulation (AON) is the mechanism that systemically controls the number of nodules per plant through communication between the shoot and the root of leguminous plants, using unknown signal molecules other than photosynthate (Ito et al. 2006a, b). The hypernodulation soybean mutant lines form profuse nodulation compared with their parent, most likely because they either lack or have decreased activity of a part of AON. AON in soybean (*Glycine max* [L.] Merr.) is controlled by a CLAVATA1 (CLV1)-like receptor kinase, GmNARK (Searle et al., 2003). However, the actual role of GmNARK in AON has not yet been clarified. The details of the growth characteristics of hypernodulation mutant lines may be important to understanding AON. Herein, the plant-growth characteristics of soybean hypernodulation mutants, NOD1-3, NOD2-4 and NOD3-7, were compared with their parent cv. Williams. Although the mutation site(s) have not been reported, evidence (Vuong et al., 1996; Vuong and Harper, 2000; Nishimura et al., 2002) suggests that the mutated locus of all NOD mutant lines is in *GmNARK*. The phenotypes of hypernodulation mutant lines showed other plant- growth features and, generally, show less vigorous growth than the parents.

Our results showed that the reduced accumulation of total dry matter by NOD mutant lines may be due to the higher consumption of photosynthate by the large number of nodules (Ito et al., 2006c, 2007). When the dry weight of each leaf was compared, NOD1-3 and NOD3-7 had different partitioning patterns to Williams and NOD2-4, with or without inoculation by *Bradyrhizobium japonicum* (Ito et al., 2007). So, we focused on the emergence rate and size of the trifoliolate leaves during the early growth stage. Both NOD1-3 and NOD3-7 had a quicker leaf-emergence, but the final size of their expanded leaves was smaller than Williams and NOD2-4. Leaf growth of NOD2-4 was similar to Williams. Surgical removal of the shoot-growing point (to prevent new leaf emergence) did not overcome the small size of the primary leaf of NOD3-7. Therefore, the small-leaf phenotype may not be due to the faster rate of new-leaf emergence.

F. D. Dakora et al. (eds.), *Biological Nitrogen Fixation: Towards Poverty Alleviation through Sustainable Agriculture*.
© Springer Science + Business Media B.V. 2008

We then investigated the difference of fully expanded leaf size at the cell level. Using light microscopy, the number of palisade cells per leaf blade of NOD1-3 and NOD3-7 was significantly lower than that of Williams and NOD2-4, in both inoculated- and uninoculated-conditions. There were no significant differences in cell size among lines. Therefore, the small size of the fully expanded leaf of NOD1-3 and NOD3-7 might be due to the lower number of leaf cells and not due to the size of individual cells.

In NOD1-3 and NOD3-7, both the number of nodules and leaves per plant increased however, their size was smaller than Williams. The gene defect in AON in NOD1-3 and NOD3-7 might be related to the control of leaf growth. In NOD2-4, however, leaf phenotype was similar to Williams. The difference may be attributed to the positional effect of the mutation in *GmNARK* among NOD mutant lines. It is known that CLV1 is involved in control of cell proliferation in shoot meristems. *GmNARK*, which encodes a CLV1-like receptor kinase, might have a role in cell proliferation of the leaf blade as well as in the control of nodule number.

References

Ito S et al. (2006a) Soil Sci. Plant Nutr. 52, 438–443.
Ito S et al. (2006b) Bull. Facul. Agric. Niigata Univ 59(1), 33–38.
Ito S et al. (2006c) Bull. Facul. Agric. Niigata Univ 59(1), 39–43.
Ito S et al. (2007) Soil Sci. Plant Nutr. 53, 66–71.
Nishimura et al. (2002) Nature 420, 426–429.
Searle et al. (2003) Science 299, 109–112.
Vuong TD and Harper JE (2000) Crop Sci. 40, 700–703.
Vuong TD et al. (1996) Crop Sci. 36, 1153–1158.

ACTINORHIZAL NODULES AND GENE EXPRESSION

V. Hocher, B. Péret, L. Laplaze, F. Auguy, H. Gherbi, S. Svistoonoff, C. Franche and D. Bogusz

Groupe Rhizogenèse Symbiotique, UMR DIAPC, Institut de Recherche pour le Développement, 911 avenue Agropolis, BP 5045, 34394 Montpellier Cedex 5, France

Actinorhizal root nodules result from the interaction between a nitrogen-fixing actinomycete called *Frankia* and roots of dicotyledonous plants belonging to 8 plant families and 25 genera (Benson and Silvester, 1993). Actinorhizal plants share common features with the exception of *Datisca*, which has herbaceous shoots. They are perennial dicots and include woody shrubs and trees such as *Alnus* (alder), *Elaeagnus* (autumn olive), *Hippophae* (sea buckthorn), and *Casuarina* (beef wood). Most actinorhizal plants are capable of high rates of nitrogen fixation comparable to those found in legumes. As a consequence, these plants are able to grow in poor and disturbed soils and are important elements in plant communities worldwide. In addition, some actinorhizal species can grow well under a range of environmental stresses, such as high salinity, extreme pH, and the presence of heavy metals. This facility for adaptation has drawn great interest to actinorhizal plants, particularly to several species of *Casuarinaceae*, such as *Casuarina glauca*, which can be used for fuelwood production, agroforestry, and land reclamation in the tropics and subtropics.

The symbiotic association between *Frankia* and actinorhizal plants is still poorly understood, although it offers striking differences to the *Rhizobium*-legume symbiosis (Obertello et al., 2003; Vessey et al., 2005). *Frankia* is a filamentous branching Gram-positive actinomycete, whereas rhizobia are Gram-negative unicellular bacteria. *Frankia* interacts with a diverse group of dicotyledonous plants, whereas rhizobia only enter symbiosis with plants from the legume family and with one non-legume, *Parasponia*. In actinorhizal plants, the formation of the nodule primordia takes place in the root pericycle and the nodule consists of multiple lobes, each representing a modified lateral root without a root cap and with infected cells present in the cortex. In indeterminate nodules formed on roots of temperate legumes, the nodule primordium starts in the root inner cortex and determinate nodule primordia are formed in the root outer cortex of tropical and subtropical legumes. Legume root nodules represent stem-like structures with peripheral vascular bundles and infected cells in the central tissue, whereas actinorhizal

F. D. Dakora et al. (eds.), *Biological Nitrogen Fixation: Towards Poverty Alleviation through Sustainable Agriculture.*
© Springer Science + Business Media B.V. 2008

nodules conserve the structure of a lateral root with a central vascular bundle and peripheral infected cortical tissue.

The understanding of regulatory events in actinorhizal nodulation at the molecular level is mainly limited by the microsymbiont *Frankia*. This actinomycete is characterized by a slow growth rate, a high G+C DNA content, and the lack of a genetic transformation system (Lavire and Cournoyer, 2003). So far, attempts to detect DNA sequences homologous to the rhizobial *nod* genes in the *Frankia* genome have failed. However, in the past decade, some progress has been made with respect to the plant genes that are expressed at different stages of actinorhizal nodule differentiation. Differential screening of nodule cDNA libraries with root and nodule cDNA has resulted in the isolation of a number of nodule-specific or nodule-enhanced plant genes in several actinorhizal plants, including *Alnus, Datisca, Eleagnus* and *Casuarina* (for reviews, see Obertello et al., 2003; Vessey et al., 2005).

Our group has concentrated on the molecular study of the plant genes involved in the interaction between *Frankia* and the tropical tree *Casuarina glauca*. ESTs from roots and nodules have been compared, providing insights into the genes expressed in the actinorhizal nodules. Tools for the functional analysis of candidate genes have been developed and they include genetic transformation procedures (based on *Agrobacterium tumefaciens* and *A. rhizogenes*) and gene silencing by RNA interference. Finally, we have analysed the role of auxin during actinorhizal nodule formation.

Search for New Early Expressed Genes: The First Genomic Platform for Actinorhizal Species

Our group has developed the first genomic platform to identify new genes involved in the symbiotic process between *Frankia* and *C. glauca*. A total of 2,028 ESTs was obtained from cDNA libraries corresponding to mRNA extracted from (i) young nodules induced by *Frankia* and (ii) non-infected roots. Their annotation, through a multi-module custom pipeline, led to 242 contigs and 1,429 singletons, giving a total of 1,616 unique genes (Figure 1). Around 650 sequences were found to be nodule specific. For half of these nodule transcripts, no similarity to previously identified genes was detected, whereas for the other half, genes of primary metabolism, protein synthesis, cell division and defence were highly represented. As expected, several nodule EST/cluster sequences corresponded to proteins previously described as actinorhizal nodulins (i.e., hemoglobin, metallothioneins, subtilisin, rubisco activase, saccharose synthase, glycine and histidine rich proteins). In order to explore the early events of *C. glauca–Frankia* symbiosis, a substractive hybridisation library (SSH) was also constructed with roots sampled 4 days after infection. Seven hundred and three SSH sequences were validated and annotated revealing a large proportion of ESTs implicated in defence, cell-wall structure and gene expression. All these results are integrated in database accessible through a web interface (http://www.mpl.ird.fr/rhizo/ – Resources; Hocher et al., 2006).

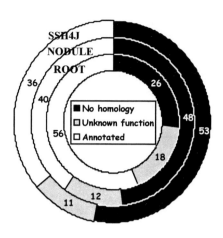

Figure 1. Classification of *Casuarina glauca* clusters and singletons from root, nodule, and substractive hybridisation library from roots sampled 4 days after infection (SSH4J). The distribution of clusters and singletons is based on E-value and top ten results of BLAST.

In order to test the usefulness of our EST database to identify symbiotic genes, 13 genes were selected to verify nodule-specific and/or nodule-enhanced expression by Quantitative Real-Time RT-PCR (qRT-PCR) on the basis of their putative involvement in nodule development and/or functioning. Differential expression was observed between roots and nodules for four genes that coded for flavonoid biosynthesis enzymes, thus suggesting a possible role of flavonoids in actinorhizal nodule development (Hocher et al., 2006). Experiments are now underway to characterize more precisely the expression of these genes at different stages of *C. glauca* nodule differentiation.

Tools for Functional Analysis of Symbiotic Genes: Gene Transfer Based on *Agrobacterium rhizogenes* and *A. tumefaciens*

The natural susceptibility of members of the *Casuarinaceae* family to *A. tumefaciens* was used to develop a gene-transfer procedure for *C. glauca* (Smouni et al., 2002). Epicotyl fragments of 2 cm in length were excised from 45-day-old plantlets and co-cultivated with the disarmed strain C58C1(pGV2260; pBIN19). After selection on kanamycin, one to three transgenic calli were observed on 26% of the epicotyls. Integration of the transgenes was further confirmed by PCR and Southern blot analyses. Transgenic plants were regenerated in approximately 9 to 10 months for *C. glauca*. The nodulation efficiency was found to be similar in transgenic Casuarinaceae and in non-transformed control plants inoculated by *Frankia*. The transgenic nodules fixed nitrogen at the same rate as those of the non-transformed control nodules. This transgenic approach has contributed to the characterization of two symbiotic genes from *C. glauca*, *Cgenod40*, a homolog of early nodulin gene *enod40* from legumes (Santi et el., 2003), and *cg12*, an actinorhizal symbiotic gene encoding a subtilisin-like serine protease (subtilases) (Svistoonoff et al., 2003).

A rapid procedure for producing composite plants of *C. glauca* is also available. It relies on the induction of a hairy root system by *A. rhizogenes*, while the aerial part of the plant remains untransformed. Young seedlings of *C. glauca* were wounded on the hypocotyl and inoculated with *A. rhizogenes* A4RS containing the chosen binary vector. After 2 weeks, highly branched roots exhibiting rapid growth were observed at the inoculation site. The normal root system was removed at the stem base and the composite plant decontaminated with cefotaxime. Co-transformation with the wild-type T-DNA and the T-DNA from the binary vector was observed in about 50% of *C. glauca* hairy roots. After inoculation with *Frankia*, nodulation was observed on 40% of the transformed roots. Using this "composite plant" approach, the pattern of expression conferred by the promoter region of a symbiotic gene can be studied in both roots and nodules of *C. glauca* within about 4 months (Diouf et al., 1998).

These composite plants are currently used to dissect symbiotic gene function by RNAi gene silencing. To demonstrate the potential of this approach for actinorhizal plants, two RNAi silencing vectors directed against the ß-glucuronidase (*GUS*) gene were introduced into transgenic plants of *Casuarina* stably transformed with the *GUS* gene under the control of the *35S* promoter. Our data established that the reporter gene was efficiently silenced in both roots and actinorhizal nodules of composite plants.

Auxin Transport and Nodule Formation in *Casuarina glauca*

Because auxin is the key signal in lateral root development and actinorhizal nodules are modified lateral roots, we are studying the role of auxin during actinorhizal nodule formation. Genes from the *AUX-LAX* family (which contains four members in *Arabidopsis thaliana*) have been identified as potential auxin influx carriers. Two of these genes (*AtAUX1* and *AtLAX3*) are involved in lateral root development. *Casuarina* homologs of these two genes, named *CgAUX1* and *CgLAX3*, have been isolated. Southern blot analysis suggests that the *Casuarina AUX-LAX* family contains only two genes. Functional complementation of *Arabidopsis aux1* and *lax3* mutants with *CgAUX1* and *CgLAX3* has shown that CgAUX1 protein is functionally equivalent to AtAUX1. *CgAUX1* and *CgLAX3* are expressed in shoots, roots and nodules as shown by RT PCR. *CgAUX1* and *CgLAX3* promoter::GUS fusion constructs have been introduced in *A. thaliana*. *CgAUX1* expression is observed in the root apex and vasculature and in leaves and *CgLAX3* expression is found in leaves and root vasculature. These constructs have also been introduced in *C. glauca* and *Allocasuarina verticillata*, a Casuarinaceae species closely related to *Casuarina*. The *AUX-LAX* gene expression patterns during lateral root and nodule formation are being analysed. These studies will help us to understand the implication of auxin transport in the development of the lateral root derived symbiotic structure.

Conclusion

In the past decade, considerable advances have been made in the identification and characterization of the plant genes involved in the development and functioning of actinorhizal nodules. However, in comparison with the progress achieved in the molecular dissection of the communication between *Rhizobium* bacteria and legumes (Crespi and Galvez, 2000), our understanding of the early steps of the interaction between *Frankia* and the actinorhizal plants lags well behind.

The genetic transformation procedures and RNA interference technology developed in the Casuarinaceae family now make it possible to perform functional analysis of the actinorhizal symbiotic genes. Moreover, emerging genomics tools may help investigate the early communication between the actinomycete and the host plant. Genoscope (France) funding will allow the sequencing of 25,000 new sequences from *Casuarina* roots and nodules in 2007. This EST collection will represent an essential resource for the understanding of actinorhizal nodule ontogenesis and for comparative genomics with the legume/*Rhizobium* and mycorrhizal symbioses.

References

Benson DR and Silvester WB (1993) Microbiol. Rev. 57, 297–319.
Crespi M and Galvez S (2000) J. Plant Growth Regul. 19, 155–166.
Diouf et al. (1998) Mol. Plant-Microbe Interact. 11, 887–894.
Hocher V et al. (2006) New Phytol. 169, 681–688.
Lavire C and Cournoyer B (2003) Plant Soil 254, 125–138.
Obertello M et al. (2003) Afr. J. Biotechnol. 2, 528–538.
Santi C et al. (2003) Mol. Plant-Microbe Interact. 16, 808–816.
Smouni A et al. (2002) Funct. Plant Biol. 29, 649–656.
Svistoonoff S et al. (2003) Mol. Plant-Microbe Interact. 16, 600–607.
Vessey JK et al. (2005) Plant Soil 274, 51–78.

PART 2B

NODULE PHYSIOLOGY AND GENETICS

CHARACTERIZATION OF GLUTATHIONE S-TRANSFERASES FROM LEGUME ROOT NODULES

D. A. Dalton, J. Scott, L. Jelinek and Z. Turner

Biology Department, Reed College, Portland, OR 97202, USA

Nitrogen fixation is an energy-intensive process. Legume root nodules, and most other N_2-fixing systems, are dependent upon aerobic metabolism because of its potential for high energy yield. However, since nitrogenase is quickly inactivated by O_2, there is a fundamental conflict. N_2-fixing systems must carefully balance the risks and benefits associated with free O_2. In nodules, the "O_2 problem" is addressed primarily by two factors: (i) an O_2 diffusion barrier that controls the entry of O_2 into the central infected zone; and (ii) an abundant supply of the O_2-buffering protein leghemoglobin. These are highly effective measures, but do not completely solve the O_2 problem in part because some O_2 is still reduced to reactive active species (ROS), such as H_2O_2 and superoxide.

Surprisingly, a major source of ROS in nodules is leghemoglobin. The O_2 that is bound to the central heme may spontaneously autoxidize, thus producing superoxide and a non-functional ferric form of leghemoglobin. The superoxide is subsequently reduced to H_2O_2, which attacks heme groups, releasing free Fe that catalyzes the formation of hydroxyl radicals through Fenton chemistry (Puppo and Halliwell, 1988). Consequently, a further major aspect of the O_2 problem in nodules is the need for strong antioxidant defenses to minimize the potential damage from ROS. These defenses include superoxide dismutase and the enzymes of the ascorbate-glutathione (ASC-GSH) cycle, namely, ascorbate peroxidase, (mono)dehydroascorbate reductase, and glutathione reduc-tase. Additional antioxidant defenses may be provided by another class of enzymes known as glutathione S-transferases (GSTs) that can also function as glutathione per-oxidases.

There is now ample evidence that the ASC-GSH cycle provides strong benefits for N_2 fixation (Becana et al., 2000). For instance, N_2 fixation in an in vitro system containing rhizobia and leghemoglobin can be enhanced four-fold by the addition of ascorbate and ascorbate peroxidase (Ross et al., 1999). A similar increase in N_2 fixation can be achieved by infusing ascorbate directly into stems of nodulated soybeans (Bashor and Dalton, 1999). In contrast, the function of GSTs in nodules is ambiguous. This uncertainty also applies to the role of GSTs in plants in general. The traditionally accepted function of GST in plants is to provide a means of detoxifying xenobiotics

F. D. Dakora et al. (eds.), *Biological Nitrogen Fixation: Towards Poverty Alleviation through Sustainable Agriculture.*
© Springer Science + Business Media B.V. 2008

(especially herbicides, such as atrazine) by conjugating potentially toxic molecules to GSH. This targets the toxin for transport to the vacuole where it can be degraded. GSTs may also function to protect plants from oxidative damage by functioning as glutathione peroxidases. In this case, the GSTs use GSH to reduce organic hydroperoxides of fatty acids and nucleic acids to the corresponding alcohol (Dixon et al., 2002). If this happens in nodules, it could provide major antioxidant benefits that have not been previously appreciated.

GSTs in plants are a complex gene superfamily with 25 recognized genes in soybean and 48 in *Arabidopsis*. These GSTs show an amazing degree of diversity with very little similarity between the different GSTs. Our lab has shown that crude extracts of soybean nodules contain substantial GST activity and that GST can be purified by affinity chromatography on GSH-agarose, but the activity is quickly lost. GST activity is usually assayed by spectrophotometry with CDNB, a substrate that forms a colored conjugate with GSH. GST from soybean nodules has an eight-fold higher activity when assayed with ethacrynic as the substrate instead of CDNB. Experiments are currently in progress to determine the peroxidase activity of the major nodule GSTs with both GSH and thioredoxin as potential reductants.

We have used Real-time PCR to determine transcript levels in nodules for all 25 of the GSTs in soybean. There were 10 GSTs that were expressed at levels substantially above those in uninfected roots, with GST9 at 91-fold (accession no. AF243364) being by far the predominant GST, followed by GST4 (accession no. AF048978) at six-fold. We have cloned the cDNA for each of the five most highly expressed nodule GSTs into *E. coli* expression vectors in order to obtain sufficient recombinant protein to produce antibodies. These antibodies will be used in immuno-staining procedures to determine the localization of GSTs in nodules. Our lab is also in the process of producing transgenic hairy roots in *Medicago truncatula* in which GSTs have been silenced using RNAi technology. Our vectors are based on the pHannibal vectors of CSIRO and will include a GFP reporter and hairpin RNAi-inverted repeats. Since the 35S cauliflower mosaic virus promoter has weak activity in alfalfa, we are incorporating a cassava virus promoter instead. Initial trials with the reporter gene alone have been successful. Future studies will address the critical question of the function of GST genes in transgenic nodules.

References

Bashor CJ and Dalton DA (1999) New Phytol. 142, 19–26.
Becana M et al. (2000) Physiol. Plant. 109, 372–381.
Dixon DP et al. (2002) Genome Biol. 3, 1–10.
Puppo A and Halliwell B (1988) Planta 173, 405–410.
Ross E et al. (1999) Phytochem. 52, 1203–1210.

OVEREXPRESSION OF FLAVODOXIN IN ALFALFA NODULES LEADS TO DELAYED SENESCENCE AND HIGH STARCH ACCUMULATION

T. C. de la Peña, F. J. Redondo, M. M. Lucas and J. J. Pueyo

Department of Plant Physiology and Ecology, Instituto de Recursos Naturales, Centro de Ciencias Medioambientales, CSIC, Serrano 115-bis, 28006 Madrid, Spain

Nodule senescence is the sequence of structural, molecular, biochemical and physiological events leading to the loss of the nitrogen-fixing activity and to cell death in the symbiotic tissue. A senescence-inducing signal from the plant causes a decrease in antioxidant levels and, thus, an increase in reactive oxygen species (ROS) up to a point of no return (Puppo et al., 2005). Nitrogen fixation is very sensitive to ROS and nitrogenase activity drastically decreases during nodule senescence. Delayed nodule senescence represents a strategy that could lead to increased nitrogen fixation and legume productivity. Delayed nodule senescence together with amelioration of sustainability under field conditions constitutes a key aim of legume improvement programs (Puppo et al., 2005).

The antioxidant bacterial protein flavodoxin contains a flavin mononucleotide group as its redox centre and transfers electrons at low potentials (Pueyo et al., 1991). Flavodoxin can function as an electron donor to nitrogenase. In its reduced state, flavodoxin can also react with ROS, such as the superoxide radical, and then revert to its original redox state in the presence of an appropriate electron source.

In the present work, transformed *Sinorhizobium meliloti* that were overexpressing *Anabaena variabilis* flavodoxin was used to nodulate *Medicago sativa* plants. Nodule natural senescence was compared and characterized in *M. sativa* plants nodulated by either the flavodoxin-expressing rhizobia or the corresponding control bacteria. Flavodoxin was expressed and accumulated in *M. sativa* nodules elicited by transformed *S. meliloti* bacteria. Nodulation kinetics, nodule number, and nodule weight were not affected by flavodoxin expression when comparing plants inoculated with flavodoxin-expressing and control bacteria.

Flavodoxin overexpression protected free-living bacteria from oxidative stress induced by oxidative agents, such as hydrogen peroxide and methylviologen. To estimate

F. D. Dakora et al. (eds.), *Biological Nitrogen Fixation: Towards Poverty Alleviation through Sustainable Agriculture.*

the antioxidant-protecting effect of flavodoxin on the senescence-associated decline of nitrogen fixation in functional symbiotic nodules, the nitrogen-fixing activity was measured by the acetylene-reduction assay for *M. sativa* plants nodulated by flavodoxin-expressing and control bacteria (Figure 1). The decline in nitrogenase activity associated with nodule natural senescence was delayed in flavodoxin-expressing nodules. Flavodoxin overexpression also reduced senescence-related structural and ultrastructural alterations in alfalfa nodules.

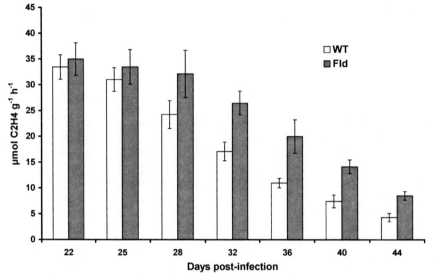

Figure 1. Nitrogenase activity per gram of fresh weight of nodule as measured by the acetylene-reduction assay in nodules of *M. sativa* elicited by wild-type *S. meliloti*, (white bars) and flavodoxin-expressing *S. meliloti* (gray bars).

Substantial changes in nodule antioxidant metabolism, involving antioxidant enzymes and ascorbate-glutathione cycle enzymes and metabolites, were observed in flavodoxin-containing nodules, suggesting that the delay in nodule senescence is likely due to the antioxidant capacity of flavodoxin. Flavodoxin over-expression led to high starch accumulation in alfalfa nodules without a net increase in plant biomass. This is probably the first evidence that indicates that an amelioration of the oxidative balance may result in delayed nodule senescence and qualitative improvements.

The authors thank the Ministerio de Educacion y Ciencia, Comunidad de Madrid and CSIC for financial support.

References

Pueyo et al. (1991) Eur. J. Biochem. 202, 1065–1071.
Puppo et al. (2005) New Phytol. 165, 683–701.

EFFECTS OF THE OVEREXPRESSION OF A SOYBEAN CYTOSOLIC GLUTAMINE SYNTHETASE GENE (*GS15*) LINKED TO ORGAN-SPECIFIC PROMOTERS ON GROWTH AND NITROGEN ACCUMULATION OF PEA PLANTS

J. K. Vessey[1], H. Fei[1], S. Chaillou[2], B. Hirel[2], P. Polowick[3] and J. D. Mahon[3]

[1]Biology Department, Saint Mary's University, Halifax, NS, B3H 3C3, Canada; [2]Unité de Nutrition Azotée des Plantes, INRA, Route de St-Cyr, 78026, Versailles cedex, France; [3]Plant Biotechnology Institute, National Research Council, 110 Gymnasium Place, Saskatoon, SK, S7N 0W9, Canada

Glutamine synthetase (GS; EC 6.3.1.2) is a critical enzyme in the assimilation of ammonia into amino acids in higher plants and enhancement of its activity in plants may increase nitrogen utilization efficiency (Miflin and Habash, 2002). Overexpression of the cytosolic GS gene (*GS1*) has lead to increases in biomass and N accumulation (Habash et al., 2001; Oliveira et al., 2002). However, other studies indicate that overexpression of *GS1* leads to nil or even negative responses in plant function and growth. For example, overexpression of a soybean *GS1* linked to a nodule specific promoter (*LBC$_3$*) in *Lotus japonicus* lead to increases in GS activity, but no effects on nodule biomass or N$_2$ fixation (Limami et al., 1999). Over-expression of a *rolD-GS1* gene in the same species actually led to decreases in root biomass (Limami et al., 1999).

Studies were pursued (Fei et al., 2003, 2006) to determine if tissue-specific overexpression of a soybean cytosolic gene could improve plant growth in pea (*Pisum sativum* L.). A GS1 gene (*GS15*) from *Glycine max*, linked to whole plant-, nodule-, or root-specific promoters (i.e., *CaMV35S-GS15*, *LBC3-GS15*, and *rolD-GS15*, respectively) was transformed into pea. The performance of transformants was tested against the wild-type parental line (cv. Greenfeast) in hydroponic culture with either NO$_3^-$ or NH$_4^+$ as the mineral N source.

Chimeric gene constructs were prepared at INRA (Limami et al., 1999). Plasmids (pGPTV) containing each of the constructs were electroporated into *A. tumefaciens* and transgenic plants were recovered from developing embryo axes of *P. sativum* cv. Greenfeast (Polowick et al., 2000). Eleven stable, homozygous transformed plant lines were generated (three *35S-GS15* lines, four *LBC3-GS15* lines and four *rolD-GS15* lines). Presence of the *GS15* gene in the transformed lines was confirmed by PCR; copy number was determined by Southern blotting; and GS protein expression was assessed by

F. D. Dakora et al. (eds.), *Biological Nitrogen Fixation: Towards Poverty Alleviation through Sustainable Agriculture.*
© Springer Science + Business Media B.V. 2008

Western blot (Table 1) using tobacco GS antibodies (Fei et al., 2003, 2006). In all, 8 of the 11 transformed lines were tested in three hydroponic experiments: (a) four single-copy *GS15* lines (# 1, 7, 8 & 9) were tested with $^{15}NO_3^-$ as mineral N source (at 0, 0.1, 1.0 or 10.0 mM); (b) the same four single-copy *GS15* lines were tested with $^{15}NH_4^+$ as mineral N source; and (c) four double-copy *GS15* lines (# 2, 4, 6 & 11) were tested with $^{15}NH_4^+$ as mineral N source. After four weeks of growth, lines were tested for GS activity, biomass accumulation, N_2 fixation (by a ^{15}N dilution technique) and N accumulation.

Table 1. A summary of the results for GS15 expression in the eight tested lines as determined by Western-blot analysis.

Tissue	Increase in GS1 expressions in transformant lines relative to cv. Greenfeast		
	35S-GS15	*LBC3-GS15*	*rolD-GS15*
Leaves	Nil/+	Nil	+
Nodules	Nil/+	+	Nil/+
Roots	Nil/+	Nil	++

Southern blot analysis showed that the *GS15* was present in all the transformants (data not shown). Increased GS1 expression was seen in all expected target tissues, except with *35S-GS15*, where increased protein levels were not consistent in all transformants. The greatest relative increase in GS1 levels were in the roots of the *rolD-GS15* transgenic lines. Despite the consistent expression of *GS15* and increased GS1 protein in many lines, GS activity was only consistently increased in the roots of the *rolD-GS15* lines (up to 10 fold). There were no increases in N_2 fixation in any lines, and increases in biomass and N accumulation were rare, however, increases of 54% in N content and 45% in biomass were reported in one of the *rolD-GS15* lines grown at 10 mM NH_4^+.

The lack of consistent increases in GS activity in target tissues, despite positive Western blot analysis, suggests that over-expressed GS1 polypeptides may not always form active octameric holoenzyme (Fei et al., 2003). Even in tissue with increased GS activity, increases in biomass and N accumulation were neither common nor consistent, suggesting that GS1 activity may not be limiting in N assimilation and growth of pea. However, it is clear that over-expression of GS1 in pea roots does not have negative effects on growth, as was observed in *L. japonicus* (Fei et al., 2003). Albeit that occurrences were infrequent, but the dramatic increases in N accumulation and biomass observed in some of the *rolD-GS15* lines indicates that this type of transformation should be the focus of future efforts in increase growth by overexpression of GS1 in pea.

References

Fei H et al. (2003) Planta 216, 467–474.
Fei H et al. (2006) Plant Physiol. Biochem. 44, 543–550.
Habash DZ et al. (2001) Ann. Appl. Biol. 138, 83–89.

Limami A et al. (1999) Planta 209, 495–502.
Miflin BJ and Habash DZ (2002) J. Exp. Bot. 53, 979–987.
Oliveira IC et al. (2002) Plant Physiol. 129, 1170–1180.
Polowick PL et al. (2000) Plant Sci. 153, 161–170.

ROLE OF PEPc ISOFORMS IN PHOSPHATE-DEFICIENT NODULES

A. Kleinert[1], M. Venter[2], J. Kossmann[1] and A. J. Valentine[3]

[1]Institute for Plant Biotechnology; [2]Department of Genetics,
University of Stellenbosch, Private Bag X1, Matieland 7602, RSA;
[3]Department of Horticulture and Landscape Technology, Cape Peninsula
University of Technology, P.O. Box 652, Cape Town 8000, RSA

In soils, the concentration of available P for plants is normally very low, because most of the P combines with iron, aluminium, and calcium to form relatively insoluble compounds (Aono et al., 2001). Phosphate (Pi) deficiency is thought to be one of the limiting factors for nitrogen fixation (Aono et al., 2001) due to the high energy requirement of plants performing nitrogen fixation for nitrogenase function (Al Niemi et al., 1997). Pi deficiency has important implications for the metabolic Pi and adenylate pools of plants, which influence respiration and nitrogen fixation (Theodorou and Plaxton, 1995). An alternative route of pyruvate supply during Pi stress has been proposed (Theodorou and Plaxton, 1995), which involves the combined activities of phosphoenolpyruvate carboxylase (PEPc), malate dehydrogenase and NAD-malic enzyme supplying pyruvate to the mitochondrion during Pi stress. Marczewski (1989) purified three isoforms of PEPc from lupin nodules and roots, with two forms being nodule specific. The aim of this project was to determine the possible roles of PEPc isoforms in *Lupinus luteus* nodules under Pi stress.

Lupinus luteus (var. Juno) seeds were inoculated with *Bradyrhizobium* sp (*lupinus*) and germinated in vermiculite. Ten days after planting, seedlings were transferred to 22-L black tanks containing modified Long Ashton solution (no nitrogen). Plants were grown on sufficient P (2 mM) for ±40 days after which the first set was harvested (control). Plants were then split into two separate treatments, one continued to receive 2 mM P (P sufficient), whereas the other received 5 µM P (P stressed). Plants from both treatments were harvested 12 days after onset of P starvation and again at 20 days and Pi concentrations and *in vitro* PEPc activity of nodules were determined. Two novel PEPc isoforms, LUP1 (AM235211) and LUP2 (AM237200), were isolated from nodules. Transcriptional analyses, using semi-quantitative real-time PCR, were performed.

Nodules under P stress had lower amounts of metabolically available Pi and, as P stress developed, the amount of Pi decreased. A typical P-stress response is higher

F. D. Dakora et al. (eds.), *Biological Nitrogen Fixation: Towards Poverty Alleviation through Sustainable Agriculture.*
© Springer Science + Business Media B.V. 2008

anaplerotic carbon fixation via PEPc. However, we found that after 12 days there was a significant increase in PEPc activity in P-stressed plants, but no difference could be discerned in PEPc activity at 20 days. Preliminary data showed that there was no change in pyruvate kinase activity for both treatments after 12 days, but after 20 days, there was an increase in pyruvate kinase activity of P-stressed plants.

The similar PEPc activities after 20 days, along with the increase in pyruvate kinase, indicate that the P-stressed nodules have adapted to P-stress and more pyruvate may be synthesized via pyruvate kinase to assist in this adaptation. Juszczuk and Rychter (2002) suggested that increased pyruvate might serve as a mechanism for oxidizing the reduce-ing equivalents, which accumulate during P stress. Therefore, it appears that, although P stress can initially stimulate the P starvation bypass reaction via PEPc, this was later restored, as more pyruvate is needed to adapt to P stress.

The semi-quantitive real-time PCR results showed a decrease in expression of both LUP1 and LUP2 after 12 and 20 days after onset of P stress compared to the control set. No significant differences were found in expression of LUP1 or LUP2 at 12 and 20 days for both P-sufficient and P-stressed plants. From these results, it appeared that an in-crease in PEPc-gene expression did not account for the increase seen in *in vitro* PEPc activity found in plants grown under similar conditions. Therefore, it is feasible that the control/modulation of PEPc activity during P stress may reside in protein expression and subsequent post-translational modifications.

References

Aono T et al. (2001) Plant Cell Physiol. 42, 1253–1264.
Al Niemi T et al. (1997) Plant Physiol. 113, 1233–1242.
Theodorou ME and Plaxton WC (1995) In: Environmental and Plant Metabolism. Flexibility and Acclimation (Smirnoff N, ed.), pp. 79–109.
Marczewski W (1989) Physiol. Plantarum 76, 539–543.
Juszczuk IM and Rycheter AM (2002) Plant Physiol. Biochem. 40, 738–788.

OVER-EXPRESSION OF THE CLASS 1 HEMOGLOBIN GENE CONTRIBUTES TO SYMBIOTIC NITROGEN FIXATION

T. Uchiumi[1], F. Sasakura[2], M. Nagata[3], Y. Shimoda[4], A. Suzuki[5], K. Kucho[1], T. Sano[1], S. Higashi[1] and M. Abe[1]

[1]Faculty of Science; [2]Frontier Science Research Center; [3]Graduate School of Science and Technology, Kagoshima University, Japan; [4]Kazusa DNA Research Institute, Kisarazu, Chiba, 292-0818, Japan; [5]Faculty of Agriculture, Saga University, Japan

Plant hemoglobins (Hbs), which were first reported as leghemoglobin in root nodules of soybean (Kubo, 1939), have been identified in various species of legumes, non-legumes, and actinorhizal plants, and are divided into three distinct types, class 1, class 2, and trun-cated Hb. The physiological functions of plant Hbs are still unclear except for that of leghemoglobins (Lb), which compose a subgroup of class 2 Hb. Lbs are crucial for nitro-gen fixation in root nodules as an O_2 transporter and regulator, but not for general plant growth and development (Ott et al., 2005). Class 1 Hbs exhibit an extremely high affinity for O_2 (Arrendondo-Peter et al., 1997) and nitric oxide (NO) (Dordas et al., 2004). Class 1 Hbs may play significant roles by modulating NO levels in various physiological situations. During the *Rhizobium*-legume interaction, the produc-tion of NO was detected both at the early stage of host-symbiont recognition (Shimoda et al., 2005) and also in the nodules (Baudouin et al., 2006). NO can itself be an inducer of the class 1 Hb gene, so that the modulation of NO by class 1 Hb might be involved in establishment and mainten-ance of symbiotic nitrogen fixation.

The genes for the class 1 Hb of *Lotus japonicus* and *Alnus firma* were cloned and are referred to as *LjHbl* and *AfHbl*, respectively (Uchiumi et al., 2002; Sasakura et al., 2006). Both genes were highly expressed in nodules, and the recombinant proteins showed a higher affinity toward NO compared to the Lb of *L. japonicus*. *Lotus* plants with both transgenic and control hairy roots were generated by *A. rhizogenes* carrying *CaMV35Spro:Hb1*. The number of nodules per unit weight on transgenic roots, which over-expressed *LjHbl* and *AfHbl*, was at least twice that on the control hairy roots. No significant increase was detected in the total nodule number per plant, indicating that *LjHbl* and *AfHbl* did not prevent autoregulation of nodulation. The acetylene-reduction activity (ARA) of nodules on hairy roots that over-expressed *LjHbl* and *AfHbl* was twice the value in the control hairy roots. The ARA of nodules of *L. japonicus* and

A. firma was strongly inhibited by the addition of an NO donor (SNAP, *S*-nitroso-*N*-acetyl-D,L-penicillamine) and enhanced by the addition of either an NO scavenger (cPTIO, carboxy-2-phyenyl-4,4,5,5-tetramethyl-imidazoline-3-oxide-1-oxyl) or an inhibitor of NO synthase (L-NAME, N^G-nitro-L-arginine methyl ester). These results indicate that class 1 Hb, when acting as an NO scavenger, might contribute to efficient nodulation and nitrogen fixation.

The *Frankia* symbiont of *Casuarina glauca* forms vesicles under free-living conditions, but does not do so in nodules. The class 2 Hb gene is strongly expressed in the nodules of *C. glauca* to create a microaerobic condition (Jacobsen-Lyon et al., 1995). In the nodules of *A. firma*, as found in free-living conditions, *Frankia* forms vesicles that are surrounded by a multilayered lipid envelope, which functions as an O_2-diffusion barrier, so that expression of class 2 Hb gene is not required. Instead of the class 2 Hb acting as an O_2 regulator, the function of the class 1 Hb can become significant in the nodules of *A. firma*. These results indicate that class 1 Hbs of host plants support the nitrogen-fixation capacity of both *Rhizobium* and *Frankia* by modulating the NO level in the nodules.

References

Arrendondo-Peter R et al. (1997) Plant Physiol. 115, 1259–1266.
Baudouin E et al. (2006) Mol. Plant-Microbe Interact. 19, 970–975.
Dordas C et al. (2004) Planta 219, 66–72.
Jacobsen-Lyon K et al. (1995) Plant Cell 7, 213–223.
Kubo H (1939) Acta Phytochimica (Toyko) 11, 195–200.
Ott T et al. (2005) Current Biol. 15, 531–535.
Sasakura F et al. (2006) Mol. Plant-Microbe Interact. 19, 441–450.
Shimoda Y et al. (2005) Plant Cell Physiol. 46, 99–107.
Uchiumi T et al. (2002) Plant Cell Physiol. 43, 1351–1358.

PART 2C

PLANT GENOMICS
AND TRANSCRIPTOMICS

STRUCTURAL AND COMPARATIVE GENOME ANALYSIS OF *LOTUS JAPONICUS*

S. Sato, T. Kaneko, Y. Nakamura, E. Asamizu, T. Kato and S. Tabata

Kazusa DNA Research Institute, 2-6-7 Kazusa-Kamatari, Kisarazu, Chiba, 292-0818, Japan

Legumes, the third largest family of flowering plants, are composed of more than 18,000 species with diverse characteristics, and many of the species are of agronomic importance. They have long been the targets of breeding programs and, in this regard, genomics may provide the greatest potential for benefit. Legume genomics in two model plants, *Lotus japonicus* and *Medicago truncatula*, has advanced rapidly over the past few years (Young et al., 2005; Sato and Tabata, 2006). *L. japonicus* is a perennial temperate pasture species with characteristics suitable for genetic and genomic studies, such as short life cycle (2–3 months), self-fertility, diploidy (n = 6), small genome size (472.1 Mb), and feasibility of genetic transformation. Genomics in *L. japonicus* was initiated in 1999 by collection of ESTs (Asamizu et al., 2004), followed by large-scale genome sequencing (Sato et al., 2001), generation of DNA markers and construction of high-density genetic linkage maps (Hayashi et al., 2001). The information and material resources accumulated during the course of these genomic analyses are useful for understanding the genetic system of *L. japonicus*. Furthermore, such resources might assist breeding programs by comparison and subsequent transfer of the knowledge on *L. japonicus* to crop legumes (Zhu et al., 2005).

Status of Genome Sequencing

Large-scale genome sequencing of *L. japonicus* began in 2000 using accession Miyakojima MG-20. Multiple seed points were chosen along the entire genome, based on the sequences of ESTs, cDNAs, and gene segments from *L. japonicus* and other legumes. Genomic libraries were constructed with transformation-competent artificial chromosome (TAC) and bacterial artificial chromosome (BAC) vectors, and screened by PCR to select clones for sequencing. The nucleotide sequence of each selected clone was determined by the shotgun strategy, which is followed by semi-automatic and manual gene modeling and annotation. In parallel, microsatellite or SNP markers were generated using the sequence information obtained and the clones were located on the genetic linkage map (Sato et al., 2001). As of January 2007, among the 2,031 clones selected

F. D. Dakora et al. (eds.), *Biological Nitrogen Fixation: Towards Poverty Alleviation through Sustainable Agriculture.*
© Springer Science + Business Media B.V. 2008

for sequencing, 1992 clones covering 190 Mbp of the genome have been sequenced including clones in draft (phase 1) stage. These accumulated sequences are compiled into 364 contigs and 623 single clones. The longest contig consists of 18 clones covering a 1.6-Mbp region. A total of 1,590 clones have been located on the genetic linkage map, using 815 microsatellite markers and 80 dCAPS markers and by overlaps with the genetically mapped clones. These DNA markers and associated sequence information provide enormous value in gene mapping and map-based cloning in *L. japonicus* and other legumes (Sato and Tabata, 2006).

To complement the ongoing sequencing effort that adopts the clone-by-clone strategy, shotgun sequencing was introduced. Two types of genomic libraries were constructed for this project. One was a whole genome shotgun library from which selected highly repetitive sequences were subtracted. The other was a shotgun library derived from a mixture of 4,603 TAC clones selected under the condition that neither one of the end sequences match previously sequenced clones or repetitive sequences (STM shotgun: selected TAC mixture shotgun), which likely covers the unsequenced euchromatic regions. Approximately 1.7 million sequence files have been accumulated from the two libraries. Forty-four percent of the obtained sequences were shown to be derived from regions covered by the sequenced genomic clones, the remaining 56% were assembled into 109,986 contigs, a total of which cover 148 Mbp of the genome. Collectively, approximately 90% of the *L. japonicus* ESTs could be assigned to currently accumulated seed clone and random genome sequences.

Gene Features and Resources of *L. japonicas*

In the 80.2 Mb sequences of the finished clones, complete structures of 6,584 potential protein-encoding genes have been deduced. The structural features of the protein-encoding genes in *L. japonicus* are broadly similar to those of *Arabidopsis thaliana*, with the exception of average intron length (371 bp in *L. japonicus* vs. 157 bp in *A. thaliana*), which results in a difference in average gene length including introns (2,883 vs. 1,918 bp). The gene density in the sequenced regions of the *L. japonicus* genome (one gene in every 12.3 kbp) is approximately half of *A. thaliana* (one gene in every 4.5 kbp), although this difference is less marked when retrotransposons are considered. Similarity searches of potential protein-encoding genes against the nr protein database indicated that approximately 40% of predicted *L. japonicus* genes were homologous to genes of known function, 40% showed similarity to hypothetical genes, and the remaining 20% exhibited no significant similarity to registered genes.

Along with the progress of the genome analysis projects of *L. japonicus*, a variety of information and material resources have been generated. The online database at http://kazusa.or.jp/lotus/ supports easy access to the information resources produced by the *Lotus* genome and cDNA sequencing projects. This web site provides nucleotide sequences of the TAC and BAC clones and the chloroplast genome with annotation of predicted genes as well as information on the DNA markers, the genetic linkage maps and genotype data of the RI lines. The material resources generated during the course of the *Lotus* genome project have been deposited into the resource center at Miyazaki University, which is supported by the Japan National Bioresource Project. Resources including seeds of wild accessions and RI lines as well as cDNA clones and genomic libraries can be accessed by following the guidance in the web database, Legume Base (http://www.shigen.nig.ac.jp/legume/legumebase/).

Comparative Genome Analysis of *L. japonicas*

The rapidly accumulating genome information for the two model legumes, *L. japonicus* and *M. truncatula*, provides a unique opportunity to explore synteny based on genome-scale sequence comparisons. This comparative analysis provides an advanced perspec- tive of comparative genomic studies. As of January 2007, microsynteny has been detected for approximately 70% of the sequenced *L. japonicus* clones, although traces of local dupli-cations, inversions, deletions, and insertions are observed. The proportion of the clones exhibiting microsynteny is expected to increase as further sequences become available for *L. japonicus* and *M. truncatula*. By combining the syntenic relations assigned at the clone level with positional information on the linkage maps, the microsyntenic relations could be expanded to macrosynteny at the level of entire or large segments of chromosome (Cannon et al., 2006). With a quite high level of synteny in both micro- and macroscale, the accumulating genome information on two model legumes will establish a robust basis for studies on the complexity and diversity of legume genomes.

References

Asamizu E et al. (2004) Plant Mol. Biol. 54, 405–414.
Cannon SB et al. (2006) Proc. Natl. Acad. Sci. USA 103, 14959–14964.
Hayashi M et al. (2001) DNA Res. 8, 301–310.
Sato S and Tabata S (2006) Curr. Opin. Plant. Biol. 9, 128–132.
Sato S et al. (2001) DNA Res. 8, 311–318.
Young N et al. (2005) Plant Physiol. 137, 1174–1181.
Zhu H et al. (2005) Plant Physiol. 137, 1189–1196.

NOVEL SYMBIOTIC REGULATORY GENES IDENTIFIED BY TRANSCRIPTOMICS IN *MEDICAGO TRUNCATULA*

J.-P. Combier, T. Vernie, S. Moreau, T. Ott, L. Godiard, A. Niebel and P. Gamas

Laboratoire des Interactions Plantes Microorganismes (LIPM), INRA CNRS, Chemin de Borde rouge, 31326 Castanet Tolosan cedex, France

Transcriptomics as a Tool to Identify Novel Regulatory Genes

Forward genetics has been extremely fruitful in the recent past to identify key genes involved in the very early stages of symbiotic signaling in model legumes. However, the number of regulatory genes, particularly transcription-factor genes, demonstrated to play a role in the control of the formation of an infected nodule remains very limited. Transcriptomics represents a complementary approach to forward genetics. It can be very powerful for identifying such genes, now that efficient reverse genetics tools have been set up for model legumes. Further, transcriptomics is likely the best method to give a global view of the range of genes that accompany a developmental program, and to compare it to other kinds of developmental program, e.g., symbiotic vs. pathogenic inter-actions or root-nodule formation vs. lateral-root formation. Transcriptomics approaches have already been used to get a view of the range of genes associated with the nodulation process. We report here on the last global approach used in the frame of the European Grain Legume Integrated Project. This approach was based on 70-mer oligonucleotide microarrays, provided by Helge Kuester, Bielefeld, representing about 16,000 *Medicago truncatula* genes.

To distinguish between various regulatory patterns, we conducted a comparative analysis of a series of nodule samples, induced by either *Sinorhizobium meliloti* wild-type or mutants. More than 2,000 up-regulated genes (in comparison to non-inoculated roots) were identified (with a false discovery rate <1% and an induction ratio >1.5) of which about 1,100 were found activated in at least two different samples. This internal consistency showed the reliability of the results. More than 100 candidate regulatory genes were identified, encoding many families of transcription factors (with the ERF-AP2 family most represented) and indicating a very complex symbiotic regulatory net-work. However, at least five regulation patterns were seen by hierarchical clustering analysis, among which one gene cluster of particular interest to us co-regulated with early nodulin genes, such as *MtENOD11*, *MtENOD40* or *MtNIN*.

F. D. Dakora et al. (eds.), *Biological Nitrogen Fixation: Towards Poverty Alleviation through Sustainable Agriculture.*
© Springer Science + Business Media B.V. 2008

Characterization of Regulatory Genes Expressed at Early Stages of Nodulation

Four genes that were strongly induced in immature nodules were characterized in more detail. Three of them are likely to be involved in the negative regulation or in the fine tuning of nodulation and infection, based on the phenotype observed following RNA interference. In two cases, the number of rhizobia recovered on plates from crushed nodules increased substantially, which correlated with a strong increase in the diameter of infection threads. In addition, the nodulation was enhanced with two RNAi constructs, whereas gene over-expression led to a strong decrease in nodulation. Possible target genes were looked for by transcriptomics for one gene, and found to be commonly associated with abiotic- or biotic-stress responses. Finally, one of these genes was found to be efficiently turned on by a bacterial root pathogen.

In contrast, the fourth gene, which encoded a protein of the HAP family, played a positive role at different stages of the nodulation process. The HAP proteins are encoded by multigene families in plants, and bind the CCAAT motif in promoters as heterotrimers of HAP2, 3 and 5 subunits. The CCAAT motif is rather common in eukaryotic promoters, but it should be underlined however that several *HAP* genes have been demonstrated to play a central role in the control of a developmental process in plants. For example *LEC1* is a key regulator of embryogenesis in *Arabidopsis thaliana* and a HAP complex mediates the interaction of Constans or Constans-like proteins with their DNA targets, thereby controlling the expression of floral integrators.

HAP2.1 belongs to a multigene family in *Medicago truncatula* and it is symbiosis-specific, based on Electronic northern. We found that knocking down *HAP.1* by RNAi led to small arrested nodules, surrounded by the endodermis, which is indicative of a non-functional meristem (Combier et al., 2006). The underlying tissues were not properly differentiated and no bacteria liberation could be observed, even if a few infection threads wre present. We established by in situ hybridisation that *HAP2.1* was expressed essentially in the nodule meristem, with some residual transcripts in the upper infection zone II. We found two potential miR169 recognition sites in the 3'-UTR of *HAP2.1* and showed that one of them was cleaved in nodules. In collaboration with M. Crespi's group (ISV CNRS Gifsur-Yvette) and J. Gouzy (LIPM Toulouse), a possible precursor gene for the miR169 was identified in the *Medicago truncatula* genome sequence, and showed to be expressed in nodules. Over-expressing this precursor led to non-functional nodules, similar to those obtained with a *HAP2* RNAi construct. A promoter:GUS fusion analysis showed that this precursor was expressed in zone II of the nodule and totally excluded from zone I. This result suggests that the miR 169 regulation helps to establish a strong gradient of HAP2.1 protein, necessary for proper nodule development.

Reference

Combier J-P et al. (2006) Genes and Dev. 20, 3084–3088.

GENOME SYNTENY OF PEA AND MODEL LEGUMES: FROM MUTATION THROUGH GENETIC MAPPING TO THE GENES

V. A. Zhukov[1], A. Y. Borisov[1], E. V. Kuznetsova[1], L. H. Madsen[2], M. D. Moffet[3], E. S. Ovchinnikova[1], A. G. Pinaev[1], S. Radutoiu[2], S. M. Rozov[4], T. S. Rychagova[1], O. Y. Shtark[1], V. E. Tsyganov[1], V. A. Voroshilova[1], J. Stougaard[2], N. F. Weeden[3] and I. A. Tikhonovich[1]

[1]All-Russia Research Institute for Agricultural Microbiology, Podbelsky schaussee 3, Saint-Petersburg Pushkin 8, 196608, Russia; [2]Department of Molecular Biology, University of Aarhus, Aarhus, DK-8000, Denmark; [3]Department of Plant Science and Plant Pathology, Montana State University, Bozeman, MT, USA; [4]Institute of Cytology and Genetics, Lavrent'ev avenue 10, Novosibirsk, 630090, Russia

Genome synteny of crop and model legumes appears to be a key approach for sequencing symbiotic genes of several agriculturally important species (Zhu et al., 2005). To exploit genome synteny, it is necessary to localize genes of interest precisely on the genetic map of pea (*Pisum sativum* L.) and then try to find candidate genes in the orthologous region of a model legume.

We concentrated on the precise location of three symbiotic genes, *sym27*, *crt* and *coch*, lying in linkage group (LG) V of pea. A set of CAPS markers, derived from genes of chromosome 7 of barrel medic (*Medicago truncatula* Gaertn.), was created for this purpose. Genetic maps of pea and barrel medic were compared and this enabled us to define the regions in chromosome 7 of *M. truncatula*, where presumed orthologs of pea symbiotic genes are located. New markers are now being designed based on *M. truncatula* genes lying between markers used earlier and closer to the target genes.

Genetic mapping of the pea symbiotic gene *Sym37* lying in LG I of pea (the same region as for *sym2*) was performed and the nucleotide sequence was established. It was assumed that this gene shared similarity with the gene *Nfr1* of *Lotus japonicus* (Regel.) Larsen, which is located in the orthologous region of chromosome I of *L. japonicus*. Screening of the pea cDNA library, using part of *Nfr1* as a probe, allowed us to identify two pea genes homologous to *Nfr1*, initially named *K1* and *K10*. Both genes encode putative Ser/Thr kinases, similar to *Nfr1*, which appears to be a Nod-factor receptor. Mutations in the coding sequence of gene *K10* were found after direct sequencing

F. D. Dakora et al. (eds.), *Biological Nitrogen Fixation: Towards Poverty Alleviation through Sustainable Agriculture*.

of pea homologues of *Nfr1* in two mutant lines (Figure 1). Inheritance of a molecular marker representing *K10* in the F2 population confirms that the gene *K10* (homologous to *Nfr1*) is the pea gene *Sym37*. Inoculation experiments with *Rhizobium leguminosarum* bv. *viciae* strain carrying the *nodX* gene demonstrated that a mutant with the Leu-to-Phe substitution in the receptor domain behaves as an "afghan" variety (the pea genotype carrying *sym2^A* allele in the homozygous state).

The structure of the putative proteins was determined by *in silico* analysis and compared to the homologous genes *Nfr1* and *LYK3* of the model legume plants *L. japonicus* and *M. truncatula*, respectively. It is suspected that both homologues of *Nfr1* in pea could work as Nod-factor receptors, probably in a complex with *Sym10*.

The results of the present study demonstrate the possibility of using the genome synteny of pea and model legumes for the subsequent nucleotide sequencing of pea symbiotic genes.

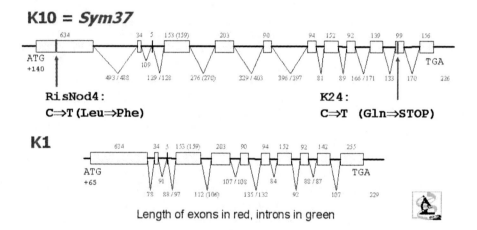

Figure 1. Pea genes homologous to *Nfr1* of *L. japonicus*.

This work was supported by State Contracts (02.445.11.7492 and 02.434.11.7122), President grant (HIII-9644.2006.4), RFBR (06-04-01856, 07-04-01171, 07-04-01558), and GLIP TTC FOOD-CT (2004-506223).

Reference

Zhu et al. (2005) Plant Physiol. 137, 1189–1196.

THE *MEDICAGO TRUNCATULA LIN* MUTANT: CAUGHT BY MAP-BASED CLONING, CHARACTERISED BY TRANSCRIPTOMICS

G. Endre[1], E. Kiss[1], A. Borbola[1], H. Tiricz[1], A. Lozsa[1],
B. Olah[1], K. Kuppusamy[2], N. Sharopova[2] and K. A. VandenBosch[2]

[1]Biological Research Center, Institute of Genetics, H-6726 Szeged, Hungary;
[2]Department of Plant Biology, University of Minnesota, St. Paul, MN 55108, USA

The *Medicago truncatula lin* (lumpy infections) mutant, which is defective in intermediate steps of nodule differentiation, has been identified previously as an EMS mutant of line A17 (Penmetsa and Cook, 2000). This mutant is characterized by a four-fold reduction in the number of infections, all of which arrest in the root epidermis, and by nodule primordia which initiate normally but fail to mature.

For genetic mapping, *lin* was crossed to *M. truncatula* A20 ecotype and an F2 mapping population was generated. The map position of *lin* was already located between CAPS markers, DSI and SCP, on Linkage Group 1 (Kuppusamy et al., 2004). These molecular markers were used to start the physical mapping and genome walking towards *lin*. The BAC clone carrying marker DSI was easily identified because clone mth2-004O04 has been assigned to genetic marker DSI on the *M. truncatula* physical mapping website (http://mtgenome.ucdavis.edu; by the Cook lab, UC-Davis). Generating and mapping a new genetic marker at one end of this BAC clone proved the real linkage of this clone to both marker DSI and *lin* mutation, and also showed in which direction the genome walking should proceed. From the other anchor genetic marker (SCP), the first BAC clone, which had been assigned to it, proved to map at a different position on Linkage Group 2. By analyzing the growing amount of sequence information, another BAC clone was identified as a potential carrier of genetic marker SCP. Again, new markers were generated at both ends of this BAC and mapped in the *lin* segregating population, so demonstrating the true linkage and determining the direction for the genome walk from this side. The next steps taken were similar from both regions. The accumulating data produced by the Medicago Genome Program (http://www.medicago.org/genome) were followed and used to gain insight for the next steps. These included the physical contig information of mth2 BAC clones, which originated from the *Hin*dIII fingerprinting

F. D. Dakora et al. (eds.), *Biological Nitrogen Fixation: Towards Poverty Alleviation through Sustainable Agriculture.*
© Springer Science + Business Media B.V. 2008

225

of random clones, and the sequence information coming from different laboratories sequencing entire BAC clones as well as BAC ends.

The indicated direction was continuously checked by both designing primers to known sequences and proving true overlap between BACs by PCR, and trying to produce genetic markers by looking for polymorphism in them. Anywhere a genetic marker was located, it was genotyped for the individuals showing recombination in the region of interest, and tested whether another recombination had been by-passed or not. For contigs that could not be connected to one another, we performed hybridization experiments to identify the following overlapping clones. In this way, we could join together four contigs of BAC clones from each side of the genome walking and identify the region carrying the *lin* gene to an ~350 kb region, where no recombination could be detected to the mutation. This region, represented by three BAC clones, has been sequenced and analyzed for its gene content. Based on their predicted function, the best candidates were selected for sequencing to determine if they carried an appropriate mutation or not. The first eight candidate genes (different transcriptional factors or genes with unknown function but with higher expression in the symbiotic tissues based on EST analysis) showed no mutation in the *lin* mutant. A mutation was found in a gene coding for a ~1,500 AA protein at a last position of an intron. Since there is no EST from any of the *Medicago* cDNA libraries for this gene, the first task was to see if this gene was active. The wild-type cDNA could be easily amplified and sequenced. In the *lin* mutant, the intron carrying the mutation is not spliced out of the mRNA and so generates an early STOP codon. Complementation experiments are in progress to prove we have identified the *LIN* gene.

To characterize LIN function, microarray experiments with both *lin* and A17 roots, following inoculation with *S. meliloti*, produced transcript profiles that showed molecular changes in *lin* roots that result in the symbiotic phenotype. A time-course experiment, including time points selected to correspond to stages when the phenotype of the mutant diverged from the wild type, revealed co-regulated genes in the wild type that were mis-regulated in the mutant. These included genes predicted to be involved in defense responses, cell cycle regulation, transport and signaling, membrane composition, and transport. We also found a set of 55 genes that were normally regulated in the mutant, suggesting that some of the machinery in the symbiotic pathway is functional in *lin*.

References

Kuppusamy KT et al. (2004) Plant Physiol. 136, 3682–3691.
Penmetsa RV and Cook DR (2000) Plant Physiol. 123, 1387–1397.

PROTEOMIC ANALYSIS OF SECRETED PROTEINS OF *GLYCINE MAX* AND *MEDICAGO SPP.*

M. A. Djordjevic[1], M. Oakes[1], L. Kusumawati[1], C. Hocart[2], C. McKinlay[2], C.-H. Hwang[3], D. Li[3] and P. M. Gresshoff[3]

Australian Research Council Centre of Excellence for Integrative Legume Research: [1]Genomic Interactions Group and [2]Mass Spectrometry Facility, Research School of Biological Sciences, Australian National University, Canberra, Australia 2601; [3]Botany Department, University of Queensland, St Lucia, Queensland, Australia 4067

Protein-based interactions occurring in plant extracellular spaces (the apoplast) have attracted increased attention because they are critical for defence and development (Boller, 2005). Important apoplast-located self- and non-self responses include the recog-nition of (i) bioactive peptides from phytobacteria and (ii) endogenous peptides. The bioactive peptides interact with membrane receptors and 100s of plant receptors are predicted to "sense the apoplast" and/or external environment. Also, long distance root-to-shoot inter-actions are mediated by unknown graft transmissible signals and the xylem is a likely conduit for this signal transmission (Searle et al., 2003; Beveridge, 2006; Suarez-Lopez, 2005). RNA expression studies suggest that root-synthesised sap proteins can be detec-ted in shoot xylem sap (Satoh, 2006). In addition, an *Arabidopsis* mutant affected in the gene encoding the apoplastic protein, lipid transfer protein, is defective in systemic acq-uired resistance (Maldonado et al., 2002).

To identify new secreted protein signalling molecules, proteomic studies of apoplast and sap proteins in *Glycine* and suspension culture secreted proteins in *Medicago* were initiated. Xylem sap (isolated from the hypocotyl of decapitated plants) and apoplastic fluid (isolated from the epicotyl) was collected from soybeans. We examined soybeans defective in autoregulation of nodulation (AON; Searle et al., 2003) at different time points post-*Bradyrhizobium* or no inoculation; these were *Nts 1007*, a super-nodulation phenotype defective in a CLAVATA1-like LRR-receptor kinase, and *nod139*, a reces-sive Nod⁻ mutant caused by a nonsense mutation in *GmNFR5α*. One- or 2-D gel electro-phoresis combined with tandem mass spectrometry (LC-MS/MS, LCQ DECA XP and MALDI-TOT/TOF, ABI 4800) were used to separate, compare, and identify proteins. Non-quantitative RT-PCR was used to examine the gene-expression pattern of the cor-responding sap/apoplast proteins. Suspension cultured cells of *M. sativa* sp. varia and *M. truncatula* (2HA) were established and secretome examined similarly.

F. D. Dakora et al. (eds.), *Biological Nitrogen Fixation: Towards Poverty Alleviation through Sustainable Agriculture.*
© Springer Science + Business Media B.V. 2008

Irrespective of the method of isolation of sap or apoplastic fluid, the genotype of the soybean plants, plant age at sap collection, treatment applied, or day post-inoculation when fluids were collected, their protein separation patterns on 1- and 2-D gels were qualitatively similar. Ten bands ranging from ~6 to ~80 kDa were present in 1-D gels. Multiple sample analyses identified 20 proteins common-to-all samples possessing predicted N-terminal secretion signals. These results suggested that none were involved directly in AON. Five cytoplasmic proteins were present as low level contaminants. About half of the soybean xylem sap proteins are found commonly in the xylem sap of other plant species and included lipid transfer proteins (LTPs), peroxidases, aspartyl proteases, β-1,3-glucanases, β-galactosidase, arabinogalactan-proteins, lipases, invertases, inhibitors and blue copper proteins (Buhtz et al., 2004; Rep et al., 2003; Alvarez et al., 2006). The more "soybean-specific" components of xylem sap included: Kunitz trypsin inhibitors (KTI), γ-glutamyl hydrolase, vegetative storage proteins, a dehydration stress/ elicitor induced protein, and a Ppr-27-like (defence-related) protein. Nq RTPCR indicated that only one LTP gene was predominantly expressed in soybean roots and the hypocotyl and therefore could move a long distance to the sap/apoplastic fluid collection points (the hypocotyl and epicotyl). If so, the well characterised hydrophobic cleft(s) of LTP (Wang et al., 2004) could carry root-produced signals to the shoot. In addition, KTI has been implicated as a signal strength modulator of defence responses (Park et al., 2001).

Secreted proteins of *M. sativa* and *M. truncatula* suspension cultures were resolved into 12 and 15 bands, respectively, using 1-D PAGE. The *M. sativa* array was not affected by adding phytohormones or other bioactive molecules. Seven proteins were detected, including SIEP1L (a Culculin-like putative mannose-binding protein), poly-galacturonase inhibitor protein (PGIP), and chitinases. In *M. truncatula*, the 16 detected proteins included chitinases, proteases, peroxidases, glucanases, glucanase and protease inhibitors, and PR proteins. PGIP was the only protein found in both suspension cul-tures. All species detected had an N-terminal secretion signal. SIEP1L and PGIP are implicated in cell adhesion and cell-to-cell signalling (Boller, 2005).

References

Alvarez S et al. (2006) J. Proteome Res. 5, 963–972.
Beveridge CA (2006) Curr. Opin. Plant Biol. 9, 35–40.
Boller T (2005) Curr. Opin. Cell Biol. 17, 116–122.
Buhtz A et al. (2004) Planta 219, 610–618.
Maldonado AM et al. (2002) Nature 419, 399–403.
Park DS et al. (2001) Physiol. Molec. Plant Pathol. 59, 265–273.
Rep M et al. (2003) FEBS Lett. 534, 82–86.
Satoh S (2006) J. Plant Res. 119, 179–187.
Searle I R et al. (2003) Science 299, 109–112.
Suarez-Lopez P (2005) Int. J. Dev. Biol. 49, 761–771.
Wang SY et al. (2004) Acta Crystallogr. D. Biol. Crystallogr. 60, 2391–2393.

PART 2D

BACTERIAL PARTNER

PROTEOMIC AND TRANSCRIPTOMIC APPROACHES TO STUDY GLOBAL GENOME EXPRESSION IN *RHIZOBIUM ETLI* AND *SINORHIZOBIUM MELILOTI*

S. Encarnación[1], E. Salazar[1], G. Martínez[1], M. Hernández[1], A. Reyes[1],
M. del C. Vargas[1], S. Contreras[1], M. Elizalde[1], R. Noguez[1], N. Meneses[1],
O. Bueno[1], R. Sánchez[2], Y. Mora[1] and J. Mora[1]

[1]Centro de Ciencias Genómicas, Universidad Nacional Autónoma de
México, Cuernavaca, Mor., México; [2]Applied Biosystems, México

Two global approaches have been used to study the free-living state of the symbiotic species, *Rhizobium etli* and *Sinorhizobium meliloti*, as well as the *R. etli-Phaseolus vulgaris* interaction. The proteome of *R. etli* was examined to determine the enzymatic reactions and cell processes that occur in the free-living state (fermentative, aerobic metabolism, and symbiosis with legume plants). All the detectable protein spots on the two-dimensional (2-D) gels between pH 3–10 were analyzed. In total, we identified 1,518 proteins. Using a combination of 2-D gel electrophoresis, peptide mass fingerprinting, and bioinformatics, our goal is to identify: (i) biofilm-, aerobic- and fermentative metabolism-, symbiosis- and stress-specific proteins; and (ii) the biochemical pathways active under different conditions tested. Using the databases of the genome sequences from *R. etli* and *S. meliloti* genomes and the software "Pathway tools", we constructed *in silico* the pathways and metabolic reactions potentially functional in both bacteria. We called these databases Sinocyc (*S. meliloti*) and Rhizocyc (*R. etli*). Using this definition, 46 pathways involved in many common anabolic and catabolic cellular processes of small molecule metabolism can be considered active in *R. etli* under the conditions examined.

In both bacterial species, approximately 30% of the proteins were present on 2-D gels in at least two distinct locations indicating charge or size variation. Using phosphor-specific stains and LC-MS, we have shown that a large number of proteins are phosphorylated, including, in both bacteria, aconitase, ATP synthase alpha and beta chains, cysteine synthase, DnaK, elongation factors P, Ts and Tu, enolase, GroEL2, GTP-binding protein (TypA), isocitrate dehydrogenase, lipoamide dehydrogenase, ribosomal protein S1, and succinyl-CoA synthetase alpha chain.

F. D. Dakora et al. (eds.), *Biological Nitrogen Fixation: Towards Poverty Alleviation through Sustainable Agriculture.*
© Springer Science + Business Media B.V. 2008

In the rhizosphere, most microbes grow as organised biofilm communities on surfaces. Biofilms can be defined as communities of microorganisms attached to a surface. It is clear that microorganisms undergo profound changes during their transition from plank-tonic (free-swimming) organisms to cells that are part of a complex, surface-attached community (sessile). With two approaches (microarray and proteome analysis), we iden-tified genes encoding proteins involved in adhesion (type 1 fimbriae and flagellum) and, in particular, autoaggregation were highly expressed in the adhered population in a man-ner consistent with current models of sessile community development. Several novel gene clusters and proteins were induced upon the transition to biofilm growth, and these included genes and proteins expressed under oxygen-limiting conditions, genes enco-ding transport proteins, putative oxidoreductases, and genes associated with enhanced antibiotic or heavy metal resistance. Of particular interest was the observation that many of the genes and proteins altered in expression have no current defined function. These genes, as well as those induced by stresses relevant to biofilm growth such as oxygen and nutrient limitation, may be important factors that trigger enhanced resistance mechanisms of sessile communities to antibiotics and hydrodynamic shear forces.

A set of 23 heat shock proteins (Hsp) was observed by subtractive two-dimensional gel electrophoresis to be induced when *R. etli* is temperature up-shifted from 30°C to 42°C. Up-regulated protein spots were identified by mass spectrometry. Five proteins in the 8-20 kDa range were identified as the small Hsp (sHsp; HspB, C, D, E and H) and three others showed strong sequence similarity to the sHsp family. Other proteins corresponded to novel proteins such as Ibp, Omp and ArgC. In *S. meliloti*, 49 heat shock proteins were induced, 19 of them being approximately 60 kDa and identified as GroEL1, GroEL2, and GroEL5. Additionally, the small Hsp were not induced in a similar way as in *R. etli*. Phosphoproteome analysis in *S. meliloti* indicated changes in phosphorylation patterns during heat shock-stress conditions. The differences observed in *R. etli* and *S. meliloti* suggest different mechanisms to respond to the same stress conditions which involve proteins that could be placed into at least three distinct regulatory groups based on the kinetics of protein appearance.

In transcriptomics, we constructed a microarray with the complete sequence of the *R. etli* genome and performed studies at different time points during nitrogen fixation with *P. vulgaris*. The results show significant differences in gene-expression pattern between the free-living state as compared to symbiosis and specific components are induced at different times during nitrogen fixation in symbiosis.

This work was supported by DGAPA-UNAM IN203003-3 grant.

CONTROL AND ACTION OF THE TRANSCRIPTIONAL REGULATOR TtsI OF *RHIZOBIUM* SP. NGR234

R. Wassem, K. Kambara, H. Kobayashi, W. J. Broughton and W. J. Deakin

LBMPS, Département de Biologie Végétale, University of Geneva, Switzerland
(Email: wassem@ufpr.br)

Development of nodules by legumes is dependent upon complex signal exchanges between the plants and symbiotic nitrogen-fixing bacteria. The primary signals exuded from the plants are flavonoids, which trigger a transcriptional activation cascade in the bacteria, leading to the production of a variety of symbiotic signals. Flavonoids are sensed by rhizobia and lead to the activation of the transcriptional regulator NodD, which, in turn, induces numerous genes responsible for the synthesis of Nod factors. A large variety of Nod factors are produced by different strains of rhizobia. Apart from flavonoid/Nod-factor signalling, proteins secreted by type-three secretion systems (T3SS) are also important for nodulation (see Broughton et al., 2000). Such secretion systems were believed to be only present in pathogenic bacteria, but have since been found in non-pathogens, such as rhizobia. The T3SS of *Rhizobium* sp. NGR234 has been extensively characterized. Mutation of T3SS leads to different effects on nodulation, depending on the host plant (Marie et al., 2003). In *Phaseolus vulgaris*, for example, knock-out of T3SS induces an increase in nodulation, whereas in *Tephrosia vogelii*, it partially inhibits nodulation. On plants like *Lotus japonicus* and *Vigna unguiculata*, however, mutations of the T3SS machinery produced no obvious effects.

In *Rhizobium* sp. NGR234, the genes encoding the T3SS machine are clustered in a 30-kb region of the symbiotic plasmid. Based upon homology with other bacteria, these genes are thought to code for proteins that form a trans-membrane pore in the bacterial membrane as well as an external needle/pilus-like structure that allows effector proteins to be directly injected into the eukaryotic host-cell cytoplasm. It has been shown that secretion of proteins by the T3SS of NGR234 is dependent on flavonoids, NodD1, and the transcriptional activator TtsI (Marie et al., 2004; Kobayashi et al., 2004). As well as inducing the T3SS, TtsI also controls (flavonoid-induced) modifications to lipo-poly-saccharides (LPS) that also affect the nodulation ability of NGR234 (Marie et al., 2004). TtsI belongs to the regulator class of two-component systems' sensor-regulator proteins, but it lacks the conserved aspartate residue, which is usually phosphorylated by the sensor component. The promoter regions of genes controlled by TtsI possess

F. D. Dakora et al. (eds.), *Biological Nitrogen Fixation: Towards Poverty Alleviation through Sustainable Agriculture.* 233
© Springer Science + Business Media B.V. 2008

conserved sequences, called *tts*-boxes (Krause et al., 2002). These are potential binding sites for TtsI and have been identified in the promoter regions of several genes of the NGR234 symbiotic plasmid, although not all of them are involved with the T3SS system. These promoters were fused to a promoter-less *lacZ* gene and their activities measured in wild type and *ttsI* mutant backgrounds, to show that their expression is dependent on TtsI. Taken together, these data allowed us to propose a regulatory cascade. After sensing flavonoids, NodD1 is the primary activator of the *ttsI* gene, which in turn activates all genes containing *tts*-boxes in their promoter regions.

To further characterise the regulatory pathway that leads to the expression of Nod-factors, T3SS proteins, and LPS, we purified the transcriptional activators NodD1, NodD2, and TtsI and tested their ability to bind to promoter regions of their candidate target genes. Electrophoretic mobility shift assays confirmed that NodD1 binds to *nod*-boxes NB3, NB6, NB8 and NB18. The ability of NodD2 to bind to these *nod*-boxes was also tested, but a slow migrating band was only observed for NB8 and NB3. DNA foot-printing assays showed, however, that both NodD1 and NodD2 are able to bind to all four *nod*-boxes tested. The ability of NodD2 to bind in the presence of NodD1 was also determined; in all the promoters tested, a band with a very low migration rate was observed. This band could be a result of strong modification of DNA conformation or a high degree of protein oligomerisation, possibly between NodD1 and NodD2. The pattern of DNaseI cleavage shows that NodD2 binding is restricted to the *nod*-box region, protecting approximately 50 bp. NodD1 on the other hand, has a much more extensive protection, covering more than 100 bp, which is characterized by changes to the DNaseI cleavage pattern both upstream and downstream of the *nod*-boxes.

Additionally, we have demonstrated for the first time binding of TtsI to *tts*-boxes by electrophoretic mobility shift assays and observed two retarded bands, suggesting TtsI forms oligomers upon binding. This observation is further supported by DNA foot-printing experiments, where a broad region, spanning the proposed *tts*-box consensus sequence, was protected from DNaseI cleavage. Point mutations in three of the conserved bases of the consensus did not impair TtsI binding, as established by both shift assays and DNaseI protection experiments. These results are in agreement with the extensive contact of TtsI with the target DNA, which covers a region of approximately 100 bp.

References

Broughton WJ et al. (2000) J. Bacteriol. 182, 5641–5652.
Kobayashi H et al. (2004) Mol. Microbiol. 51, 335–347.
Krause A et al. (2002) Mol. Plant-Microbe Interact. 15, 1228–1235.
Marie C et al. (2003) Mol. Plant-Microbe Interact. 16, 743–751.
Marie C et al. (2004) Mol. Plant-Microbe Interact. 17, 958–966.

SYMBIOTIC ROLES AND TRANSCRIPTIONAL ANALYSIS OF THE TYPE III SECRETION SYSTEM IN *MESORHIZOBIUM LOTI*

S. Okazaki[1], S. Okabe[2], S. Zehner[1], M. Göttfert[1] and K. Saeki[2]

[1]Institute of Genetics, Dresden University of Technology, Dresden, Germany; [2]Department of Biological Science, Faculty of Science, Nara Women's University, Kitauoyanishimachi 630-8506, Japan
(Email: Shin.Okazaki@mailbox.tu-dresden.de)

Type III secretion systems (T3SS) have been identified in a number of rhizobia and were shown to affect symbiosis with legumes. For further exploitation of rhizobial T3SS, we analyzed *Mesorhizobium loti* MAFF303099, a symbiont for *Lotus japonicus*.

T3SS mutants were constructed and symbiotic properties were examined on about 25 species of *Lotus*. Different symbiotic phenotypes were observed between wild type and T3SS mutants dependent on host plant used. For example, the T3SS mutant formed nodules more effectively than the wild type on *L. halophilus* but the opposite result was observed on *L. corniculatus* subsp. *Frondos*. These results suggest that T3SS have either a positive or a negative effect on symbiosis with different subsets of host plants.

To study the expression of the *tts* genes, translational fusions of *lacZ* with *ttsI* and *nopX* were constructed and were integrated into the chromosome of the wild type and a *ttsI* mutant. Expression of *ttsI* was induced by naringenin and the *nodD* product from *Rhizobium leguminosarum* (Lopez-Lara et al., 1995). TtsI was found to be involved in the translational regulation of *nopX*.

To examine the expression of *ttsI* and *nopX* during the nodulation process, we inoculated *L. japonicus* MG-20 with an *M. loti* strain carrying transcriptional fusions of *gusA* with *ttsI* and *nopX*. The results showed that both *nopX* and *ttsI* were expressed in nodules. Furthermore, the expression of *nopX* was observed in both the infection thread and the mature nodule. This result suggests that the T3SS might have functions both in the early and the later steps of nodule development.

Reference

Lopez-Lara et al. (1995) Mol. Microbiol. 15, 627–638.

F. D. Dakora et al. (eds.), *Biological Nitrogen Fixation: Towards Poverty Alleviation through Sustainable Agriculture*.
© Springer Science + Business Media B.V. 2008

TOXIN-ANTITOXIN MODULES AND SYMBIOSIS

M. Bodogai[1], Sz. Ferenczi[2], S. P. Miclea[1], P. Papp[2] and I. Dusha[1]

[1]Institute of Genetics, Biological Research Center, Szeged, Hungary;
[2]Institute of Genetics, Agricultural Research Center, Gödöllő, Hungary

The recent expansion of microbial DNA and protein databases that followed the sequencing of a large number of prokaryotic genomes has promoted the identification of numerous toxin-antitoxin (TA) modules present the bacterial plasmids and chromosomes. The first TA modules were identified on plasmids acting as post-segregational killing systems. Their function was to prevent the proliferation of plasmid-free progeny. Further TA loci were also found in chromosomes and were considered to be associated with the modulation of the global level of translation under various stress conditions.

A typical TA module consists of two small genes that form an operon, in which the first gene determines an unstable antitoxin and the second gene, a stable toxin protein. The two proteins form a complex, thus, the antitoxin prevents the lethal effect of the toxin. Under stress conditions, the antitoxin is degraded by proteases and the activity of the free toxin results in the inhibition of translation (Gerdes et al., 2005). Seven TA gene families have been described (Gerdes et al., 2005). One family is the most abundant *vapBC* family present in Gram-positive and Gram-negative bacteria as well as in Archaea. The antitoxin protein of this family is an AbrB/MazE homolog and the toxin partner belongs to the PIN domain family.

Based on protein homologies, domain architectures, and gene neighborhood analysis, we could identify 17 TA modules in the genome of *Sinorhizobium meliloti* strain 1021. Ten of these 17 modules are members of the *vapBC* gene family and all of them are localized on the bacterial chromosome.

The previously identified *ntrPR* operon in *Sinorhizobium meliloti* is one of the 17 *vapBC*-type TA modules. As demonstrated earlier, a Tn5 insertion in the *ntrR* gene resulted in increased transcription of nodulation and nitrogen-fixation genes as compared to that of the wild-type strain, and this effect was more pronounced in the presence of an external ammonium source (Dusha et al., 1989; Oláh et al., 2001). We

F. D. Dakora et al. (eds.), *Biological Nitrogen Fixation: Towards Poverty Alleviation through Sustainable Agriculture.*
© Springer Science + Business Media B.V. 2008

supposed that the NtrR protein has a role in the nitrogen regulation of the *nod* and *nif* genes. However, when the gene expression patterns of the entire genomes of the wild type and *ntrR* mutant strains were compared, an unexpectedly large number of genes exhibited altered expression in the mutant strain, suggesting a more general function for NtrR (Puskás et al., 2004).

In order to test the hypothesis that the *ntrPR* genes form a functional TA module, we examined the possible toxic function of NtrR. We tested the growth and viability of *E. coli* derivatives carrying plasmids with *ntrR* or *ntrP* or both genes controlled by the arabinose inducible promoter. NtrR overexpression resulted in the inhibition of cell growth and colony formation, and this effect was counteracted by the presence of the antitoxin NtrP. We have shown that the autoregulatory functions of the *ntrPR* operon are in accordance with those of other TA systems. The antitoxin NtrP is able to recognize a DNA segment in the promoter region of the *ntrPR* operon, but its binding is weak. The toxin component alone is not able to bind to the same DNA region, but the complex of NtrP and NtrR strongly binds to the promoter region and results in negative autoregulation. The N-terminal part of NtrP is responsible for the interaction with the promoter DNA, whereas the C-terminal part is required for protein-protein interactions (Bodogai et al., 2006).

What is the reason for carrying such a high number of *vapBC* modules and why is it only the chromosome that encodes complete modules, whereas the plasmids encode only solitary and probably inactive toxins? Considering the possible physiological role of these modules in stress management, their presence in high numbers in the genome of *Sinorhizobium meliloti* may not be surprising. Rhizobia, as symbiotic bacteria, have to face various stresses, from different environmental conditions (in the soil and inside the root nodule) to different nutritional circumstances in the free-living and symbiotic states. The TA systems may be involved in helping bacteria cope with these transitions.

These results and the observed higher symbiotic efficiency of the *ntrR* mutant strain with the host plant alfalfa suggest the involvement of the *ntrPR* module in the adaptation to the symbiotic state and open new perspectives on how *S. meliloti* manages to adjust its metabolic processes to the ever-varying environmental requirements.

References

Bodogai M et al. (2006) Mol. Plant-Microbe Interact. 19, 811–822.
Dusha I et al. (1989) Mol. Gen. Genet. 219, 89–96.
Gerdes K et al. (2005) Nature Rev. Microbiol. 3, 371–382.
Oláh B et al. (2001) Mol. Plant-Microbe Interact. 14, 887–894.
Puskás L et al. (2004) Mol. Gen. Genomics 272, 275–289.

COMPARATIVE ANALYSIS OF ArgC PROTEIN EXPRESSION IN *RHIZOBIUM* SPECIES

M. del C. Vargas, M. A. Villalobos, R. Díaz, M. A. Flores, S. Contreras,
S. Encarnación, Y. Mora, M. F. Dunn, L. Girard and J. Mora

Universidad Nacional Autónoma de México, Centro de Ciencias
Genómicas, Programa Genómica Funcional de Procariotes, Apto. Postal
565-A, Cuernavaca, Mor., México
(Email: cvargas@ccg.unam.mx)

Comparative genomics in bacteria belonging to the Rhizobiaceae family, such as *Sinorhizobium meliloti*, *Agrobacterium tumefaciens*, *Mesorhizobium loti*, *Brucella melitensis* and *Rhizobium etli*, has led to the understanding that syntenic genes participate in 80–90% of the essential functions for growth and cellular division. In this way, synteny has an important biological significance in these organisms (Guerrero et al., 2005). Proteins encoded by syntenic genes had a high degree of identity, although they displayed some differences in their amino acid composition. These changes may result from an evolutionary adaptation to their specific environments.

To document this hypothesis, we focused our analysis on the ArgC protein encoded by a syntenic gene, *argC*, which is involved in the arginine synthesis pathway, an essential amino acid for the cellular metabolism. ArgC polypeptides display high identity levels among the species analyzed with a molecular weight in the same range, but with some differences in the amino acid sequences. To understand the role of these differences in the Rhizobiaceae speciation, we constructed an *argC* mutant strain of *S. meliloti*. The *argC* mutant is an arginine auxotroph that cannot grow in minimal medium. We also designed specific oligonucleotides to generate different fragments containing *speB-argC* genes of each species of *Rhizobium* by PCR amplification. The same strategy was used to amplify and subsequently clone fragments containing the *argC* gene of each species transcribed under control of the *S. meliloti argC* promoter as well as the *argC* gene of each species under a constitutive promoter. The resulting plasmids were then mobilized into each species studied as well as into the *S. meliloti* mutant.

Several carbon and nitrogen sources were assayed to determine the best medium on which to grow the different strains. The minimal medium (MM) plus mannose and nitrate was the best condition for growth and to establish significant differences between the species. The strains were grown in MM-succinate plus ammonia and MM-mannose plus

F. D. Dakora et al. (eds.), *Biological Nitrogen Fixation: Towards Poverty Alleviation through Sustainable Agriculture.*
© Springer Science + Business Media B.V. 2008

nitrate. Under these conditions, the *S. meliloti argC* mutant accumulates and excretes great amounts of glutamate, glycine and α-ketoglutarate (αKG).

The *argC* mutant of *S. meliloti*, which contained the *speB-argC* genes of each species transcribed under their own promoter, showed differences in growth, cellular excretion of metabolites, and ArgC activity. Only the strain containing the *argC* gene of *M. loti* was unable to grow, excreted high levels of metabolites, and had a reduced ArgC activity.

When the *argC* gene of each species was under the control of the *S. meliloti* promoter, clear differences were observed in the growth and excretion of metabolites, however, only minor differences were detected in ArgC activity, except in the *argC* mutant with the *argC* gene of *M. loti*, which showed reduced activity. In the *S. meliloti argC* mutants containing the *argC* gene of each species of *Rhizobium* under a constitutive promoter, growth was similar in all strains including the *argC* mutant with the *argC* gene of *M. loti*. In contrast, the *S. meliloti argC* mutant with the *argC* gene of *Escherichia coli* under the same constitutive promoter had ArgC activity similar to wild type strain, although its growth was poor.

The strains containing the *argC* genes belonging to different *Rhizobium* species showed differences in the growth, cellular metabolic flow, and ArgC activity, which indicated a particular behavior of each protein in the same genetic background, referred here as "the species signature".

Reference

Guerrero G et al. (2005) BMC Evolutionary Biology 5, 55.

GENISTEIN-DEPENDENT REGULATORY CIRCUITS IN *BRADYRHIZOBIUM JAPONICUM*

K. Lang[1], A. Lindemann[2], F. Hauser[2], H. Hennecke[2], S. Zehner[1] and M. Göttfert[1]

[1]Institute of Genetics, Technical University Dresden, Dresden, Germany; [2]Institute of Microbiology, ETHZ, Zürich, Switzerland

The *B. japonicum* genome consists of a single circular chromosome (Kaneko et al., 2002). The regulation of genes directly involved in nodulation (*nod*) is well studied (Figure 1; Loh and Stacey, 2003). Genistein is a potent signal molecule of soybean. In *B. japonicum*, two regulatory systems are required for efficient symbiosis. NodD1, a LysR-type regulator, is a well-conserved activator protein in rhizobia. NodVW, belonging to the two-component regulatory family, are essential for the nodulation of several host plants (Göttfert et al., 1990). *nwsB*, if overexpressed, could repress the Nod⁻ phenotype of the *nodW* (Grob et al., 1993). *nwsB* is also involved in nodulation-density-dependent repression of *nod* genes (Loh et al., 2002). NolA and NodD2 mediate this effect (Loh et al., 2001).

Besides the presence of a common set of *nod* genes, rhizobia have developed additional strategies for their interaction with host plants. One example is the type three secretion system (T3SS), which is present in several rhizobia including *B. japonicum* (Marie et al., 2001). Depending on the host, the T3SS improves, impairs, or does not affect symbiotic efficiency. *ttsI*, which is preceded by a *nod*-box promoter, is the activator of genes encoding the T3SS (Krause et al., 2002).

Little is known about the effect of genistein on the expression of other genes. We studied this question by whole-genome microarray analysis. The Affymetrix chip used in the analysis covers coding and intergenic regions. RNA was isolated from cultures 8 hours after addition of genistein. This analysis revealed that about 100 genes are induced by genistein by at least two-fold, including all nodulation genes, most of which were not known previously to be inducible by genistein. This result prompted us to study regulatory mutants. We found that *nodW* is essential for detectable activation of most of these genes, whereas *nodD1* was less important. In addition, there were also a few genes that were induced by genistein independent of *nodW* or *nodD1*. Examples are the genes *bll6621* and *bll6622*. A database search revealed that they might encode a transporter. It is unknown how these genes are activated. A putative regulator, encoded

F. D. Dakora et al. (eds.), *Biological Nitrogen Fixation: Towards Poverty Alleviation through Sustainable Agriculture.*
© Springer Science + Business Media B.V. 2008

by the neighbouring gene *blr6623*, might be involved. Data obtained after over-expression of the *nodD2/nolA* region support its role in repression. However, a few other genes were overexpressed within the same strain, albeit independent of genistein. To our surprise, little or no induction was observed for genes encoding the T3SS. Therefore, we constructed reporter gene fusions. For blr1806, we could show that the gene is inducible by genistein and expressed in soybean nodules.

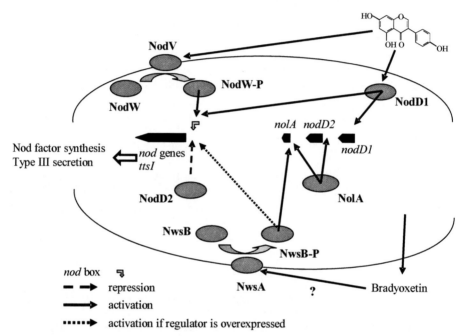

Figure 1. Regulatory model of genistein-inducible genes in *B. japonicum*. Microarray analysis revealed that this model has to be extended to genes not preceded by a *nod* box (see text).

References

Göttfert M et al. (1990) Proc. Natl. Acad. Sci. USA 87, 2680–2684.
Grob P et al. (1993) Mol. Gen. Genet. 241, 531–541.
Kaneko T et al. (2002) DNA Res. 9, 189–197.
Krause A et al. (2002) Mol. Plant-Microbe Interact. 15, 1228–1235.
Loh J and Stacey G (2003) Appl. Environ. Microbiol. 69, 10–17.
Loh J et al. (2002) J. Bacteriol. 184, 1759–1766.
Loh J et al. (2001) Mol. Microbiol. 42, 37–46.
Marie C et al. (2001) Curr. Opin. Plant Biol. 4, 336–342.

RHIZOBIUM LEGUMINOSARUM MODIFICATION OF ITS LIPOPOLY-SACCHARIDE DURING SYMBIOTIC BACTEROID DEVELOPMENT

E. L. Kannenberg[1,2], T. Härtner[2], L. S. Forsberg[1] and R. W. Carlson[1]

[1]Complex Carbohydrate Research Center, University of Georgia, Athens, GA, USA; [2]Department of Microbiology/Biotechnology, University of Tübingen, Tübingen, Germany

During pea nodule development, rhizobial lipopolysaccharides (LPSs) undergo host-specific modifications, likely of critical importance to nodule development. Previously, we analyzed the LPSs from *Rhizobium leguminosarum (Rl)* 3,841 bacteroids and found modifications to the LPS O-chain and lipid A that result in more hydrophobic LPSs. Consequently, the bacteroid LPS partitions mostly into phenol rather than water during hot phenol/water extraction (see Figure 1; Kannenberg and Carlson, 2001).

To investigate whether similar LPS modifications occur during bean nodule development, we isolated LPS from *Rl* B625 bean nodule bacteria. Nodule bacteria were isolated from both pink N_2-fixing nodules and immature white non-fixing nodules. From a free-living culture, the bulk of the LPS partitioned into the water phase. In contrast, LPS from bacteria from immature nodules partitioned exclusively into phenol, whereas the LPS from bacteroids of mature pink nodules partitioned into both the water and phenol phases (these results were also corroborated using LPS-specific mAbs).

These changes in LPS hydrophobicity for both *Rl* 3841 and *Rl* B625 bacteroids correlate with changes in overall bacteroid cell-surface properties: Pea *Rl* 3841 bacteroids display hydrophobic cell-surface characteristics, whereas *Rl* 3841 free-living bacteria show hydrophilic cell-surface properties. In the bean-specific *Rl* B625 (Table 1), we found a transient increase in the cell-surface hydrophobicity of bacteria from immature non-fixing nodules. No water-soluble LPSs were detected in these bacteria (Figure 1). Mature N_2-fixing bacteroids from pink nodules displayed hydrophilic cell surface properties and this correlated with detectable amounts of water-soluble LPSs.

The correlation between hydrophobic LPSs in the pea and bean nodule bacteria with their cell-surface hydrophobicity suggests LPS as a major factor in determining their biophysical cell-surface properties. If confirmed, this result could be important for LPS structure/function relationships and give insight into the fundamental differences

F. D. Dakora et al. (eds.), *Biological Nitrogen Fixation: Towards Poverty Alleviation through Sustainable Agriculture.*
© Springer Science + Business Media B.V. 2008

between pea and bean symbiosomes. In pea, individual bacteroids are tightly wrapped by the peribacteroid membrane, whereas in bean, several bacteroids are less tightly packaged. Working out the molecular mechanisms of these *Rhizobium*-legume interactions could lead to understanding the role of LPS in pathogen-host interactions, particularly pathogens that form chronic intracellular infection (e.g., *Brucella*).

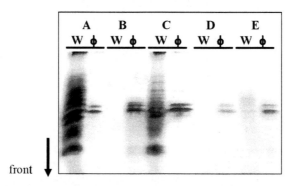

Figure 1. Gel banding patterns of *Rl* LPSs. LPSs from: **A**, *Rl* 3841 lab culture; **B**, *Rl* 3841 pink pea nodules; **C**, *Rl* B625 lab culture; **D**, *Rl* B625 white immature non-fixing bean nodules; **E**, *Rl* B625 mature pink bean nodules. Strain *Rl* B625 is derived from *Rl* 3841 by placing a bean-specific symbiotic plasmid in a mutant of *Rl* 3841 cured of the pea-specific symbiotic gene region (Sindhu et al., 1990). LPSs were obtained from rhizobial cells by: hot phenol (Φ) or water (W) extraction; separation by 18% polyacrylamide gel electrophoresis; and visualization by carbohydrate-specific alcian blue/silver stain.

Table 1. Overall cell surface hydrophobicity in rhizobia established using a microbial adhesion to hydrocarbons [MATH] procedure (Kannenberg and Carlson, 2001).

Strain	Origin	Rel. hydrophobicity [%]	Plant host
Rl 3841	Free-living	~10	
	Pink nodules/ bacteroids	90–1000	Pea
Rl B625	Free-living	~10	
	White non-fixing nodules	60–65	Bean
	Pink nodules/ bacteroids	15–20	

References

Kannenberg EL and Carlson RW (2001) Mol. Microbiol. 39, 379–391.
Sindhu et al. (1990) J. Bacteriol. 172, 1804–1813.

ALTERED FORMATIONS OF SPECIAL EUBACTERIAL MEMBRANE LIPIDS AFFECT BACTERIAL INTERACTIONS WITH EUKARYOTIC HOSTS

O. Geiger, I. M. López-Lara and C. Sohlenkamp

Centro de Ciencias Genómicas, Universidad Nacional Autónoma
de México, Cuernavaca, Mor., Mexico
(Email: otto@ccg.unam.mx)

Since our review of the biosyntheses and possible functions of membrane lipids in plant-associated bacteria (López-Lara et al., 2003), some new and important findings have occurred. A major surprise was the recent finding of the PlsX/Y pathway (Lu et al., 2006). All bacterial membrane phospholipids are derived from the central precursor CDP-diacylglyceride, which is formed from glycerol-3-phosphate by two subsequent acylations that yield phosphatidic acid and a final activation to give CDP-diacylglyceride. Although in *Escherichia coli*, acylation at the 1-position of glycerol-3-phosphate is carried out by PlsB, the majority of bacteria lack a *plsB* gene. Instead, PlsX produces a uniquely activated fatty acid by catalyzing the synthesis of acyl-phosphate from acyl-acyl carrier protein, and then PlsY transfers the fatty acid from acyl phosphate to the 1-position of glycerol-3-phosphate (Lu et al., 2006). The PlsX/Y pathway is more widespread than PlsB and constitutes the initial acylation of glycerol-3-phosphate in rhizobia.

Phosphatidylglycerol (PG), cardiolipin, and phosphatidylethanolamine are the major membrane phospholipids found in most eubacteria. However, phosphatidylcholine (PC), a major lipid in eukaryotes, is formed in both rhizobia and many other bacteria that interact with eukaryotic hosts in a benefical (symbiotic) or harmful (pathogenic) way. Although many symbiotic bacteria form PC by either the phospholipid *N*-methyltransferase pathway or the PC synthase pathway (Sohlenkamp et al., 2003), pathogens, such as *Brucella*, *Borrelia*, *Pseudomonas*, or *Legionella*, depend on choline provided by the eukaryotic host to form PC (via the PC synthase pathway). In the case of *Legionella*, bacterial PC is clearly required for virulence.

Lysyl-PG (LPG) is a major membrane lipid in Gram-positive staphylococci and it is thought to provide resistance against cationic peptides produced by the innate immune defense. In *Staphylococcus aureus*, the gene product of *mprF* is responsible for LPG

F. D. Dakora et al. (eds.), *Biological Nitrogen Fixation: Towards Poverty Alleviation through Sustainable Agriculture.*
© Springer Science + Business Media B.V. 2008

formation. Low pH-inducible genes, termed *lpiA*, have been identified in the Gram-negative alpha-proteobacteria, *Rhizobium tropici* (Vinuesa et al., 2003) and *Sinorhizobium medicae* (Reeve et al., 2006), in screens for acid-sensitive mutants and they encode homologues of MprF. Expression of *lpiA* from *R. tropici* in the heterologous hosts *E. coli* and *Sinorhizobium meliloti* causes formation of LPG. A wild-type strain of *R. tropici* forms about 1% LPG (of total lipids) when the cells are grown in minimal medium at pH 4.5, but not when grown in minimal medium at neutral pH and not in complex TY medium at either pH. LPG biosynthesis does not occur when *lpiA* is deleted and is restored upon complementation of *lpiA*-deficient mutants with a functional copy of the *lpiA* gene. When grown in the low-pH medium, *lpiA*-deficient rhizobial mutants are greater than four-times more susceptible to the cationic peptide polymyxin B than the wild type. An analysis of the membrane lipid composition of *R. tropici* bacteroids isolated from bean nodules shows that LPG is not formed in larger amounts in bacteroids. This is the first report that a Gram-negative bacterium, *Rhizobium tropici*, which is able to form nitrogen-fixing root nodules with bean, can form LPG.

Under phosphate-limiting conditions of growth, some eubacteria replace their membrane phospholipids by lipids not containing any phosphorus. In *S. meliloti*, these phosphorus-free lipids are sulfoquinovosyl diacylglycerol, ornithine-containing lipids (OL), and diacylglyceryl trimethylhomoserine (Geiger et al., 1999). These phosphorus-free membrane lipids are not required for the symbiosis with alfalfa, but contribute to increased cell yields under phosphorus-limiting conditions of growth (López-Lara et al., 2005). Recently, we have characterized the biosynthetic pathway for OL (Gao et al., 2004), which stimulate the mammalian immune system; these are therefore bioactive lipids that are encountered in many eubacteria. In different bacteria, modifications can occur in different parts of the OL molecule, thereby altering its properties. For example, in *R. tropici*, hydroxylations of either the δ-amino group of the ornithine residue or the piggy-back fatty acyl residue of OL can occur (Rojas-Jiménez et al., 2005). This latter hydroxylation, caused by the *olsC* gene, affects symbiotic efficiency and acid tolerance of *R. tropici*.

References

Gao J-L et al. (2004) Mol. Microbiol. 53, 1757–1770.
Geiger O et al. (1999) Mol. Microbiol. 32, 63–73.
López-Lara IM et al. (2003) Mol. Plant-Microbe Interact. 16, 567–579.
López-Lara IM et al. (2005) Mol. Plant-Microbe Interact. 18, 973–982.
Lu Y-J et al. (2006) Mol. Cell 23, 765–772.
Reeve WG et al. (2006) Microbiology 152, 3049–3059.
Rojas-Jiménez K et al. (2005) Mol. Plant-Microbe Interact. 18, 1175–1185.
Sohlenkamp C et al. (2003) Prog. Lipid Res. 42, 115–162.
Vinuesa P et al. (2003) Mol. Plant-Microbe Interact. 16,159–168.

RHIZOBIUM LEGUMINOSARUM BV *VICIAE* STRAIN LC-31: ANALYSIS OF NOVEL BACTERIOCIN AND ACC DEAMINASE GENE(S)

F. Y. Hafeez, Z. Hassan, F. Naeem, A. Bashir, A. Kiran, S. A. Khan and K. A. Malik

National Institute for Biotechnology and Genetic Engineering (NIBGE), P.O. Box 577, Jhang Road, Faisalabad-38000, Pakistan (Email: fauzia@nibge.org)

Rhizobium spp., in addition to symbiotic nitrogen fixation, utilizes a variety of mechanisms, both direct and indirect, to stimulate the growth of plants and/or to compete in nodulation. Ethylene is a gaseous phytohormone produced by the plant during normal growth conditions. But, under biotic and abiotic stresses, the synthesis of ethylene increases and results in stunted root growth and low nodulation in legumes. The bacteria with ACC-deaminase gene activity convert the precursor of ethylene, ACC (1-Amino cyclopropane-1-carboxylic acid), into NH_3 and alpha-ketobutyric acid.

The amplified ACC-deaminase gene from bacterial strain LC-31 showed 99% homology with *Rhizobium leguminosarum* bv. *viciae* (Wenbo et al., 2003). A probe based on the LC-31 ACC-deaminase gene was synthesized and used for screening 15 rhizobial strains by the dot-blot technique. All but one strain showed the presence of the ACC-deaminase gene on the chromosomal DNA, whereas two strains showed its presence on both plasmid and chromosomal DNA. Expression profiling of the ACC-deaminase gene was studied by RT-PCR and differential display PCR. The gene was up-regulated at 1 mM ACC. In another experiment, LC-31 was supplemented with ACC (1 mM) and IAA (5–50 mM) to determine any correlation between IAA and ACC-deaminase gene activity. The ACC-deaminase gene is down regulated at 50 mM. *R. leguminosarum* LC-31 exhibited both properties, i.e., IAA and ACC-deaminase activity. The regulation of different genes, after ACC treatment of LC-31, is being investigated by differential PCR. cDNA was constructed by random hexamer primers using cDNA single strand kit. Differentially expressed genes were cloned and sequenced and data analysis is in progress.

In addition to ACC-deaminase activity, *R. leguminosarum* bv. *viciae* strain LC-31 also showed a typical narrow spectrum activity. It was more effective against the most closely related *R. leguminosarum* strains, but less effective against *Agrobacterium* and *Bradyrhizobium* strains. A 40-kDa protein band was sequenced and submitted to Swiss Prot Database under the accession number P84703. MALDI-TOF analysis of the 40-kDa

F. D. Dakora et al. (eds.), *Biological Nitrogen Fixation: Towards Poverty Alleviation through Sustainable Agriculture*.
© Springer Science + Business Media B.V. 2008

protein gave a molecular weight of 37.6 kDa, which correlated with SDS-PAGE results (Hafeez et al., 2005). The bacteriocin gene has three encoding regions; *rzcA*, for bacteriocin, and *rzcB and rzcD*, which are required for bacteriocin secretion. The sequence of accession number AF273216 was retrieved from gene bank and used to design random primers. An 82-bp *rzcA* fragment of the bacteriocin gene was amplified. The protein sequence of the *rzcA* fragment showed 88% homology with the protein sequence of *rzcA* from the gene bank. It also showed 58% similarity with protease-associated peptidase of *Bacillus cereus* sub sp. *cytotoxis* NVH, which has a molecular weight of 2,961.3 Da.

R. leguminosarum bv *viciae* LC-31 is an efficient nodulating N_2-fixing (182 ± 9.50 nmole C_2H_4 produced $h^{-1}g^{-1}$ nodule dry weight) and IAA-producing strain (3.40 mg L^{-1}). Moreover, it is potent bacteriocin producer and has ACC-deaminase activity. Therefore, LC-31 is a novel *Rhizobium* strain having competitive ability that can be used as a good inoculant.

References

Hafeez FY et al. (2005) Environ. Exp. Bot. 54, 142–147.
Wenbo M et al. (2003) Appl. Environ. Microbiol. 69, 4396–4402.

INFECTION OF LEGUMES BY BETA-RHIZOBIA

E. K. James[1], G. N. Elliott[1], W.-M. Chen[2], C. Bontemps[3], J. P. W. Young[3], S. M. de Faria[4], F. B. dos Reis, Jr.[5], M. F. Simon[6], E. Gross[7], M. F. Loureiro[8], V. M. Reis[4], L. Perin[4], R. M. Boddey[4], C. E. Hughes[6], L. Moulin[9], A. R. Prescott[1] and J. I. Sprent[1]

[1]College of Life Sciences, University of Dundee, Dundee DD1 5EH, UK; [2]National Kaohsiung Marine University, Kaohsiung City 811, Taiwan; [3]Department of Biology, University of York, York YO10 5YW, UK; [4]EMBPRAPA-Agrobiologia, km 47, Seropédica, 23851-970, RJ, Brazil; [5]EMBRAPA-Cedrrados, Planaltina, Brasília, 73301-970, DF Brazil; [6]Department of Plant Science, University of Oxford, Oxford, OXI 3RB, UK; [7]Depto de Ciências Agrárias e Ambientais, UESC, km 16, Ilhéus 45662-000 BA, Brazil; [8]Faculdade de Agronomia, UFMT, Cuiabá, 78060-900, MT, Brazil; [9]LSTM, IRD/CIRAD/INRA/AGROM/UMII, 34398 Montpellier, France

It is now well established that many species and strains in the large genus *Burkholderia* have the ability to fix nitrogen in free-living culture, particularly those, such as *B. tropica*, *B. unamae* and *B. Vietnamiensis*, that are associated with (mainly tropical) gramineous plants (Reis et al., 2004). Many *Burkholderia* strains have also been found within nodules of tropical legumes (Moulin et al., 2001), particularly in nodules on *Mimosa* spp. (Barrett and Parker, 2005, 2006; Chen et al., 2005a, b). Some of these strains have now been described as novel species of *Burkholderia*, including *B. mimosarum* and *B. nodosa* isolated from *Mimosa* spp. (Chen et al., 2006, 2007), *B. phymatium* isolated from *Machaerium lunatum* (Vandamme et al., 2002), and *B. tuberum* isolated from *Aspalathus carnosa* (Vandamme et al., 2002). These *Burkholderia* strains possess *nod* genes and, together with strains of a newly described species of *Ralstonia, R. taiwanensis* (now renamed *Cupriavidus taiwanensis* and also isolated from *Mimosa* nodules; Chen et al., 2001, 2003a; Verna et al., 2002), are collectively termed "beta-rhizobia". Recent studies with strains from South America and Taiwan have confirmed that both *C. taiwanensis* and *Burkholderia* beta-rhizobia isolated from *Mimosa* spp. (including *B. mimosarum* and *B. nodosa*) are effective symbionts of plants in this genus (Chen et al., 2003b, 2005a, b). More surprising is the recent discovery that *B. phymatum* is also a highly effective symbiont of several *Mimosa spp.* and that it has a broader host range in the genus *Mimosa* than *C. taiwanensis* (Elliott et al., 2007).

F. D. Dakora et al. (eds.), *Biological Nitrogen Fixation: Towards Poverty Alleviation through Sustainable Agriculture.*
© Springer Science + Business Media B.V. 2008

So far, all attempts to nodulate *Machaerium* spp. with *B. phymatum* have been unsuccessful (Elliott et al., 2007) and, indeed, there has been very little evidence published of effective nodulation by beta-rhizobia of legumes in any genera other than *Minosa*. However, we have recently obtained evidence for effective nodulation by *B. phymatum* STM815 of other Mimosoid genera, including *Acacia seyal*, *Leucaena leu-cocephala*, *Piptadenia gonoacantha*, *P. oblique*, *P. stipulacea* and *Pitecellobium dulce* (G.N. Elliott et al., unpublished data, 2006). Interestingly, although *Piptadenia* spp. are close taxonomically to *Mimosa* and, therefore, it might not be considered surprising that they would be nodulated by a *Mimosa* symbiont, the other species, in particular *P. dulce*, are not. These results suggest that beta-rhizobia are widespread within the sub-family Mimosoideae, but are not universally symbiotic within it. We are collaborating closely with legume taxonomists to determine the depth of the relationship between Mimosoid legumes and beta-rhizobia.

With regard to nodulation of papilionoid legumes by Beta-rhizobia, with the exception of the ineffective nodulation of the promiscuous legume, *Macroptilium atropurpureum*, by *B. phymatum* and *B. tuberum* (Moulin et al., 2001), to date there have been no published reports of a genuinely symbiotic relationship with plants in this sub-family. However, Elliott et al. (this volume) have recently reported effective nodulation of the South African endemic papilionoid legumes, *Cyclopia galioides*, *C. genistoides* and *C. pubescens*, by *B. tuberum* STM678.

STM678 appears to be unique among the known beta-rhizobia in having a *nod*A gene very separate in phylogenetic terms from *Mimosa*-nodulating bacteria. Further, it can nodulated neither any *Mimosa* spp. nor any *Aspalathus* spp., although its original host, *A. carnosa*, has not yet been tested. Regardless of whether it can or cannot nodulate *A. carnosa*, our very strong evidence that *B. tuberum* can nodualte *Cyclopia* spp. is the first confirmed report of nodualtion by beta-rhizobia in the sub-family Papilionoideae.

References

Barrett CF and Parker M (2005) Syst. Appl. Microbiol. 28, 57–65.
Barrett CF and Parker M (2006) Appl. Environ. Microbiol. 72, 1198–1206.
Chen W-M et al. (2001) Int. J. Syst. Evol. Microbiol. 51, 1729–1735.
Chen W-M et al. (2003a) J. Bacteriol. 185, 7266–7272.
Chen W-M et al. (2003b) Mol. Plant-Microbe Interact. 16, 1051–1061.
Chen W-M et al. (2005a) Appl. Environ. Microbiol. 71, 7461–7471.
Chen W-M et al. (2005b) New Phytol. 168, 661–675.
Chen W-M et al. (2006) Int. J. Syst. Evol. Microbiol. 56, 1847–1851.
Chen W-M et al. (2007) Int. J. Syst. Evol. Microbiol. 57, 1055–1059.
Elliott GN et al. (2007) New Phytol. 173, 168–180.
Moulin L et al. (2001) Nature 411, 948–950.
Reis VM et al. (2004) Int. J. Syst. Evol. Microbiol. 54, 2155–2162.
Vandamme P et al. (2002) Int. J. Syst. Evol. Microbiol. 25, 507–512.
Verna et al. (2002) Can. J. Microbiol. 50, 313–322.

IN SEARCH OF BETA-RHIZOBIA: EXPLORING THE SYMBIONTS OF *MIMOSA* IN BRAZIL

C. Bontemps[1], G. Elliott[2], E. James[2], J. Sprent[2], M. Simon[3], C. Hughes[3] and J. P. W. Young[1]

[1]Department of Biology, University of York, York, UK; [2]CHIPs Complex, University of Dundee, Dundee, UK; [3]Department of Plant Science, University of Oxford, Oxford, UK

In 2001, the first identification of β-proteobacteria, which belong to the *Burkholderia* and *Ralstonia* genera and are able to nodulate legumes, has changed the long-held perception that legumes can only form nitrogen-fixing symbioses with bacteria in the α-proteobacteria subgroup (Moulin et al., 2001). The terms α-rhizobia and β-rhizobia are now used to designate these two subgroups of symbionts. β-rhizobia are real rhizobia that fix nitrogen in symbiosis (Chen et al., 2003a, 2005a), but are taxonomically far from the "classical" α-rhizobia. Nevertheless, both α- and β-rhizobia share the same symbiotic genes (*nod* and *nif*) that allow interaction with their hosts. These genes have probably moved between these two bacterial lineages through several gene transfers (Chen et al., 2003b). The β-rhizobia form a poorly described symbiotic group in comparison to the α-rhizobia. In consequence, the aims of this study were: (i) to obtain a large collection of new β-rhizobia isolates; and (ii) to provide insight into their evolution.

Because previous studies have suggested that the *Mimosa* genus may have a special affinity for the β-rhizobia (Chen et al., 2003b, 2005b), we have prospected for symbionts of 69 different *Mimosa* species. Sampling expeditions were undertaken in the Cerrado and Caatinga regions of central Brazil, which are the diversification centres of this plant genus. Just a small part of our samples (33/131 isolates from 20 different *Mimosa* species) have been studied so far by sequencing two housekeeping genes (the 16S rDNA and *recA* genes) and two symbiosis-related genes (the *nifH* and *nodC* genes).

Preliminary sequencing of the 16S rDNA and *recA* genes indicates that all isolates but one belong to the *Burkholderia* genus and suggests that *Burkholderia* are the main symbionts of *Mimosa* in Brazil. The *nifH* sequences obtained for a large part of these samples suggest diazotrophic capability for the isolates. As previously shown, phylogenetic analysis of these *nifH* sequences confirm a closer relationship with free-living diazotrophic β-proteobacteria *nifH* sequences rather than with those of α-rhizobia. The ancestor of the Brazilian *Mimosa* β-rhizobia was probably already a diazotrophic bacterium when it acquired symbiotic functions. *NodC* sequences have been obtained for most of the tested strains and confirm the symbiotic character of our samples.

16S rDNA, *recA*, and *nodC* gene trees all show the same repartition of our strains into several well supported clusters. The strong phylogenetic congruence between housekeeping and symbiosis-related genes suggests: (i) that symbiosis is an old and stable character of the *Burkholderia* lineage; and (ii) a limitation on gene-transfer events between major groups of nodulating *Burkholderia*. Interestingly, the observed clusters seem linked with the phytogeographic repartition of the plant samples. Specific association with the host may have restricted lateral gene transfer among strains which could have shaped these conserved clade topologies.

In conclusion, our study suggests that *Burkholderia* are the dominant symbionts of *Mimosa* in Brazil and a strong relationship (and maybe coevolution) seems to exist between these two symbiotic partners.

References

Chen et al. (2003a) Mol. Plant-Microbe Interact. 16, 1051–1061.
Chen et al. (2003b) J. Bacteriol. 185, 7266–7272.
Chen et al. (2005a) New Phytol. 168, 661–675.
Chen et al. (2005b) Appl. Environ. Microbiol. 71, 7461–7471.
Moulin et al. (2001) Nature 411, 948–950.

EFFECT OF PHOSPHOGLYCERATE MUTASE DEFICIENCY ON THE FREE-LIVING AND SYMBIOTIC LIFE OF *BURKHOLDERIA PHYMATUM*

S.-Y. Sheu[2], D.-S. Sheu[2], C.-Z. Ke[2], K.-Y. Lin[1], J.-H. Chou[1], J. Prell[3], E. K. James[4], G. N. Elliott[4], J. I. Sprent[4] and W.-M. Chen[1]

[1]Laboratory of Microbiology, Department of Seafood Science; [2]Department of Marine Biotechnology, National Kaohsiung Marine University, Kaohsiung, Taiwan; [3]School of Biological Sciences, University of Reading, Reading RG6 6AJ, England, UK; [4]School of Life Sciences, University of Dundee, Dundee DD1 5EH, Scotland, UK

Burkholderia phymatum KM16-22 is a Tn5-induced mutant of the beta-rhizobium *Burkholderia phymatum* STM815. The mutant KM16-22 is unable to form root nodules on *Mimosa pudica*, but still causes root-hair deformation, which is the early step of infection. The Tn5-interrupted site of gene in KM16-22 has been determined (Figure 1) and the sequences strongly resemble the deduced amino-acid sequences of the genes encoding 2,3-bisphosphoglycerate-dependent phosphoglycerate mutase (PGM) (87–93% identity to *Burkholderia* spp., 78–81% identity to *Cupriavidus/Ralstonia* spp., and 66% identity to *Escherichia coli*). PGM is an important enzyme for both glycolysis and gluconeogenesis.

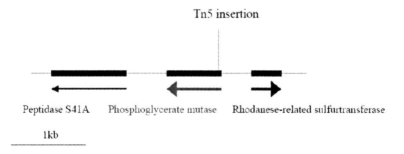

Figure 1. Location of ORFs in the *pgm* region of the *B. phymatum* chromosome. "Tn5 insertion" indicates the position of the Tn5 insertion resulting in the interruption of the gene encoding PGM.

F. D. Dakora et al. (eds.), *Biological Nitrogen Fixation: Towards Poverty Alleviation through Sustainable Agriculture*.
© Springer Science + Business Media B.V. 2008

KM16-22 grew efficiently on complex medium, however, it could not grow on minimal medium containing sugars or glycolytic intermediates (Table 1). Enzyme assays showed that the KM16-22 mutant had lost PGM activity, whereas it is normally expressed in both STM815 free-living bacteria and symbiotic bacteroids. The putative *pgm* gene was cloned and overexpressed in *E. coli* using the pET expression vector. Characterization of the recombinant protein confirmed that it is 2,3-bisphosphoglycerate- dependent and inhibited by vanadate. Additionally, mutant KM16-22 recovered its enzyme activity after introduction of a wild-type *pgm* gene, which also restored its ability to use hexoses and related compounds as carbon source, and to form root nodules. PGM activity appears essential for both growth and symbiosis of *Burkholderia phymatum* on its plant host.

Table 1. Growth of *Burkholderia phymatum* STM815 and KM16-22 on complex medium and minimal medium containing various carbon sources.

Medium	Growth of strain	
	STM815	KM16-22
Yeast Extract Mannitol	+	+
Luria-Bertani	+	+
Minimal medium with compound:		
Glucose, fructose, gluconate, mannitol, pyruvate, succinate, fumarate, malate, or glutamate	+	– (+)[a]
Minimal medium with compound mixture:		
Glucose and pyruvate	+	+
Glucose and malate	+	+
Glucose and glutamate	+	+
Mannitol and pyruvate	+	+
Mannitol and malate	+	+
Mannitol and glutamate	+	+
Pyruvate and malate	+	–
Pyruvate and glutamate	+	–
Glucose and gluconate	+	–
Glucose and mannitol	+	–

[a]Results in parentheses were obtained with KM16-22/*pgm* strain harboring the wild-type *pgm* gene

LIVING LARGE: ELUCIDATION OF THE *FRANKIA* EAN1pec GENOME
SEQUENCE SHOWS GENE EXPANSION AND METABOLIC VERSATILITY

L. S. Tisa[1], D. R. Benson[2], G. B. Smejkal[4], P. Lapierre[2], J. P. Gogarten[2],
P. Normand[5], M. P. Francino[3] and P. Richardson[3]

[1]Department of Microbiology, University of New Hampshire, Durham,
NH, USA; [2]Department of Molecular and Cellular Biology, University
of Connecticut, Storrs, CT, USA; [3]Joint Genome Institute, Walnut Creek,
CA, USA; [4]Pressure Biosciences, Inc, Bridgewater, MA, USA;
[5]Ecologie Microbienne, UMR CNRS 5557, Université Lyon,
Villeurbanne, France

Frankia are nitrogen-fixing actinomycetes that form root nodules with dicotyledonous plants in eight families of distantly related angiosperms. The paucity of genetic information stemming from the absence of standard genetic tools has been a major obstacle in our understanding of actinorhizal symbiosis. Although we are trying to resolve this situation by developing genetic tools, we have also pursued new genomic approaches.

We sequenced the genome of *Frankia* strain EAN1pec, a broad host-range strain that infects plants from three families and represents one of the three major lineages (Group 3) within the genus. The genome is circular with an average G+C content of 71%. The genome encoded 3 rRNA operons, 46 tRNA, and 7492 CDS. The genome is larger than that of *Frankia* ACN14a, a medium host range *Alnus* strain, and CcI3, a narrow host range *Casuarina* strain. These genomes represent two of the three major lineages and vary in size from 5.43 Mb for *Frankia* sp. CcI3 to 7.50 Mb for *Frankia alni* ACN14a to 9.1 Mb for EAN1pec. On a total DNA basis, this is the largest range known for free-living prokaryotes related at the 98–99% 16S rRNA sequence level.

Some features of the *Frankia* EAN1pec genome immediately stand out. For example, EAN1pec has the highest total number and ranks third in percentage of secondary metabolite-related genes of the 395 bacterial genomes listed in the IMG database (395, 6.4%). The positions of genes annotated as transposases, phage integrases, and other mobile genetic element-related genes indicate a high degree of genome plasticity in the recent past. Genes associated with symbiosis include those for nitrogen fixation and regulation, but none similar to rhizobial common *nod* genes were found. These genome sequences allow 2-D gel electrophoresis to be used to identify development-specific proteins and to identify vesicle-specific proteins (the site of nitrogen fixation).

F. D. Dakora et al. (eds.), *Biological Nitrogen Fixation: Towards Poverty Alleviation
through Sustainable Agriculture.*
© Springer Science + Business Media B.V. 2008

PART 2E

SUMMARY PRESENTATIONS

PHOSPHATE-DEPENDENT GENE EXPRESSION IN FREE-LIVING AND SYMBIOTIC *SINORHIZOBIUM MELILOTI*

R. Zaheer, Z. Yuan, R. Morton and T. M. Finan

Center for Environmental Genomics, McMaster University, Hamilton, Ontario, Canada

Growth of *Sinorhizobium meliloti* under Pi-deficient conditions induces many genes involved in both P acquisition and in a general response to P-limitation. PhoB regulates genes induced by P limitation by binding to Pho-regulon promoters with a Pho-Box. PhoR is the histidine-protein kinase that activates PhoB by phosphorylation. Pi-PhoB interacts with the $\sigma70$ subunit of RNA polymerase. Using a whole genome *in silico* analysis of known Pho-box sequences, we identified 34 genes as putative members of the Pho regulon. Thirty-one genes were induced and three repressed by P-starvation in a PhoB-dependent manner (Yuan et al., 2006a; Krol and Becker, 2004), including genes with no obvious relation to P-metabolism, e.g., for iron transporters and *katA*, which encodes catalase and has two independent promoters regulated by OxyR and PhoB (Yuan et al., 2005). Pho regulon members in *S.meliloti* include the *pstSCAB* and *phoCDET* operons, encoding ABC-type high-affinity Pi-transporters (Bardin et al., 1998; Voegele et al., 1997; Yuan et al., 2006b), and the *orfA-pit* operon, which encodes a low-affinity Pi-transport system negatively regulated by PhoB (Bardin et al., 1998). Expression of these *S. meliloti* Pi transporters in root nodules, using promoter-*gusA* fusion plasmids in wild type and *phoB⁻ S. meliloti*, showed that the *orfA-pit* system was highly expressed in nodules, whereas very little *phoCDET* or *pstSCAB* expression was detected in either strain. There was no zone-specific expression of *orfA-pit::gusA*, and neither *phoC* nor *pstS* expression was detected in any nodule zone. These and other (Yuan et al., 2005, 2006b) data suggest that bacteroid metabolism in alfalfa nodules is not Pi limited.

References

Bardin SD and Finan TM (1998) Genetics 148, 1689–1700.
Bardin SD et al. (1998) J. Bacteriol. 180, 4219–4226.
Krol E and Becker A (2004) Mol. Gene Genomics 272, 1–17.
Voegele RT et al. (1997) J. Bacteriol. 179, 7226–7232.
Yuan ZC et al. (2005) Mol. Microbiol. 58, 877–894.
Yuan ZC et al. (2006a) Nucleic Acids Res. 34, 2686–2697.
Yuan ZC et al. (2006b) J. Bacteriol. 188, 1089–1102.

F. D. Dakora et al. (eds.), *Biological Nitrogen Fixation: Towards Poverty Alleviation through Sustainable Agriculture.*
© Springer Science + Business Media B.V. 2008

CHARACTERIZATION OF *BRADYRHIZOBIUM ELKANII* TYPE III SECRETION SYSTEM

S. Okazaki, S. Zehner, K. Lang and M. Göttfert

Institute of Genetics, Dresden University of Technology, Dresden, Germany

Type III secretion systems (T3SS) have been identified in a number of rhizobia and were shown to affect symbiosis with legumes. The genes encoding the conserved core of the T3SS (*rhc; rhizobium conserved*) are highly conserved in all cases, although differences exist in the secreted proteins. In our preliminary study, the presence of *rhcC2* and *rhcJ* in *Bradyrhizobium elkanii* was verified by PCR and sequencing of the fragments. In this study, we characterized the T3SS of *B. elkanii* USDA61 by DNA sequencing, mutagenesis, RT-PCR and extracellular protein analysis.

The DNA sequence of the T3SS gene cluster of *B. elkanii* USDA61 revealed the presence of all of the 9 *rhc* genes. Putative secreted proteins are also encoded in the region. The corresponding genes contain the *tts box* motif in the putative promoter region. Mutations in T3SS genes resulted in host specific symbiotic phenotype as reported for other rhizobia. The *rhcJ* mutant lost the ability to secrete several proteins. These results suggest that *B. elkanii* encodes a T3SS that functions in symbiosis and secretes proteins.

F. D. Dakora et al. (eds.), *Biological Nitrogen Fixation: Towards Poverty Alleviation through Sustainable Agriculture.*
© Springer Science + Business Media B.V. 2008

SEARCH FOR NODULATION GENE INDUCER FOR *MESORHIZOBIUM LOTI* SECRETED FROM ROOTS OF *LOTUS CORNICULATUS*

K. Kojima[1], T. Yokoyama[1], M. Itakura[2], K. Minamisawa[2] and Y. Arima[1]

[1]Tokyo University of Agriculture and Technology, Tokyo, Japan; [2]Graduate School of Life Sciences, Tohoku University, Japan
(Email: 50005951004@st.tuat.ac.jp)

It has been established that an inducer secreted from roots of host leguminous plants is needed for the expression of the *nod* genes. Several types of compounds, e.g., flavonoids, betains and aldonic acids, have been identified as inducers. In the particular case of *nod*-gene expression in *Mesorhizobium loti*, which makes root nodules on the *Lotus* plants, tetronic acid was reported to act as an effective inducer at a high (mM) concentration. However, there is no evidence that tetronic acid is secreted from *Lotus* plant roots. This suggests the existence of another inducer. Therefore, we searched to identify which compound was secreted from seeds and was in intact roots exudates of *Lotus corniculatus*.

The *M. loti* strain ML001, carrying *nod-lacZ* fusion, was exposed to a tetronic acid solution or to a concentrated plant culture solution, which contained seed and root exudates of *L. corniculatus* seedlings. The *nod* gene-inducer activity of the tested solutions was determined by measuring the β-galactosidase activity. Results were as follows. First, seed and root exudates showed significant *nod* gene-inducing activities for ML001 under poor nutrient conditions (with $CaSO_4$), but not under rich nutrient conditions (YEM). Second, and in contrast, inducing activity of tetronic acid was found for ML001 only under rich nutrient, but not poor nutrient, conditions. This strategy should enable us to identify the physiological *nod* gene-inducers, which differ between tetronic acid treatment and substances produced by the host plant under natural conditions.

F. D. Dakora et al. (eds.), *Biological Nitrogen Fixation: Towards Poverty Alleviation through Sustainable Agriculture.*
© Springer Science + Business Media B.V. 2008

IDENTIFICATION OF BetX, A PERIPLASMIC PROTEIN INVOLVED IN BINDING AND UPTAKE OF PROLINE BETAINE AND GLYCINE BETAINE IN *SINORHIZOBIUM MELILOTI*

B. Sagot, G. Alloing, D. Hérouart, D. Le Rudulier and L. Dupont

Unité Mixte de Recherche "Interactions Plantes-Microorganismes et santé végétale", n°6192 CNRS-INRA-University of Nice-Sophia Antipolis, Centre Agrobiotech, 06903 Sophia Antipolis cédex, France (Email: sagot@antibes.inra.fr)

An osmotic stress, such as excessive salinity or drought, has deleterious effects on the establishment and maintenance of symbiosis between *Sinorhizobium meliloti* and its host plant alfalfa. Bacteria protect themselves against high external osmolarity by accumulating osmoprotective organic compounds in their cytoplasm, the so-called compatible solutes. Although glycine betaine (GB) is one of the most effective osmoprotectant used by bacteria, proline betaine (PB) is the major betaine produced by alfalfa. PB is secreted by germinating seedlings of many *Medicago* species and is found in roots, nodules, and bacteroids. *S. meliloti* has the capacity to use betaines as compatible solutes but also as carbon and nitrogen sources. Three different transport systems (two ABC transporters and one BCCT transporter) are involved in the uptake of PB and GB in *S. meliloti*. BetX, a periplasmic binding protein encoded by a gene located in a PB catabolic locus, has been characterized. Although the *betX* gene does not belong to an ABC transporter operon, BetX contributes to the binding and uptake of PB and GB. Further, BetX is not associated with the already characterised betaine ABC transporters and the protein-membrane complex with which it is associated is unknown. Expression analysis of a *betX-lacZ* fusion revealed induction by PB, mono-methyl-proline (MMP, an intermediate in PB catabolism), and salt stress, suggesting that BetX has a catabolic and osmo-protection role. BetR, a TetR-like regulator is the *betX* expression repressor in the absence of ligand, and the repression is abolished by addition of substrate (MMP or PB). However, *betX* induction by osmotic stress is independent of BetR, suggesting an additional level of regulation. A *betX* mutant shows a defect in growth capacity but osmoprotection by PB and GB was unchanged due to the presence of other betaine transporters in *S. meliloti*. Nodulation capacity and nitrogen-fixation activity of a *betX* mutant was not affected.

F. D. Dakora et al. (eds.), *Biological Nitrogen Fixation: Towards Poverty Alleviation through Sustainable Agriculture.*
© Springer Science + Business Media B.V. 2008

H_2O_2 IS REQUIRED FOR OPTIMAL INFECTION-THREAD FORMATION DURING *SINORHIZOBIUM MELILOTI* SYMBIOSIS

A. Jamet[1], K. Mandon[1], A. Chéron[2], C. Coste-Maehrel[1], D. Le Rudulier[1], F. Barloy-Hubler[2], A. Puppo[1] and D. Hérouart[1]

[1]UMR INRA-CNRS-UNSA IPMSV, 400 Route des Chappes, 06903 Sophia-Antipolis Cedex, France; [2]UMR CNRS 6061, Université de Rennes I, Bat. 13, Campus de Beaulieu, 35042 Rennes Cedex, France (Email: mandon@unice.fr)

The involvement of reactive oxygen species in the establishment and maintenance of the legume-*Rhizobium* symbiosis has been under emphasized. Hydrogen peroxide (H_2O_2) accumulates all around bacteria during the infection process. To cope with H_2O_2, *S. meliloti* contains three catalases, which are differentially regulated, both in culture and during symbiosis. To better assess the role of H_2O_2 in the early steps of symbiosis, an *S. meliloti* strain overexpressing the constitutive catalase KatB was constructed. In culture, the Rm*katB*[++] resulting strain displayed a total catalase activity three-fold higher than the control strain, and a two to five fold decrease in intracellular H_2O_2 concentration. In addition, the mutant could be considered as a sink for exo-genous H_2O_2 because the Rm*katB*[++] cells were clearly more resistant to toxic levels of this compound. Microarray studies from exponentially growing cells revealed that more than 60 genes, mainly involved in cell process, transport and metabolism, were up-regulated in the Rm*katB*[++] mutant. None of the genes deregulated in the mutant appeared directly involved in oxidative defence mechanisms. These results strongly suggest that, at low concentration, H_2O_2 may have a signaling role in *S. meliloti*.

In planta, the Rm*katB*[++] mutant displayed a delayed nodulation phenotype with a significant reduction in the number of nodules one week post-inoculation. The initial response of the plant was not perturbed following inoculation with the mutant, but a significant decrease in infection-thread elongation was observed. This result was in agreement with ultrastructural analysis which showed that the infection threads induced by the Rm*katB*[++] mutant were enlarged and disorganized, with hardly any trace of H_2O_2 detectable around them. These results show that a minimal H_2O_2 threshold is required for optimal nodule development. Two hypothesis (not mutually exclusive) might explain the Rm*katB*[++] phenotype: (i) the absence of H_2O_2 leads to a structural perturbation of the infection thread; and (ii) H_2O_2 is involved in a signaling process.

MOLECULAR APPROACHES FOR THE CHARACTERIZATION OF *BRADYRHIZOBIUM* STRAINS

S. B. Campos, A. Giongo, A. Ambrosini, A. Beneduzi, N. Cobalchini, M. H. Bodanese-Zanettini and L. M. P. Passaglia

Núcleo de Microbiologia Agrícola, Departamento de Genética, Universidade Federal do Rio Grande do Sul, RS, Brazil C. P. 15053, 91501-970, Porto Alegre, RS
(Email: lpassaglia@terra.com.br)

Conventional phenotypic methods based on serotyping have been widely applied to differentiate species of the family Rhizobiaceae (Sprent, 1997) and DNA-DNA homology, based on quantitative DNA-DNA hybridization, was considered the standard method for species designation (Graham et al., 1991). The amplification of the 16S rRNA gene sequence by the polymerase chain reaction (PCR) has been used to provide additional information for studying the occurrence, distribution, and relationship of different rhizobia species (Willems et al., 2003). Nevertheless, the species of the genus *Bradyrhizobium* present low genetic variability in their rRNA gene sequences (Chueire et al., 2003), which renders their discrimination difficult. *B. japonicum* and *B. elkanii* have been largely used together as inoculant in Brazilian's soybean fields. In some cases, it is practically impossible to classify them. Here, we report the development of a fast low-cost PCR-based assay of the 16S rRNA gene to discriminate between these two related *Bradyrhizobium* species from 11 reference strains and 37 strains isolated from fields cultivated with soybean.

These bacteria also have a type III secretion system (TTSS) that mediates the interaction with the host. Although the TTSS of *B. japonicum* strain USDA110 has already been described (Göttfert et al., 2001), few studies have shown the presence of TTSS in *B. elkanii*. It seems that the sequences of this system in these two species are structurally different. We report the presence of genes and proteins related to TTSS in *B. japonicum* and *B. elkanii*. Besides confirming the presence of TTSS-related genes in the genome of *B. elkanii* strains, the hybridization study has confirmed the different organization between the genomes of the two *Bradyrhizobium* species.

F. D. Dakora et al. (eds.), *Biological Nitrogen Fixation: Towards Poverty Alleviation through Sustainable Agriculture.*

References

Chueire LMO et al. (2003) Rev. Bras. Ciên. Solo 27, 833–840.
Göttfert M et al. (2001) J. Bact. 183, 1405–1412.
Graham PH et al. (1991) Int. J. Syst. Bacteriol. 41, 582–587.
Sprent JI (1997) Plant Soil 188, 65–75.
Willems A et al. (2003) Syst. Appl. Microbiol. 26, 203–210.

GENOTYPIC AND PHENOTYPIC VARIATIONS AMONG RHIZOBIA NODULATING RED CLOVER IN SOILS OF NORTHERN SCANDINAVIA

S. Duodu[1], G. Carlsson[2], K. Huss-Danell[2] and M. M. Svenning[1]

[1]Department of Biology, University of Tromsø, N-9037 Tromsø, Norway; [2]Department of Agricultural Research for Northern Sweden, Crop Science Section, Swedish University of Agricultural Sciences, S-904 03 Umeå, Sweden

Indigenous populations of *Rhizobium leguminosarum* bv. *trifolii* were sampled from red clover (*Trifolium pratense* L.) nodules at Tromsø in northern Norway and Umeå in northern Sweden, during summer and autumn. Four hundred thirty-one nodule isolates were characterised genetically by targeting both chromosomal and symbiotic genes. Enterobacterial Repetitive Intergenic Consensus (ERIC)-PCR fingerprinting of chromosomal DNA revealed 114 different ERIC types, with only one found at both sites. In contrast, RFLP analysis of the *nodEF* genes on the symbiotic plasmid revealed a high proportion of *nod* types common to both sites (Figure 1). The symbiotic efficiency of eight isolates (both dominating and rare *nod* types) was measured in a greenhouse using ^{15}N dilution. All but one isolate (a very rare genotype) showed high N_2 fixation rates in symbiosis (Figure 2). Effective N_2-fixing strains of *R. l.* bv. *trifolii* nodulating red clover are thus common and genetically diverse in the sampled soils (Duodu et al., 2007).

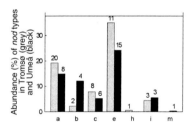

Figure 1. Relative abundance of seven *nod* types at the two sampling sites. Numbers on top of the bars show the number of ERIC types represented by each *nod* type.

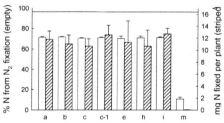

Figure 2. N_2 fixation in red clover plants inoculated with *R. l.* bv. *trifolii* isolates representing different *nod* types. Mean $\pm$ SD, n = 5. *Nod* types c and c-1 are identical, but represent different ERIC types.

Reference

Duodu S et al. (2007) J. Appl. Microbiol. 1002, 1625–1635.

F. D. Dakora et al. (eds.), *Biological Nitrogen Fixation: Towards Poverty Alleviation through Sustainable Agriculture.*
© Springer Science + Business Media B.V. 2008

NH$_4^+$ NUTRITION AFFECTS THE PHOTOSYNTHETIC AND RESPIRATORY C SINKS IN THE DUAL SYMBIOSIS OF A MYCORRHIZAL LEGUME

P. E. Mortimer[1], M. A. Pérez-Fernández[2] and A. J. Valentine[1]

[1]Plant Physiology Group, South African Herbal Science Institute, University of the Western Cape, Private Bag X17, Belleville 7535, South Africa; [2]Area de Ecología, Facultad de Ciencias Experimentales, University Pablo de Olavide, Cartera a Utrera, Km 1, 41013, Seville, Spain
(Email: alexvalentine@mac.com)

We investigated the effects of NH$_4^+$ supply on the AM and nodular development, C sink strengths, and the relative contributions of the two symbionts on host N nutrition. Legumes can form a tripartite symbiosis with both arbuscular mycorrhizal (AM) fungi and rhizobia. N-fixation is an energy-intensive process with rhizobial nodules requiring relatively high amounts of P (Almeida et al., 2000). The reliance of the host on these symbionts is most pronounced under N- and P-limiting conditions. However, both symbionts require photosynthetically derived C from the host, thus acting as C sinks (Harris et al., 1985). AM fungi can improve NH$_4^+$ nutrition (Toussaint et al., 2004), which may diminish host reliance on nodular N. *Phaseolus vulgaris*, inoculated with *Rhizobium leguminosarum* and *Glomus mosseae*, were grown for 30 days in sterile sand culture with a 2 μM P Long Ashton solution with either 1 mM NH$_4^+$ or a N-free control.

In the absence of NH$_4^+$, AM colonization reduced the ‰Ndfa in nodulated plants. This reduction of ‰Ndfa was even more pronounced during NH$_4^+$ addition to nodulated roots, possibly due to AM contributing to NH$_4^+$ uptake. With the decline in AM colonization, the reliance on NH$_4^+$ uptake decreased. This effect of AM in reducing ‰Ndfa may be related to the C-sink effect of AM or the symbiont's contribution to NH$_4^+$ uptake. The double symbiosis resulted in a greater drain on host C as reflected in the increased root O$_2$ consumption and the photosynthetic (Pm) rates. However, as AM colonization declined at 31 days, the Pm stimulation became less pronounced. Only the nodulated AM roots with NH$_4^+$ supply maintained a higher Pm rate, which may be related to the AM improvement of NH$_4^+$ nutrition. NH$_4^+$ nutrition reduced the ‰Ndfa in nodulated AM plants. The synergistic effect of the double symbiosis was an increased root respiratory sink, which was balanced by improved photosynthetic rates.

F. D. Dakora et al. (eds.), *Biological Nitrogen Fixation: Towards Poverty Alleviation through Sustainable Agriculture.*
© Springer Science + Business Media B.V. 2008

References

Almeida et al. (2000) J. Exp. Bot. 51, 1289–1297.
Harris et al. (1985) New Phytol. 101, 427–440.
Toussaint et al. (2004) Can. J. Microbiol. 50, 251–260.

THE EFFECT OF P NUTRITION ON AM COLONIZATION, NODULE DEVELOPMENT AND THE N AND C METABOLISM OF *PHASEOLUS VULGARIS*

P. E. Mortimer[1], M. A. Pérez-Fernández[2] and A. J. Valentine[1]

[1]Plant Physiology Group, South African Herbal Science Institute, University of the Western Cape, Private Bag X17, Belleville 7535, South Africa; [2]Area de Ecología, Facultad de Ciencias Experimentales. University Pablo de Olavide, Cartera a Utrera, Km 1, 41013, Seville, Spain (Email: alexvalentine@mac.com)

The objective is to investigate the effects of high and low P on nodule and AM performance and the subsequent influence on host N and C metabolism. Legumes can form a tripartite symbiosis with both arbuscular mycorrhizal (AM) fungi and rhizobia. N-fixation is an energy intensive process, resulting in the rhizobial nodules requiring relatively high amounts of P (Almeida et al., 2000). Low soil P can, therefore, diminish the benefits from the rhizobia. Under these conditions, the host plant will be more dependent on the AM fungi for P supply. Both symbioses require photosynthetically derived C from the host, thus acting as C sinks (Harris et al., 1985). *Phaseolus vulgaris*, inoculated with *Rhizobium leguminosarum* and *Glomus mosseae* were grown for 10 days in vermiculite. Thereafter seedlings were placed in a N-free hydroponic culture. The plants (AM and non-AM) were exposed to high (HP, 2 mM) and low P (LP, 2 µM).

AM colonization peaked after 17 days and then declined. In spite of this decline, AM symbionts improved P uptake for the nodules. The roots of the double-symbiosis hosts had greater nodular growth, which agrees with the higher specific phosphate utilization rates of AM roots under LP, however, nodulated AM roots had no improvement in N nutrition, as shown by similar plant N content and lower ‰ N derived from atmosphere. In the absence of additional N supply to AM roots, this decline may be due to the sink effect of the AM symbiont. This is supported by the effect of P supply to AM roots, where HP reduced the AM sink. A consequence of the double symbioses is that, during the declining phase of AM colonization, the nodules reached their highest biomass, indicating that AM symbionts may be stronger sinks for host C under LP conditions. The AM sink effect on nodulated roots was enhanced under LP conditions, where AM activity and colonization are higher. The sink strength of the double symbiosis was evidenced by the higher photosynthetic rates, O_2 consumption and construction costs.

F. D. Dakora et al. (eds.), *Biological Nitrogen Fixation: Towards Poverty Alleviation through Sustainable Agriculture.*
© Springer Science + Business Media B.V. 2008

References

Almeida et al. (2000) J. Exp. Bot. 51, 1289–1297.
Harris et al. (1985) New Phytol. 101, 427–440.

EFFECTS OF Al^{3+} TOXICITY ON PHOTOSYNTHETIC AND RESPIRATORY RESPONSES OF *PHASEOLUS VULGARIS* UNDER P DEFICIENCY

R. Hugo[1], M. A. Perez-Fernandez[2] and A. J. Valentine[1]

[1]Plant Physiology Group, South African Herbal Science Institute, University of the Western Cape, Private Bag X17, Belleville 7535, South Africa; [2]Area de Ecología, Facultad de Ciencias Experimentales, University Pablo de Olavide, Cartera a Utrera, Km 1, 41013, Seville, Spain (Email: alexvalentine@mac.com)

About 40% of the world's arable land is considered acidic and Al^{3+} toxicity poses a serious agricultural problem (Zheng et al., 1998). In acid soils, P deficiency is common together with Al toxicity (Ligaba et al., 2004). The effects of Al^{3+} toxicity on organic-acid exudation has been reported in legumes (Nian et al., 2003). However, the photosynthetic and respiratory costs of Al^{3+} toxicity in P-deficient legumes remain unclear. Here, we investigate Al^{3+} toxicity on P-deficient legumes, using *Phaseolus vulgaris*, inoculated with *Rhizobium leguminosarum*, grown for 10 days in vermiculite and then placed in a N-free hydroponic culture and grown with high P (HP, 2 mM) and low P (LP, 2 μM) for 3 weeks. Al^{3+} toxicity was induced by weekly additions of 50 μM $AlCl_3$ (pH 4.5). Biomass, photosynthesis, respiration and N_2 fixation were measured.

Although HP supply prevented the decrease in nodular growth during Al^{3+} toxicity, this high level of P nutrition did not lead to any significant advantage during the exposure to Al^{3+}. This was evident in the reduction of N_2 fixation at both P levels, but with a more pronounced decline during HP nutrition. In addition, the more marked reduction in photosynthetic rate and O_2 consumption of HP plants during Al^{3+} exposure may indicate that P-sufficient plants are less adapted than P-deficient plants. Furthermore, the relatively higher growth respiration during Al^{3+} toxicity at high P supply than at low P supply suggest that the costs of replacing Al^{3+}-damaged tissues are lower in P-deficient plants. These apparent physiological benefits of low P supply during Al^{3+} toxicity may be based on the high levels of organic acids normally associated with P deficiency (Nian et al., 2003), which can complex and detoxify Al^{3+} ions.

References

Ligaba et al. (2004) Physiol. Plantarum 120, 575–584.
Nian et al. (2003) Physiol. Plantarum 117, 229–236.
Zheng et al. (1998) Physiol. Plantarum 103, 209–214.

F. D. Dakora et al. (eds.), *Biological Nitrogen Fixation: Towards Poverty Alleviation through Sustainable Agriculture.*
© Springer Science + Business Media B.V. 2008

ASSIMILATE ALLOCATION BETWEEN THE ORGANIC ACID AND AMINO ACID POOLS IN *LUPINUS ANGUSTIFOLIUS* UNDER P STRESS

M. R. Le Roux[1], S. Khan[2] and A. J. Valentine[1]

[1]Plant Physiology Group, South African Herbal Science Institute, University of the Western Cape, Private Bag X17, Belleville 7535, South Africa; [2]Department of Health Sciences, Cape Peninsula University of Technology, Cape Town, South Africa
(Email: alexvalentine@mac.com)

The dark assimilation of inorganic C (DIC) into amino-acid and organic-acid pools of P-deficient nodules was studied. The major exchange of nutrients between symbiotic partners is reduced-C from the plant, to fuel nitrogenase and bacteroid respiration, and fixed-N from the bacteroid to the plant (Udvardi and Day, 1997). Pi limitation can increase DIC utilization via PEPc dark fixation (Theodorou and Plaxton, 1993). Under P limitation, plants circumvent the adenylate and Pi-dependant reactions of respiration and the combined activities of PEPc, malate dehydrogenase and NAD-malic enzyme may function as an alternative pyruvate supply to the mitochondrion. *Phaseolus vulgaris*, inoculated with *Rhizobium leguminosarum*, were grown for 10 days in vermiculite and then placed in N-free hydroponic culture with high (HP, 2 mM) or low P (LP, 2 μM) for 3 weeks. Dark $^{14}CO_2$ incorporation and assimilation, nodular respiration, amino acid- and organic acid-synthesizing enzymes, and N compounds were measured.

Nodule mass, N_2 fixation and metabolic Pi levels decreased under LP supply. Respiratory costs associated with maintaining a high degree of functionality declined with lower Pi levels. Prolonged P deprivation results in an increased organic-acid synthesis at the expense of amino-acid assimilation. The enhanced uptake and assimilation of dark CO_2 under P deficiency and the subsequent preferential incorporation into organic acids concurs with the engagement of the alternative route of PEP metabolism via PEPc. The inhibition of N_2 fixation under P deficiency cannot be ascribed to metabolite feedback inhibition, because there was a reduction in both amino-acid and NH_4^+ levels. Instead, the elevated malate levels may cause inhibition of N_2 fixation. The increased capacity for synthesis of organic acids over amino acids may allow nodules to survive long-term P stress, but this might impair the protein quality of the crop, since less N compounds are synthesized.

F. D. Dakora et al. (eds.), *Biological Nitrogen Fixation: Towards Poverty Alleviation through Sustainable Agriculture*.
© Springer Science + Business Media B.V. 2008

References

Theodorou ME and Plaxton WC (1993) Plant Physiol. 101, 339–344.
Udvardi M and Day D (1997) Ann. Rev. Plant Physiol. Plant Mol. Biol. 48, 493–523.

AFRICAN SPECIES BELONGING TO THE TRIBE PSORALEEAE PRODUCE UREIDES, A SYMBIOTIC TRAIT CHARACTERISTIC OF THE PHASEOLEAE

S. Kanu[1], S. B. Chimpango[2], J. Sprent[3] and F. D. Dakora[1]

[1]Cape Peninsula University of Technology, Cape Town 8000, South Africa; [2]Department of Botany, University of Cape Town, Rondebosch 7701, South Africa; [3]University of Dundee, Dundee, Scotland, UK

Modern studies in plant physiology have revealed the importance of nodulation and symbiotic traits in the taxonomy of legumes (Sprent, 2001). A recent phylogenetic tree showed a close relationship between the tribes Phaseoleae and Psoraleeae. This study aimed to assess the symbiotic traits of eight *Psoralea* species, which occur uniquely as endemics in the Cape region of South Africa and to compare them to Phaseoleae. Eight *Psoralea* species were collected from six sites, namely, *P. pinnata, P. repens, P. aphylla, P. asarina, P. monophylla, P. aculeata, P. restioides and P. laxa*. All nodules had a spherical shape resembling that of cowpea nodules (Phaseoleae). Nodule anatomy and ultrastructure of *P. pinnata* revealed an internal arrangement similar to that of Phaseoleae (Dakora and Atkins, 1989; Dakora 2000). Ureides were present in all plant organs of all eight species (assayed according to Dakora et al., 1992). The concentration of ureide was found to be significantly different ($p < 0.5$) at the species and organ levels. *P. pinnata* had the highest ureides concentration. This ureide production by all eight species supports their relatedness to the tribe Phaseoleae. Interestingly, ureide biosynthesis and the formation of round determinate nodules with an internal anatomy similar to cowpea and soybean nodules are unique symbiotic features of the tribe Phaseoleae (Dakora 2000). Based on these shared symbiotic traits of nodule morphology, nodule anatomy and ureide biogenesis, we propose that *Psoralea* species be re-classified in the tribe Phaseoleae.

Support from NRF and Cape Peninsula University of Technology is acknowledged.

References

Dakora FD (2000) Aus. J. Plant Physiol. 27, 885–892.
Dakora FD and Atkins C (1989) Aus. J. Plant Physiol. 16, 131–138.
Dakora FD et al. (1992) Plant Soil 140, 255–262.
Sprent JI (2001) Nodulation in Legumes. Cromwell Press, London, pp. 66–87.

NEW LOOK AT OLD ROOT-NODULE BACTERIA: MOLECULAR TECHNIQUES UNCOVER NOVEL ISOLATES

J. K. Ardley[1], R. J. Yates[1,2], K. Nandasena[1], W. G. Reeve[1], I. J. Law[3], L. Brau[1], G. W. O'Hara[1] and J. G. Howieson[1,2]

[1]Centre for *Rhizobium* Studies, Murdoch University, Perth, W.A. 6150, Australia; [2]Department of Agriculture Western Australia, Baron-Hay Court, South Perth, W.A. 6151, Australia; [3]ARC-Plant Protection Research Institute, Private Bag X134, Queenswood 0121, South Africa

Exotic pasture legumes and their associated microsymbionts are important in providing biological nitrogen fixation in Australian agricultural systems. Southern African species of *Lotononis* from the *Listia* section can potentially provide sustainable agricultural productivity in systems affected by increasing dryland salinity and climate change. There are eight species in the *Listia* section: *L. angolensis*, *L. bainesii*, *L. macrocarpa*, *L. marlothii*, *L. minima*, *L. subulata* and *L. solitudinis* (Van Wyk, 1991). They are perennial, stoloniferous and collar-nodulated. The root-nodule bacteria (RNB) isolated from several of these species are pigmented and the symbiosis between these RNB and their hosts is highly specific (Yates et al., 2007). Pioneering work on *L. angolensis*, *L. bainesii* and *L. listii* isolates was performed in Africa in the 1950–60s by Botha (Kenya), Sandman (Zimbabwe) and Verboom (Zambia) and in Australia (Norris, 1958).

Modern molecular techniques have now enabled further characterization of these *Lotononis* isolates. Jaftha et al. (2002) determined that *L. bainesii* isolates were phylogenetically related to *Methylobacterium*. The South African *L. bainesii*, *L. listii* and *L. solitudinis* isolates are slow-growing, alkali-producing, dark pink and dry and form a cross-inoculation group. They induce ineffective nodulation on *L. angolensis*. Although 16S rRNA gene sequencing shows they are closely related to *Methylobacterium nodulans*, they are unable to utilise methanol as a sole carbon source. *L. angolensis* isolates (collected in Zambia) are fast growers, pale orange-pink and mucilaginous and do not nodulate *L. bainesii*, *L. listii* or *L. solitudinis*. They are also unable to utilize methanol. Sequencing of the 16S rRNA gene shows <94% sequence similarity to any other α-proteobacterial strains. These isolates therefore constitute a novel group of RNB.

F. D. Dakora et al. (eds.), *Biological Nitrogen Fixation: Towards Poverty Alleviation through Sustainable Agriculture.*
© Springer Science + Business Media B.V. 2008

References

Jaftha JB et al. (2002) Syst. App. Micro. 25, 440–449.
Norris DO (1958) Aust. J. Ag. Res. 9, 629–632.
Van Wyk B-E (1991) A Synopsis of the Genus *Lotononis*, Rustica Press, Cape Town.
Yates RJ et al. (2007) Soil Biol. Biochem. 39, 1680–1688.

CHARACTERIZATION OF A NOVEL NON-RHIZOBIAL SYMBIONT FROM *CICER ARIETINUM*

A. Aslam, K. A. Malik and F. Y. Hafeez

National Institute for Biotechnology and Genetic Engineering (NIBGE),
P.O. Box 577, Jhang Road, Faisalabad, 38000-Pakistan
(Email: fauzia@nibge.org)

Since 2001, there have been reports of effective legume nodulation by non-rhizobial bacteria (Ngom et al., 2004, Verma et al., 2004) of which that by *Ochrobactrum* has been reported recently (Lebuhn et al., 2000; Trujillo et al., 2005). One Gram-negative fast-growing strain, Ca-34, was isolated from nodules of *Cicer arietinum*. Its colonies are white, circular, isolated, smooth, and mucoid with a 2–3 mm diameter. The strain is nitrogen-fixing, it is high salt, acid, and temperature tolerant, it has the ability to utilize a wide range of carbon sources, and is identified as *Ochrobactrum* sp. on the BIOLOG system. It is resistant to many antibiotics, including aztreonam (β-lactams), which is a characteristic of the genus *Ochrobactrum*. Fatty acid similarity index with the *Ochrobactrum* type strain is 0.6. It shows a unique RAPD pattern and ARDRA profiles that are not similar to those of reported chickpea-specific rhizobia. It has one mega plasmid of 590 Kb that might be involved in symbiosis. Partial (AY499125) and full-length (DQ647056) 16S rDNA sequencing show 99% homology with *O. intermedium* and uncultured *Ochrobactrum* sp. One hundred thirty-five nucleotides are different from *Rhizobium* and 247 different from *Bradyrhizobium* along the entire length of 16S rDNA molecule. A segment of 42 nucleotides is highly variable and has no similarity to the 16S rDNA molecules of *Brady* (*rhizobium*), *Agrobacterium*, and nodulating *Ochrobactrum* spp. Phylogenetically, it is distinct from rhizobia and other diazotrophs, and is more closely related to *Ochrobactrum* sp. reported from *Acacia*. These distinctive characteristics suggest that Ca-34 does not belong to the reported *Ochrobactrum* spp. and should be assigned the status of new specie within genus *Ochrobactrum*. This is first report of effective symbiotic association of *Ochrobactrum* sp. with root nodules of chickpea.

References

Lebuhn et al. (2000) Int. J. Syst. Evol. Microbiol. 50, 2207–2223.
Ngom et al. (2004) J. Gen. Appl. Microbiol. 50, 17–27.
Trujillo et al. (2005) Appl. Environ. Microbiol. 71, 1318–1327.
Verma et al. (2004) Biotechnol. Lett. 26, 425–429.

F. D. Dakora et al. (eds.), *Biological Nitrogen Fixation: Towards Poverty Alleviation through Sustainable Agriculture.*
© Springer Science + Business Media B.V. 2008

BURKHOLDERIA TUBERUM EFFECTIVELY NODULATES *CYCLOPIA* SPP., BUT NOT *ASPALATHUS* SPP.

G. N. Elliott[1], W.-M. Chen[2], C. Bontemps[3], F. D. Dakora[4], J. P. W. Young[3], J. I. Sprent[1] and E. K. James[1]

[1]School of Life Sciences, University of Dundee, Dundee, UK; [2]Laboratory of Microbiology, Department of Seafood Science, National Kaohsiung Marine University, Kaohsiung City 811, Taiwan; [3]Department of Biology, University of York, P.O. Box 373, York, UK; [4]Cape Peninsula University of Technology, P.O. Box 652, Cape Town 8000, South Africa

Recently, many ß-proteobacteria have joined the established α-proteobacteria as capable of effectively nodulating legumes. However, despite their isolation from nodules of papilionoid and other mimosoid legumes (Moulin et al., 2001; Barrett and Parker, 2005), there is little evidence that these ß-proteobacteria effectively nodulate any legume outside the sub-family *Mimosoideae*. So, the ability of the ß-proteobacterium *Burkholderia tuberum* STM678 (isolated from nodules of *Aspalathus carnosa*) to effectively nodulate the South African endemic legumes *Aspalathus* and *Cyclopia* (sub-family *Papilionoideae*) was compared with that of *B. tuberum* strain DUS833 (isolated from *Aspalathus callosa*), *Mesorhizobium* sp. DUS835 (from *A. linearis*), *Sinorhizo-bium* sp. NGR234 (from *Lablab purpureus*), and *Methylobacterium nodulans* ORS2060 (from *Crotalaria podocarpa*), an isolate known to contain a closely related *nodA* gene to STM678. STM678 formed nodules with all five species of *Cyclopia*, four of which were effective, whereas DUS833 formed nodules on three *Cyclopia* spp, two of which were effective. No nodules formed on any *Aspalathus* species inoculated with either *B. tuberum*, and all plants subsequently died from apparent N-limitation. No other ino-culum formed nodules with any plant. *C. genistoides* showed GFP or GUS activity with marked strains of STM678. Nodules formed by both STM678 and DUS833 on *C. geni-stoides* were effective in appearance, and immunogold labelling of bacteroids confirmed them as *Burkholderia* containing effective symbiotic structures. Further studies showed that *C. genistoides* was infected via the root hairs. Although originally isolated from *Aspalathus carnosa*, *Burkholderia tuberum* now appears more likely to form effective *in vivo* symbioses with *Cyclopia* spp. *Burkholderia tuberum* is the first effective ß-proteobacterium symbiont of legumes from the sub-family Papilionoideae.

F. D. Dakora et al. (eds.), *Biological Nitrogen Fixation: Towards Poverty Alleviation through Sustainable Agriculture*.
© Springer Science + Business Media B.V. 2008

References

Barrett CF and Parker MA (2005) Syst. Appl. Microbiol. 28, 57–65.
Moulin L et al. (2001) Nature 411, 948–950.

ENDOSYMBIOSIS FORMATION BETWEEN A NITROGEN-FIXING MIX OF BACTERIA (*XANTHOMONAS SP. + ARTHROBACTER SP.*) AND WHEAT ROOT CELLS

B. Abbas

Agricultural Faculty, Gorgan University of Agricultural Sciences
and Natural Resources, Iran
(Email: abs346@yahoo.com)

This research attempted to produce para-nodules on wheat roots and to study them by SEM and TEM. The auxin-like 2,4-D (at 0.1 µg/mL) was used as an abiological nodulation agent. A nitrogen-fixing mixed culture (*Arthrobacter sp + Xanthomonas sp*) was used to inoculate wheat (*Triticum aestivum var. Pameiti Fedina*) seedlings that were isolated from typical Sierozem in Turkmen. SEM showed bacterial colonization of the wheat root surface (Figure 1). Mucus creates a microclimate for nitrogen-fixing bacteria that may lead to bacterial penetration of plant tissues and eventually to colonization. One of the dominant bacterial forms was V-shaped cells, characteristic of coryneform bacteria that were components of the mixed culture used. In ultra-thin sections of the para-nodules, bacterial cells were located both inter-cellularly (Figure 2) and inside the para-nodule cells (Figure 3), where a microaerophilic condition exists. Most microorganisms (bacteria) were oriented along the plant cell wall (Figure 4), which maximizes their contact area the plant cell and increases nitrogen-fixation activity. Bacteria in the intercellular spaces were alive and actively dividing. These figures show that non-leguminous plants can be made capable of forming artificial nitrogen-fixing symbioses through inducing para-nodules.

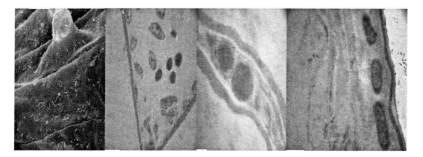

Figures 1–4. See text for details.

F. D. Dakora et al. (eds.), *Biological Nitrogen Fixation: Towards Poverty Alleviation through Sustainable Agriculture.*
© Springer Science + Business Media B.V. 2008

SECTION 3

MICROBIOLOGY OF NITROGEN FIXATION

PART 3A

GENOMICS, PHYLOGENY, AND EVOLUTION

INTEGRATING *SINORHIZOBIUM MELILOTI* GENOMICS: AN EXPRESSION LIBRARY OF THE GENOME

A. Cowie, J. Cheng, B. Poduska, A. MacLean, R. Zaheer, R. Morton and T. Finan

Centre for Environmental Genomics, McMaster University, Hamilton, Ontario, Canada
(Email: finan@mcmaster.ca)

To investigate gene function, we developed a robust transcription fusion reporter vector to measure gene expression in bacteria. The vector, pTH1522, was used to construct a random insert library for the *Sinorhizobium meliloti* genome (Cowie et al., 2006). Homologous recombination of the DNA fragments cloned in pTH1522 into the *S. meliloti* genome generates transcriptional fusions to either the reporter genes *gfp+* and *lacZ* or *gusA* and *rfp*, depending on the orientation of the cloned fragment. Over 12,000 fusion junctions in 6,298 clones were identified by DNA sequence analysis and the plasmid clones were recombined into *S. meliloti*. Reporter enzyme activities following growth of these recombinants in complex media (LBmc) and in minimal medium with glucose or succinate as sole carbon sources, allowed the identification of genes highly expressed in one or more growth conditions and those expressed at very low to background levels.

Of the 6,204 predicted ORFs in the *S. meliloti* genome, 2,480 are annotated as "conserved hypothetical", "hypothetical" or "unknown". The *S. meliloti*::pTH1522 reporter gene fusion library contains transcriptional fusions to 951 of these PUFs (Proteins of Unknown Function). Analysis of the expression data for fusion strains to just these genes reveals that approximately 60% are not expressed in complex medium (LBmc) or in minimal medium with either glucose or succinate as sole carbon source. The remaining 40% show varying degrees of expression in the different test media. Many are constitutively expressed in all three media, others in only one media type and some in different combinations of two media types (Figure 1). For the genes that showed little/no expression in minimal media, we are now screening them with a battery of carbon and nitrogen compounds to find inducers to further elucidate function.

F. D. Dakora et al. (eds.), *Biological Nitrogen Fixation: Towards Poverty Alleviation through Sustainable Agriculture.*
© Springer Science + Business Media B.V. 2008

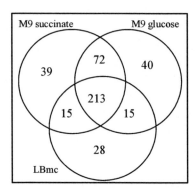

Figure 1. Numbers of fusion strains and the media on which they grow.

A major advantage of this approach is that it allows the identification of essential genes. When the *S. meliloti* DNA gene fragment cloned in pTH1522 is internal to a gene or operon, recombination will disrupt the genomic copy and lead to a loss-of-function mutation. If the gene is essential for growth in rich medium, then recombinant colonies will not be recovered. During transfer of the pTH1522 library clones to *S. meliloti*, we found 101 clones that did not generate recombinant strains. Many genes identified as essential were expected, e.g., amino-acyl tRNA synthetases, DNA and RNA polymerase subunits, ribosomal proteins, and genes involved in replication, transcription, translation, cell division, and protein translocation, all of which map to the *S. meliloti* chromosome. There were also 19 clones representing 16 genes of unknown function (PUFs), mostly located on the chromosome (Table 1).

Table 1. The sixteen essential genes of unknown function identified in this study.

Gene	Product description	Gene	Product description
SMc00074	Hyp. trans-membrane signal peptide protein	SMc02802	Conserved hypothetical protein
SMc00471	Put. sensor histidine kinase	SMc03872	Hypothetical protein
SMc00532	Conserved hypothetical protein	SMc04010	Hypothetical protein
SMc00611	Hyp. trans-membrane protein	SMc04347	Conserved hypothetical protein
SMc00722	Trans-membrane protein		
SMc00951	Trans-membrane protein	SMa2253	Conserved hypothetical
SMc01721	Inner membrane protein	SMb20057	ABC-type transporter permease
SMc02090	Conserved hypothetical protein	SMb21182	Conserved membrane anchor protein
SMc02756	Put. sensor histidine kinase		

Interestingly, there was one gene on the pSymA megaplasmid, SMa2253, for which several different clones existed, but each failed to produce recombinants. This result was unexpected because the whole of pSymA has been deleted previously without loss of viability (Oresnik et al., 2000). Whether this is because the gene is essential or changes in expression are lethal, such as in a toxin-antitoxin system, remains to be determined as these clones contained the 5' upstream region of this gene and so should not have caused loss of function when recombined into the genome, but may have resulted in loss of regulation. Clones representing two genes on the pSymB megaplasmid also failed to produce recombinants. For SMb21182, the predicted gene product shows homology to acyl-CoA transferases and lies in an operon with SMb21181 annotated as a putative glutaryl-CoA dehydrogenase but little else is known. SMb20057 is annotated as an ABC transporter permease protein and the clone representing this gene also contains an internal fragment which would cause loss of function of the transporter following recombination (Mauchline et al., 2006). We have confirmed that this gene is essential for growth on complex medium and a more detailed analysis of this transport system is currently underway.

References

Cowie A et al. (2006) Appl. Environ. Microbiol. 72, 7156–7167.
Mauchline T et al. (2006) Proc. Natl. Acad. Sci. USA 103, 17933–17938.
Oresnik IJ et al. (2000) J. Bacteriol. 182, 3582–3586.

FULL GENOME SEQUENCE OF *AZOTOBACTER VINELANDII*: PRELIMINARY ANALYSIS

C. Kennedy[1], D. R. Dean[2], B. Goodner[3], B. Goldman[4], J. Setubal[5], S. Slater[6] and D. Wood[7]

[1]Department of Plant Pathology and Microbiology, University of Arizona, Tucson, AZ, USA; [2]Department of Biochemistry, Virginia Tech, Blacksburg, VA, USA; [3]Hiram College, Hiram, OH, USA; [4]Monsanto Company, St. Louis, MO, USA; [5]Virginia Bioinformatics Institute, Virginia Tech, Blacksburg, VA, USA; [6]Department of Applied Biological Sciences, Arizona State University, Tempe, AZ, USA; [7]Department of Biology, Seattle Pacific University, Seattle, WA, USA

One of the most versatile and widely studied diazotrophs is *Azotobacter vinelandii*, which has a diversity of features that are of interest to several scientific communities. In addition to having three distinct nitrogenases encoded by different genes, *A. vinelandii* is the most aerotolerant of free-living (non-plant associated) diazotrophs. Members of this species also produce alginates, siderophores, and can form dormant-stage cysts (for review, see Kennedy et al., 2005). *A. vinelandii* is a member of the gamma-proteobacteria and is therefore related to *Escherichia coli*. It is found in soils throughout the world and has been studied for >100 years with respect to its physiology, bio-chemistry, and genetics of nitrogen fixation and related aspects of metabolism (Kennedy and Bishop, 2004). In addition, the organism is easily genetically manipulated, easily mutagenized, and grows quickly in air in liquid culture or on plates, forming single well-defined colonies.

Scientists with interests in each of these topics took part in an annotation workshop held at the University of Arizona in July 2002, after a rough draft sequence had been produced at the Joint Genome Institute of the Department of Energy in Walnut Creek, CA. Completion of the sequence occurred in 2006 under the direction of Derek Wood at Seattle Pacific University (also affiliated with The University of Washington), working with the other co-authors. Optical mapping helped to complete and close the genome.

The current genome is a single 5.37 megabase (Mb) circular chromosome probably encoding 5,221 proteins, with six ribosomal operons (Figure 1). The circular nature of the chromosome was established earlier by genetic methods (Blanco et al., 1990). Relatedness to *Pseudomonas* species is remarkable (as expected, see Rediers et al., 2004). The sequence will be released in the spring of 2007 via the project website (http: //www.azotobacter.org). A second genome annotation workshop will be held at the University of Seattle in 2007 with members of the *Azotobacter* scientific community taking part. Undergraduates at Hiram College and Seattle Pacific University are generating sequence-indexed transposon libraries, screening these libraries for essential genes under various physiological conditions, and developing an *in vivo* transposition system to facilitate future genetic analyses.

CK thanks Baomin Wang for help with the figure. This work was supported by grant 0523357 from the National Science Foundation.

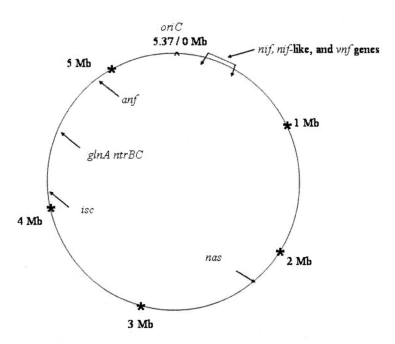

Figure 1. Location of genes related to nitrogen fixation or metabolism on the *A. vinelandii* chromosome. *nif* and *vnf* genes encode Mo and V nitrogenases; *nas* genes encode nitrate and nitrite reductases; *isc* genes encode enymes involved in iron-sulfur cluster formation; *glnA* encodes glutamine synthetase; *ntrBC* encode nitrogen-regulatory genes; and *anf* genes encode Fe nitrogenase.

References

Blanco G et al. (1990) Mol. Gen. Genetics 224, 241–247.
Kennedy C and Bishop P (2004) In Klipp et al. (eds) Genetics and Regulation of Nitrogen Fixation in Free-Living bacteria, 27–52.
Kennedy C et al. (2005) Bergey's Manual of Systematic Bacteriology, Vol. III, 384–402.
Rediers H et al. (2004) Microbiology 150, 1117–1119.

TRANSCRIPTIONAL PROFILE OF *BRADYRHIZOBIUM ELKANII* SEMIA 587 IN SYMBIOSIS WITH SOYBEAN (*GLYCINE MAX* L. Merrill) ANALYZED BY DNA MICROARRAY

J. Marcondes, M. Cantão, L. M. C. Alves and E. G. M. Lemos

Department of Technology, UNESP, Jaboticabal, Brazil
(Email: jackson@fcav.unesp.br)

Bradyrhizobium elkanii SEMIA 587 (Rumjanek et al., 1993) is one of two unique species of rhizobia officially specified for soybean inoculant composition in Brazil and other places in the world. This soybean endosymbiotic bacterium optimizes symbiotic nitrogen fixation and results in more vigorous plant growth and higher soybean crop yield because nitrogen is the nutrient required in the highest amount for soybean crops. Recently, knowledge of the genome compositions of rhizobia have arisen from complete DNA sequencing of either entire genomes or specific replicons, like plasmids or symbiotic islands, as shown by RhizoBase. Although we know about much about the symbiotic process, advances in rhizobial genomics have given us better knowledge of rhizobia-legume interactions at molecular level.

We have sequenced the DNA of *B. elkanii* from genomic shotgun libraries. Next, we have applied the Optimal Clone Identifier tool (Cantão et al., 2007) to choose clones containing an unique ORF. Selected clones had their inserts amplified by PCR, using universal primers, and amplicons higher than 300 bp were used to generate a DNA microarray for *B. elkanii*. This array contains 2,654 genes and was used in transcriptional analysis looking for bacteroid genes that were differentially expressed during symbiosis with soybean plants at three developmental stages, at 13, 28 and 48 days after inoculation. For each stage, we found 51, 46 and 26 differentially expressed genes, respectively, based on a fold change of 2.0 or 1.5.

In bacteroids, both the number and level of expression of differentially expressed genes required for amino-acid biosynthesis decreased with increasing nodule age. In *B. japonicum*, the energy required for the fast bacterial growth seems to be directed toward the reduction of nitrogen during the symbiotic state, when cellular division ceases (Karr and Emerich, 1996). At different bacteroid stages, the analyses detected genes related to nodulation and nitrogen-fixation regulation more than structural genes (see Table 1). Furthermore, organic nitrogen recycling might be involved in such regulatory processes in bacteroids based on the detection of a significant number of genes involved

F. D. Dakora et al. (eds.), *Biological Nitrogen Fixation: Towards Poverty Alleviation through Sustainable Agriculture.*
© Springer Science + Business Media B.V. 2008

in amino-acid metabolism (see Table 1). Roles for the products of genes for amino-acid metabolism during nitrogen fixation in *Rhizobium leguminosarum* bacteroids have been suggested previously (Lodwig et al., 2003).

Table 1. Some differentially expressed genes in bacteroids of *B. elkanii*.

d.a.i.	Nodulation	Nitrogen fixation	Amino acid biosynthesis
13	Nodulation protein N (***nodN***)	–	3-isopropylmalate dehydrogenase (***leuB***)cysteine synthase (***cysK***)dihydrodipicolinate reductase (***dapB***)
28	Rhizopine catabolism (***moc***)	Iron-sulfur cofactor synthesis protein (***nifZ***)short-chain alcohol dehydrogenase (***fixR***)	Dihydrodipicolinate synthase (***dapA***)O-acetylhomoserine sulfhydrylase (***cysD***)glutamine synthetase I (***glnA***)nitrogen regulatory protein P_{II} (***glnB***)urocanase (***hutU***)prephenate dehydratase (***pheA***)
	–	E1-E2 type cation ATPase (***fixI***)ferredoxin (***fer3***)	Tryptophan synthase beta subunit (***trpB***)
13/28	Bifunctional enzyme (***nodQ***)beta (1–6) glucans synthase (***ndvC***)superoxide dismutase (***sodF***)	–	–
28/48	–	Cytochrome-c oxidase (***fixN***)	–

References

Cantão M et al. (2007) Genet. Mol. Res. 6, 743–755.
Karr DB and Emerich DW (1996) Appl. Environ. Microbiol. 62, 3757–3761.
Lodwig EM et al. (2003) Nature 422, 722–726.
Rumjanek NG et al. (1993) Appl. Environ. Microbiol. 59, 4371–4373.

CHRONOS, CHAIROS AND CO-EVOLUTION IN *RHIZOBIUM*-LEGUME SYMBIOSES

K. Lindström

Department of Applied Chemistry and Microbiology, Viikki Biocenter 1, University of Helsinki, Finland
(Email: Kristina.Lindstrom@Helsinki.Fi)

The evolution of micro-organisms can be viewed conceptually from two different perspectives, which are termed '*Chronos*' and '*Chairos*' evolution. Chronos evolution is characterized through its long-term perspective; slowly evolving genes and genomes leading to speciation. Evolutionary events can be fixed on a time scale in chronological order. In contrast, Chairos evolution is fast, takes place under favourable, often non-selective conditions, leads to niché adaptation and cannot be used to measure evolutionary time. The hypothesis is that Chronos evolution occurs with all DNA sequences (core and accessory genomes) within an organism, whereas Chairos evolution impacts most effectively the accessory genome. Speciation in rhizobia is mainly seen as Chronos evolution of the core genome, whereas symbiotic interactions with plants are largely determined by accessory genes that are subjected to both Chronos and Chairos evolution (Figure 1).

In the legume–*Rhizobium* symbioses, the phylogeny of the bacteria, as determined by their taxonomic status, is not congruent with that of their host plants. However, the rhizobial symbiotic genes (accessory genes) seem to track the legumes hosts in evolution. Using the *Rhizobium galegae–Galega* sp. system, we find that rhizobial *nod* genes in this very specific symbiosis evolved under host constraint, an indication of Chairos evolution. The rhizobial *nod* genes encode a signal molecule, a substituted lipo-chito-oligosaccharide (Nod Factor, NF, LCO), which is perceived by the plant. The structure of several Nod factors representing several rhizobial taxa has been determined and critical chemical substitutions contributing to host specificity have been proposed. However, the range of symbioses that have so far been studied from an evolutionary and phylogenetic perspective is narrow. The focus has so far been mainly on traditionally important agronomical crop plants and recently also on model legumes.

In my laboratory, we have broadened the scope by focussing on the biogeographic patterns of two legume genera, namely *Calliandra* sp. (*C. calothyrsus*) and *Galega* sp.

F. D. Dakora et al. (eds.), *Biological Nitrogen Fixation: Towards Poverty Alleviation through Sustainable Agriculture.*
© Springer Science + Business Media B.V. 2008

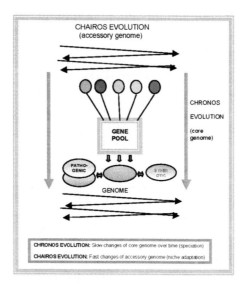

Figure 1. The concepts of Chronos and Chairos evolution.

(*G. orientalis* and *G. officinalis*). The former is a tropical woody legume with broad host range, the latter a perennial forage plant of temperate origin. By phylogenetic analysis of core and accessory genes of rhizobia isolated from these plants, we have demonstrated that the host plant is an important determinant that directs the evolution of rhizobial symbiotic genes (Suominen et al., 2001). The diversity of the rhizobia (genomic, core and accessory genes) displayed the greatest diversity in the gene center (center of origin) of their host plants (Dresler-Nurmi et al., 2003; Andronov et al., 2003) and so did the genomic fingerprinting patterns (AFLP) of *Galega* plant accessions collected from the gene center.

Recently, plant receptors receiving and transmitting the rhizobial signals have been identified, allowing comparisons of signal (i.e., ligand) and receptor structures. Partial gene sequences of two proteins involved in Nod-factor perception (SymRK of the LRR family, and NFR5 of the LysM family) from selected *Galega* plant accessions were also determined. They displayed diversity within the *Galega* genus. The *Galega* sequences were distinct from other sequenced plant genes and belonged to the same clade as those from other galegoid plants. In order to prove or disprove real co-evolution, we need to test whether the legume hosts also evolve to respond to the rhizobial signals. Thus, the next step is to look for signs of adaptive evolution in the receptor proteins.

References

Andronov EE et al. (2003) Appl. Environ. Microbiol. 69, 1067–1074.
Dresler-Nurmi A et al. (2003) Report from EU project IC18-CT97-0194.
Suominen L et al. (2001) Mol. Biol. Evol. 18, 906–916.

GENOMICS AND EVOLUTION OF BACTERIA: NODULATION, NITROGEN FIXATION AND THE REST OF THE ACCESSORY GENOME

J. P. W. Young

Department of Biology, University of York, P.O. Box 373, York YO10 5YW, UK

The alpha-proteobacteria display a variety of conspicuous properties. There are N_2-fixing root-nodule bacteria in such genera as *Rhizobium*, *Mesorhizobium*, and *Bradyrhizobium*; *Agrobacterium* causes plant tumours; *Bartonella* and *Brucella* are animal pathogens; *Methylobacterium* consumes methanol; and *Rhodobacter* is photosynthetic. The distribution of these diverse capabilities is not explained in any simple way by the relatedness of the bacteria; indeed, it is actually more complicated because many bacteria have more than one of these properties. Some root nodule bacteria are *Agro-bacterium* and some tumour-inducers are *Rhizobium*; there are photosynthetic *Brady-rhizobium* and nodulating *Methylobacterium* strains. It is clear that characters such as these cannot provide an unambiguous way to classify bacteria.

The reason for this complex pattern is that these properties are conferred by the accessory genome. Bacterial genomes are compound and can be thought of as comprising two parts, the core and accessory. The core genome is mostly carried on the main chromosome; it is stable in organisation and carries essential genes that are rarely transferred between distantly related organisms. In contrast, the accessory genome is carried on plasmids, islands, transposons, or phages; it is prone to rearrangement and horizontal transfer between species; and it confers functions that provide special adaptations that are valuable in certain environments but not needed in others. Accessory genes usually have a lower G+C composition than the core genes. As a consequence of these differences, core genes can provide a consistent basis for defining species and the phylogenetic relationships among species, whereas the accessory genome may vary substantially in size and gene content even between strains of the same species.

Complete genome sequences are now available for many bacteria, providing a rich source of data that confirms these generalisations. Our own contribution has been the genome sequence of *Rhizobium leguminosarum* biovar *viciae* strain 3841 and considerable discussion of the accessory genome concept and relevant analyses can be found in the corresponding publication (Young et al., 2006).

F. D. Dakora et al. (eds.), *Biological Nitrogen Fixation: Towards Poverty Alleviation through Sustainable Agriculture.*
© Springer Science + Business Media B.V. 2008

Recent work in our laboratory has addressed the extent to which the different species within a bacterial genus share part of the same accessory gene pool. We used Suppressive Subtraction Hybridisation to isolate a sample of the genes present in the type strain of each of nine species in the genus *Sinorhizobium* but absent from the genome of *S. meliloti* 1021, which has been completely sequenced. We also carried out a similar search for *Mesorhizobium* accessory genes using M. *huakuii* biovar *loti* MAFF303099 as the reference. Overall, these studies gave us sequences of parts of over 1,000 accessory genes. The DNA composition of these genes was strikingly lower in G+C than that of housekeeping genes in the corresponding species. In part, this is a consequence of the SSH technique, which tends to favour A+T-rich sequences, but it is also consistent with the compositional bias of accessory genes found in complete genome sequences.

For a subset of 355 of these SSH-derived sequences, matching 70-mer oligonucleotides were designed and synthesised. These were arrayed on glass slides, and these microarrays were hybridised with total genomic DNA of each of 22 strains of *Sinorhizobium* and *Mesorhizobium*, including the source strains and other representatives of their species. After appropriate normalisation, the hybridisation data indicated that each sequence was either present or absent in each of the strains, with relatively few ambiguities. A clustering of the strains based on the shared presence of accessory genes reflected, in most cases, the known phylogenetic relationships of the strains based on housekeeping gene sequences. However, there was a striking exception in the case of *S. arboris*. The phylogenetic position of this species is close to *S. meliloti* and *S. medicae* but its accessory genome resembles that of *S. kostiense*, with which it shares its host range (*Acacia* and *Prosopis*) and geographic origin (Sudan). This result is consistent with the idea that the core genome provides general housekeeping functions, whereas the accessory genome provides much of the ecological specialisation and may be shared by horizontal gene transfer among bacteria inhabiting a particular niche.

This work was supported by the NERC Environmental Genomics programme and by the EC FP5 Bacdivers project.

Reference

Young JPW et al. (2006) Genome Biology 7, R34.

EVOLUTION OF FUNCTIONAL DIVERSITY IN NITROGENASE HOMOLOGS

R. E. Blankenship[1], J. Raymond[2], C. Staples[3] and B. Mukhopadhyay[4]

[1]Departments of Biology and Chemistry, Washington University, St. Louis, MO 63130, USA; [2]Lawrence Livermore National Laboratory, 7000 East Avenue, Livermore, CA 94550, USA; [3]Department of Chemistry and Biochemistry, Arizona State University, Tempe, AZ 85287 USA; [4]Virginia Bioinformatics Institute and Departments of Biochemistry and Biology, Virginia Polytechnic Institute and State University, Blacksburg, VA 24061, USA

We have investigated the origin and distribution of nitrogen fixation using phylogenetic tools, making use of all publicly available complete bacterial and archaeal genome sequences to understand how the core components of nitrogenase, namely the NifH, NifD and NifK proteins, originated and have evolved. These genes are universal in nitrogen-fixing organisms and have remarkably congruent phylogenetic histories. Additional clues to the early origins of this system are available from two distinct clades of nitrogenase paralogs: (i) a group of Nif-like proteins coded for by genes involved in photosynthetic pigment biosynthesis (*bchLNB*, *bchXYZ*); and (ii) a group of previously uncharacterized *nif*-like genes present in methanogens and in some photosynthetic bacteria (*nflH*, *nflD*). We explore the complex genetic history of the nitrogenase family, which is replete with gene duplication, recruitment, fusion, and horizontal gene transfer and discuss these events with respect to where nitrogen fixation may have originated and how it came to have its current complex phylogenetic distribution.

Figure 1 shows an evolutionary tree of Nif-like proteins from analysis of 473 complete genomes. This tree is an updated version of the one published earlier (Raymond et al., 2004) and includes additional sequences from thermophilic taxa, including the recent hyperthermophilic Archaeon described by Mehta and Baross (2006). The tree shows five groups of nitrogenases and Nif-like (Nfl) proteins. The catalytic activity of the Group IV Nfl proteins is not known and may contain more than one type of enzyme.

We have also carried out preliminary physiological and biochemical analysis of the Nif-like Nfl proteins to determine their function and expression pattern. Biochemical evidence in *Methanocaldococcus jannaschii* indicates that the NflH and NflD gene products are constitutively expressed under a variety of growth conditions and physically

F. D. Dakora et al. (eds.), *Biological Nitrogen Fixation: Towards Poverty Alleviation through Sustainable Agriculture.*
© Springer Science + Business Media B.V. 2008

associate with each other. Figure 2 shows Western blots using antibodies raised against NflH and NflD. We suggest that these proteins may have a role in methanogen cofactor (F430) biosynthesis, similar to the role of the BchLNB and BchlXYX protein complexes in bacteriochlorophyll biosynthesis.

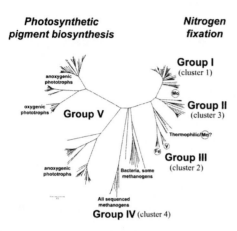 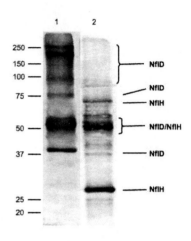

Figure 1. Neighbor-joining phylogenetic tree of sequences of NifD and related proteins. Groups I and II are Mo-dependent nitrogenases; Group III is alternative nitrogenases (V and Fe); Group IV are Nif-like proteins (Nfl); and Group V are light-independent (proto)chlorophyllide reductase enzymes involved in (bacterio)chlorophyll biosynthesis.

Figure 2. Western blots of proteins isolated from *M. jannaschii*. Anti-NflD and Anti-NflH Western blots of the elution buffer fraction of the crosslinked Anti-NflD IgG/protein-A agarose column. MW markers (kDa) shown at left. Lane 1: Anti-NflD (1:20,000 dilution) primary/Novagen goat anti-rabbit IgG-AP conjugate. Lane 2: Anti-NflH (1:1,000 dilution) primary/Novagen goat anti-rabbit IgG-AP conjugate.

References

Mehta MP and Baross JA (2006) Science 314, 1783–1786.
Raymond J et al. (2004) Molec. Biol. & Evol. 21, 541–554.

PART 3B

RHIZOSPHERE ASSOCIATIONS

DIVERSITY OF NITROGEN FIXING BACTERIAL COMMUNITY ASSESSED BY MOLECULAR AND MICROBIOLOGICAL TECHNIQUES

L. F. W. Roesch[1], R. A. Soares[2], G. Zanatta[2], F. A. de O. Camargo[1] and L. M. P. Passaglia[2]

[1]Department of Soil Science, Federal University of Rio Grande do Sul, 7712 Bento Gonçalves Avenue, 91540-000 Porto Alegre, Brazil; [2]Department of Genetics, Federal University of Rio Grande do Sul, 9500 Bento Gonçalves Avenue, 91540-000 Porto Alegre, Brazil

It is now accepted that the diversity of microorganisms in natural environments is higher than that detected by traditional methods (Ueda et al., 1995; Hugenholtz and Pace, 1996; Hugenholtz et al., 1998). Within the past decade, the number of identifiable bacterial divisions has been increased to more than 40 through culture-independent phylogenetic surveys of environmental microbial communities (Hugenholtz et al., 1998). Although these molecular strategies have opened a window into the world of non-culturable bacteria, they have the disadvantage that organisms whose genes have been isolated by one method cannot be studied for any other trait (Ueda et al., 1995).

In recent years, agricultural systems have changed to both improve environmental quality and avoid environmental degradation. Soil management and exploration of the biodiversity of cropping systems has been used to optimize high crop yields and improve the sustainability of the ecosystem. In this regard, endophytic diazotrophic bacterial associations with crops represent one of the most promising alternatives for plant growth promoters and management of soil and environment quality. In attempts to evaluate the diversity of nitrogen-fixing bacteria in various ecosystems, the *nifH* gene has been studied, mainly by culture-independent approaches. The results have provided a more complete picture of the diazotrophic community than culture-based approaches.

Molecular and microbiological techniques were used to evaluate the nitrogen-fixing bacterial community associated with oat (*Avena sativa*) in different soil management systems of Rio Grande do Sul State, Brazil. The conventional microbiological approach used semi-solid nitrogen-free and solid selective media to isolate some of these nitrogen-fixing bacteria. For the molecular approach, a PCR-RFLP strategy was used to amplify a segment of the *nifH* gene, which encodes nitrogenase reductase, from DNA samples extracted from rhizosphere soils, roots and leaves of oat (Soares et al., 2006).

F. D. Dakora et al. (eds.), *Biological Nitrogen Fixation: Towards Poverty Alleviation through Sustainable Agriculture.*
© Springer Science + Business Media B.V. 2008

The amplified *nifH* fragments were cleaved with *TaqI* and *HaeIII* endonucleases and the products analyzed in polyacrylamide gels. One sample was analyzed in more detail and both the rhizosphere-soil and leaf PCR fragments were cloned into pUC18, generating 55 *nifH*-positive clones, from which 19 different RFLP patterns were obtained. These patterns were compared with RFLP patterns generated from DNA obtained from pure strains or by theoretical digestion data. To identify and confirm the RFLP patterns obtained, at least one clone of each was sequenced and the nucleotide sequences were compared with the GenBank nucleotide database. As the partial *nifH* sequences analyzed had shown homologies with several known and unknown diazotrophs, we attempted to identify the diazotrophs through the positioning of the clones in the NifH partial sequence-derived tree. Through this molecular biological analysis, we were able to identify a reasonable occurrence of diazotrophs in the rhizosphere soil and leaf samples analyzed, 9 and 13 different genera, respectively. Of these, two genera, *Azospirillum* and *Herbaspirillum*, were particularly abundant. These two genera were also the most abundant when isolated by conventional microbiological techniques. Both approaches indicated that the nitrogen-fixing communities were different in soils under cultivation and permanent pasture from these under cultivation without live stock. Our results showed that whereas the molecular approach gives us a general view of the variability of nitrogen fixers in the community, the conventional methodology allows us to isolate those bacteria and to select some of them that have showed the highest nitrogen-fixation capacity and auxin production.

Using this PCR-RFLP approach, the maize endophytic diversity was also characterized at sites that vary by climate and soil in the Rio Grande do Sul State. Sequence analysis of *nifH* cluster I clone libraries were used to assess diversity in maize plants. Using these data, diversity indices were calculated to estimate the diazotroph diversity as well as estimates of the nucleotide diversity and the average sequence divergence ($\theta(\pi)$) to estimate gene diversity. To assess population differences, we performed the F_{ST} test. Much larger variation was detected between, rather than within, regions, particularly among communities from different soil types, water regimes, and geographical regions. The Shannon-Weaver index indicated a high difference in the diversity of species among communities. The communities located in the North region, with a higher rain distribution and higher clay content, tend to show higher diversity than those from the South region, with lower rain distribution and lower clay content. The Evenness index indicates that there is a dominant species within each of the communities analyzed. All clones grouped into *nifH* gene cluster I, a phylogenetic group containing sequences from the Mo-containing *nifH* and some *vnfH*. We recovered *nifH* sequences types from alpha, beta and gamma proteobacteria. These results demonstrate that there is a large diversity of nitrogen-fixing bacteria able to colonize the interior of maize and different environment conditions are correlated with *nifH* diversity.

References

Hugenholtz P and Pace NR (1996) Tib. Tech. 14, 190–197.
Hugenholtz P et al. (1998) J. Bacteriol. 180, 4765–4774.
Soares R et al. (2006) Appl. Soil Ecol. 30, 221–234.
Ueda T et al. (1995) J. Bacteriol. 177, 1414–1417.

QUORUM SENSING IN ROOT-ASSOCIATED AND ENDOPYTIC DIAZOTROPHS

M. Schmid[1], M. Rothballer[1], B. Hai[1], A. Fekete[2], X. Li[2], M. Englmann[2], M. Frommberger[2], P. Schmitt-Kopplin[2] and A. Hartmann[1]

[1]Department Microbe-Plant Interactions, GSF-National Research Center for Environment and Health, Neuherberg/Munich, Germany; [2]Institute of Ecological Chemistry, GSF-National Research Center for Environment and Health, D-85764 Neuherberg/Munich, Germany

Root-associated or endophytic diazotrophs, like *Azospirillum*, *Herbaspirillum* or *Burkholderia* spp., are of key interest in plant-growth promotion and biological nitrogen-fixation research. Recently, new species of the genera *Herbaspirillum* and *Burkholderia* were described (Reis et al., 2004; Rothballer et al., 2006). To analyze the diazotrophic or endophytic life style of bacteria, culture dependent methods, like isolation of bacteria from surface sterilized roots in semisolid N-free media, are not sufficient. Most important for the *in situ* characterization of the endophytic life style are microscopic techniques. For this purpose, species-specific phylogenetic fluorescently labeled oligo-nucleotide probes (Stoffels et al., 2001) during FISH analysis and the use of GFP-tagged strains in combination with confocal laser scanning microscopy (CLSM) were used. With orthogonal views of confocal image stacks prepared with digital-image ana-lysis tools, the endophytic colonization of plant cells can easily be followed (Rothballer et al., 2003). TEM-based investigations after staining with monospecific polyclonal antibodies were also successfully used (Rothballer et al., unpublished data, 2007).

These interactions with plants are of key importance for successful establishment of the bacteria in roots. Plants recognize and respond to bacterial signal molecules produced in the rhizosphere. Response to small signal molecules of the N-acylhomo-serine lactone (AHL) type, which regulate gene expression in many Gram-negative bacteria in a cell density dependent manner, is known as "quorum sensing". We could show the production of AHL-signal molecules on the root surface of tomato plants using bacterial in situ reporter constructs (Steidle et al., 2001). With the help of geo-statistical software tools the in situ "calling distance" of bacteria in the rhizosphere could be analyzed. Gradients of AHLs, which can reach out far beyond the immediate cell surrounding, exist in the rhizosphere (Gantner et al., 2006). In the case of tomato, the plant is responding to AHL-production with both pathogen defense (e.g., against

F. D. Dakora et al. (eds.), *Biological Nitrogen Fixation: Towards Poverty Alleviation through Sustainable Agriculture.*
© Springer Science + Business Media B.V. 2008

the leaf pathogenic fungus *Alternaria alternata*) and induced systemic resistance in the presence of AHL-producing *Serratia liquefaciens* in the rhizosphere (Schuhegger et al., 2006).

A detailed analysis of AHL production in bacteria is therefore key to understanding the basis of bacteria-plant interactions. Classical methods for AHL signal molecule isolation, e.g., from culture supernatants with dichlormethane and ethylacetate, and analysis are cross-streak assays with biosensors that respond to AHL, e.g., *gfp*- or *lux*-tagged sensor strains. Reversed Phase Liquid Chromatography is also a commonly used technique for AHL detection. However, both methods have their limitations. Firstly, different sensitivities of biosensor strains exist towards long- or short-chain and different substitutions in AHL molecules and the risk of failed detection of certain AHLs is high. Secondly, the TLC method suffers from its very poor capacity to separate long chain AHL molecules. Therefore, a highly sensitive and reliable chemical analysis strategy for a specific characterization of AHL compounds, using nano-LC/ESI-ion trap MS method, was developed.

We could determine the production of additional AHL signal molecules (C_{10}-HL, d-C_{10}-HL, OH-C_{10}-HL) by a diazotrophic *Burkholderia* strain originally isolated from rice plants in contrast to TLC analysis, which only indicated C6-HL and C8-HL (Frommberger et al., 2004). We further developed a high resolution technique using a 12 Tesla FT-ICR-MS. However, even such a highly sophisticated method like FT-ICR-MS showed limitations because molecules belonging to the diketopiperazines ($C_{10}H_{20}N_4O_2$) have exactly the same mass as C_8-HSL [(OHL), $C_{12}H_{22}NO_3$] up to the fourth decimal place and these molecules are present in standard LB or other complex media. Therefore, an UPLC-based analytic tool is now used in combination with FT-ICR-MS for a confident and reliable AHL analysis (Fekete et al., 2006).

The production of AHL by a number of plant-associated diazotrophic bacteria has been examined, using classical biosensor and TLC-based methods as well as our newly developed instrumental analytical tools. Although in some phylogenetic groups of bacteria, like *Burkholderia* spp., AHL production was frequently found, it was rather infrequent or absent in other genera. For example, only in *Azospirillum lipoferum* DSM 1691 could we find the production of C_6-HL, OH-C_8-HL, O-C_8-HL, C_8-HL and OH-C_{10}-HL. In *Herbaspirillum frisingense* DSM 13128, only OH-C14-HL was detectable.

References

Fekete A et al. (2006) Anal. Bioanal. Chem. 387, 455–467.
Frommberger M et al. (2004) Anal. Bioanal. Chem. 378, 1014–1020.
Gantner S et al. (2006) FEMS Microbiol. Ecol. 56, 188–194.
Reis V et al. (2004) Int. J. Syst. Evol. Microbiol. 54, 2155–2162.
Rothballer M et al. (2003) Symbiosis 34, 261–279.
Rothballer M et al. (2006) Int. J. Syst. Evol. Microbiol. 56, 1341–1348

Schuhegger R et al. (2006) Plant Cell Environ. 29, 909-918.
Steidle A et al. (2001) Appl. Environ. Microbiol. 67, 5761-5770.
Stoffels M et al. (2001) Syst. Appl. Microbiol. 24, 83-97.

CHARACTERIZATION OF *CHSA*, A NEW GENE CONTROLLING THE CHEMOTACTIC RESPONSE IN *AZOSPIRILLUM BRASILENSE* SP7

R. Carreño-Lopez[1], A. Sánchez[1], N. Camargo[1], C. Elmerich[2] and B. E. Baca[1]

[1]Centro de Investigaciones Microbiológicas, Universidad Autónoma de Puebla, 72 000, Puebla, México; [2]Département de Microbiologie, BMGE, Institut Pasteur, Paris, France

Nitrogen-fixing bacteria of the genus *Azospirillum* are able to colonize the plant rhizosphere (Elmerich and Newton, 2007). Effective colonization includes survival in the soil, motility and chemotactic response to root exudates, and mechanisms of attachment to the root system through flagella and capsular polysaccharides as well as phytohormone production. *Azospirillum* possesses a single polar flagellum and multiple lateral flagella. The polar flagellum is responsible for the swimming motility in liquid media, whereas the lateral flagella are responsible for spreading on solid or semi-solid surfaces. Chemotaxis towards root exudates and to a variety of oxidizable substrates (including sugars, amino acids, and organic acids) as well as to O_2 and redox molecules has been reported for *Azospirillum*. Only metabolizable substrates can serve as chemoattractants or repellents in *Azospirillum* (Alexandre et al., 2000). Knowledge of the chemotaxis machinery in *Azospirillum* is still limited. A cluster of 5 genes with high similarity to *cheA*, *-W*, *-Y*, *-B* and *-R* (encoding the central transduction pathway for chemotaxis) and a gene encoding a chemoreceptor, *tlp1*, have now been identified in *A. brasilense* (Hauwaerts et al., 2002; Greer-Philips et al., 2004; Stephens et al., 2006).

Identification of a New Gene Involved in the Chemotactic Response

The *A. brasilense* Sp7S strain was found to be impaired in flocculation and in motility when grown in semi-solid agar. The flocculation defect was due to a mutation in *flcA*, encoding a transcriptional regulator of the LuxR family (Pereg-Gerk et al., 1998). It was first hypothesized that the motility defect was due to the lack of lateral flagella (Pereg-Gerk et al., 2000), but we observed that the plasmid pAB7115 that restored motility to Sp7S did not carry the corresponding structural gene *laf1*.

By Tn5 insertion in pAB7115, it was possible to localize the DNA region responsible for the complementation of the motility defect of strain Sp7S. Nucleotide sequencing

F. D. Dakora et al. (eds.), *Biological Nitrogen Fixation: Towards Poverty Alleviation through Sustainable Agriculture.*
© Springer Science + Business Media B.V. 2008

317

led to the identification of a new gene, *chsA*. The deduced translation product of 64 kDa, designated ChsA, shows characteristics typical of cytoplasmic signalling proteins. It contains a PAS sensory domain and an EAL domain. The EAL domain was described in proteins involved in the hydrolysis of cyclic-di-GMP, a compound known as a secondary messenger in a broad spectrum of cellular processes, including motility (D'Argenio and Miller, 2004; Römling et al., 2005; Jenal and Malone, 2006).

A *chsA*-Tn5 insertion mutant derivative of the wild type strain Sp7 was constructed. The resulting strain Sp74031 had a defect in motility. Flagella purified from the wild-type Sp7 grown in liquid and semi-solid cultures were used to raise specific antibodies. Further, SDS PAGE and Western blot analysis showed that Sp7S714031 has polar as well as lateral flagella. The mutant strain displayed a decrease in the chemotactic response compared to the wild-type Sp7, as well as reduced colonization of rice. Chemotactic ability was restored in strain Sp74031 by introducing the wild-type *chsA* gene.

Discussion

Data reported here suggest that ChsA plays a role in the chemotactic response and that it is a component of the signalling pathway that governs chemotaxis in *Azospirillum*. Indeed, inactivation of *chsA* only partially reduced chemotaxis, but this was also the case for *cheB* and *cheR* mutants, suggesting multiple chemotaxis systems in *Azospirillum* (Stephens et al., 2006). The actual role of the ChsA protein in the signalling pathway remains unknown, including the respective functions of the EAL and PAS domains as well as the identity of a small molecule that could act as the cytoplasmic sensory signal.

This work was partially supported by a grant of VIEP-SEP. We thank Ma. Luisa Xiqui for technical assistance.

References

Alexandre G et al. (2000) J. Bacteriol. 182, 56042–6048.
D'Argenio DA and Miller SI (2004) Microbiology 150, 2497–2502.
Elmerich C and Newton WE (2007) Associative and Endophytic Nitrogen-fixing Bacteria and Cyanobacterial Associations, Springer, Dordrecht, The Netherlands.
Greer-Philips SE et al. (2004) J. Bacteriol. 186, 6595–6604.
Hauwaerts D et al. (2002) FEMS Microbiol. Lett. 208, 61–67.
Jenal U and Malone J (2006) Annu. Rev. Genet. 40, 285–407.
Pereg-Gerk L et al. (1998) Mol. Plant-Microbe Interact. 11, 177–187.
Pereg-Gerk L et al. (2000) Appl. Environ. Microbiol. 66, 2175–2184.
Römling U et al. (2005) Mol. Microbiol. 57, 629–639.
Stephens BB et al. (2006) J. Bacteriol. 188, 4759–4768.

EXOPOLYSACCHARIDE PRODUCTION AND CELL AGGREGATION IN *AZOSPIRILLUM BRASILENSE*

A. Valverde, S. Castro-Sowinski, A. Lerner, S. Fibach, O. Matan, S. Burdman and Y. Okon

The Hebrew University of Jerusalem, Faculty of Agricultural, Food and Environmental Quality Sciences, Rehovot Campus, 76100 Israel (Email: okon@agri.huji.ac.il)

Azospirillum brasilense is a free-living, nitrogen-fixing, plant growth-promoting rhizobacterium that belongs to the alpha-proteobacteria. It closely associates with a large number of plants of agronomic importance, including grain and forage legumes, and cereal and forage grasses (Okon, 1994). *A. brasilense* strains are highly pleiomorphic and versatile in their metabolic activities in response to environmental changes. Under unfavourable conditions, such as high oxygen partial pressure (Nur et al., 1981), desiccation and nutrient limitation, azospirilla can convert into ovoid, non-motile, encapsulated cyst-like forms (Sadasivan and Neyra, 1985). Similar changes occur when cells are grown in liquid minimal media supplemented with certain carbon sources, such as fructose or β-hydroxybutyrate (Bleakley et al., 1988). Under these conditions, cells aggregate and flocculate in a matrix of polysaccharide material, forming large macroscopic clumps. Cell aggregation is a phenomenon of great interest for the production, storage, survival and adsorption to roots of bacterial inoculants for agricultural application.

Cell aggregation in different *A. brasilense* strains depends on both the major outer membrane protein and the concentration and composition of the exopolysaccharide (EPS) (Burdman et al., 1998, 2000). Bahat-Samet et al. (2004) showed that the monomer composition of the *A. brasilense* (EPS) varies during the growth phase. The wild-type strain Sp7 produces a glucose-rich EPS during the exponential growth phase and an arabinose-rich EPS during the stationary growth phase. Furthermore, cell aggregation is likely dependent on an arabinose-rich EPS (Burdman et al., 2000; Bahat-Samet et al., 2004). A mutant strain defective in lipopolysaccharide (LPS) production (probably in dTDP 4-rhamnose reductase) produced a glucose-rich EPS and showed no aggregation (Jofre et al., 2004; Bahat-Samet et al., 2004).

F. D. Dakora et al. (eds.), *Biological Nitrogen Fixation: Towards Poverty Alleviation through Sustainable Agriculture.*
© Springer Science + Business Media B.V. 2008

The response regulatory gene *flcA* controls the differentiation process of *A. brasilense* from the vegetative state to cyst-like forms, both in culture and in association with plants. In contrast to the wild type strain Sp7, strain Sp72002, a Tn5-induced *flcA* mutant, does not aggregate, does not differentiate from motile, vibroid cells into non-motile cyst-like forms, and lacks most of the EPS material on its cell surface under all tested conditions. The mutant strain also colonizes the root surface to a lesser extent than the wild-type strain (Pereg-Gerk et al., 1998). Using a cDNA-AFLP approach, Valverde et al. (2006) were able to identify differentially expressed genes, whose expression could be modulated by *flcA*. Eighty-one transcript-derived fragments (TDFs) were detected, showing differential expression between strains Sp7 and Sp72002 during the exponential growth phase in an aggregation-inducing medium. The fragments were sequenced and analyzed. Among them were genes homologous to: *nodQ*, involved in sulfation; *narK*, involved in nitrite/nitrate transport; *flp*, involved in autoaggregation; and genes encoding a biopolymer-transport protein and the signal-recognition particle (SRP). This study demonstrates the usefulness of the cDNA-AFLP approach to reveal genes, which are differentially expressed during aggregation in *A. brasilense*, and provides new insights into the aggregation process of this bacterium.

Finally, the above-mentioned genes as well as others involved in EPS and LPS biosynthesis (Van Bleu et al., 2004), for example, *noeJ* (coding for mannose isomerase), *noeL* (for GDP-mannose 4,6 dehydratase), and *glgP* (for glycogen phosphorylase), are being characterized by mutagenesis and phenotypic analysis. The latter includes glycosyl composition and the structure of EPS, cell aggregation, survival under environmental stresses, cell morphology (by SEM and TEM), adsorption of cells to roots and root colonization, plant-growth promotion, nitrogen fixation, carotenoid pigments, and polyhydroxyalkanoate (PHA) production.

References

Bahat-Samet E et al. (2004) FEMS Microbiol. Lett. 237, 195–203.
Bleakley BH et al. (1988) Appl. Environ. Microbiol. 54, 2986–2995.
Burdman S et al. (1998) Microbiology 144, 1989–1999.
Burdman S et al. (2000) FEMS Microbiol. Lett. 189, 259–264.
Jofre E et al. (2004) FEMS Microbiol. Lett. 231, 267–275.
Nur I et al. (1981) J. Gen. Microbiol. 122, 27–32.
Okon Y (1994) Azospirillum/Plant Associations. CRC Press, Boca Raton, FL.
Pereg-Gerk L et al. (1998) Mol. Plant-Microbe Interact. 11, 177–187.
Sadasivan L and Neyra CA (1985) J. Bacteriol. 163, 716–723.
Valverde A et al. (2006). FEMS Microbiol. Lett. 265, 186–194.
Van Bleu E et al. (2004). FEMS Microbiol. Lett. 232, 165–172.

GLUCONACETOBACTER DIAZOTROPHICUS PAL5 STRAIN: SELECTION AND CHARACTERIZATION OF MUTANTS DEFICIENT IN NITROGEN-FIXATION ABILITY

H. V. Guedes, L. F. M. Rouws, A. L. M. Oliveira, J. L. Simões-Araújo, K. R. S. Teixeira and J. I. Baldani

Embrapa Agrobiologia, BR 465, km 07, 23851-970 – Seropédica, Rio de Janeiro, Brazil

Gluconacetobacter diazotrophicus species has been considered one of the most important endophytic nitrogen-fixing bacteria associated with sugar cane plants (Baldani and Baldani, 2005). Although many physiological studies have already been carried out on *G. diazotrophicus* (Reis et al., 2002), there are still many important aspects of the bacteria that need clarification. The complete genome sequence of this bacterium was recently finished and the annotation process is partially completed (P. Ferreira, personal communication, 2006). The knowledge of the complete genome opens new perspective for understanding the function of various *Gluconacetobacter diazotrophicus* genes when in association with sugarcane plants.

In this work, random mutants were constructed (Rouws et al., 2006), using a commercial transposon kit, and selected in semi-solid LGI medium using the ampicillin-enrichment method (Fitzgerald and Williams, 1975). The mutants were characterized with regard to nitrogenase activity in comparison to the wild type strain PAL5. In addition, the production of indolic acid substances (IAA) was also measured. The position of the transposon insertion into the genome of the mutants was determined by inverse PCR methodology (Rouws et al., 2006), followed by sequencing the PCR products, and comparison with the available database of *G. diazotrophicus*.

Sixteen mutants were selected of which eight showed a similar nitrogenase activity to the PAL5 strain. The other seven had rates that were much lower and differed statistically (p 0.005) from that of the wild-type strain. However, one mutant showed nitrogenase activity 2.5-times higher than the control strain. The measurement of the IAA content in those mutants with the lowest nitrogenase activities showed four mutants with IAA production similar to the PAL5 (29.3 μM/mL protein), two mutants with double the concentration, and one mutant with very high IAA production (173.7 μM/mL protein).

F. D. Dakora et al. (eds.), *Biological Nitrogen Fixation: Towards Poverty Alleviation through Sustainable Agriculture.*
© Springer Science + Business Media B.V. 2008

321

The sequencing within the transposon-insertion region of the four low nitrogenase-activity mutants (#09, 20, 29 and 30) showed that no common nitrogen-fixation (*nif*) genes were mutated. The mutated sequence of mutant #20 showed similarity to a putative oxidoreductase of *Gluconobacter oxydans* and also to dehydrogenases with different specificities (related to short-chain alcohol dehydrogenases) in *Nostoc puncti-forme*. The short-chain alcohol dehydrogenase (SCAD) superfamily is a phylogene-tically related group of enzymes that act on substrates as diverse as compounds involved in nitrogen metabolism (Krozowski, 1994).

In mutant #30, the insertion was within a putative membrane-protein region. Trans-port systems allow the uptake of essential nutrients and ions, excretion of end products of metabolism and deleterious substances, and also communication between cells and the environment.

In contrast, the insertion in mutant #29 occurred within the region of the *G. diazotrophicus* genome homologous to a putative transduction histidine kinase that likely work as a sensor. A region that codes for a transcriptional regulator present in *Gluconobacter oxydans* was also found upstream of this putative histidine kinase. Genes coding for chemotaxis, motility, and quorum sensing exist in *G. diazotrophicus*.

In mutant #9, the insertion occurred within the region related to the hydrolase nudix family of genes. According to Bessman et al. (1996), genes coding for the nudix hyro-lases can be considered "housecleaning" genes, whose function is to cleanse the cells of potentially deleterious endogenous metabolites.

So far, the results did not provide a clear picture about the functions of the mutated genes. Additional studies, including site-directed mutagenesis, will be carried out to determine their functions in *G. diazotrophicus*, especially those involving the nitrogen-fixation process. The response of micropropagated sugar cane plants to inoculation with the mutants should provide a better understanding of their role in this unique endophytic association.

Supported by CNPq (Project Genoma Functional, n° 506355/04-7); CNPq/FAPERJ (Project Pronex , n° E-26/171.208/2003). J. I. Baldani also has a Research Fellowship from CNPq – 501864/2004-0.

References

Baldani JI and Baldani VLD (2005) Ann. Acad. Bras. Ciênc. 77, 549–579.
Bessman MJ et al. (1996) J. Biol. Chem. 271, 25059–25062.
Fitzgerald G and Williams LS (1975) J. Bacteriol. 122, 345–346.
Krozowski ZS (1994) J. Steroid Biochem. Mol. Biol. 51, 125–130.
Reis VM et al. (2002) Crit. Rev. Plant Sci. 19, 227–247.
Rouws LFM et al. (2006) In: Proc. 7th European nitrogen fixation conference, Aarhus, Denmark.

PART 3C

ENZYMOLOGY AND GENETICS

THE ROLE OF NIF PROTEINS IN NITROGENASE MATURATION

L. M. Rubio, J. A. Hernández, B. Soboh, D. Zhao, R. Y. Igarashi, L. Curatti
and P. W. Ludden

Department of Plant and Microbial Biology, University of
California-Berkeley, Berkeley, CA 94720, USA

The majority of the N_2 fixed biologically is catalyzed by the molybdenum nitrogenase,
which is composed of two O_2-labile metalloproteins; dinitrogenase (also referred to as
MoFe protein) and dinitrogenase reductase (also referred to as Fe protein). Dinitrogenase is a 230-kDa $\alpha_2\beta_2$ tetramer of the *nifD* and *nifK* gene products (Kim and Rees,
1992). Each $\alpha\beta$ dinitrogenase dimer contains an iron-molybdenum cofactor (FeMo-co)
and a P-cluster. The FeMo-co, located at the active site of dinitrogenase, is one of the
most complex metalloclusters known in biology, and is composed of 7 Fe, 9 S, 1 Mo, 1
homocitrate, and 1 light atom proposed to be either N, O, or C (Einsle et al., 2002; see
Figure 1). Dinitrogenase reductase is a dimer of the *nifH* gene product that contains a
single [4Fe-4S] center coordinated between the two subunits.

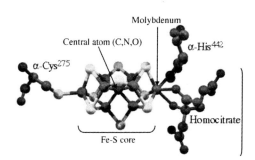

The structural genes for dinitrogenase,
nifD and *nifK*, appear not to be required
for the biosynthesis of FeMo-co. FeMo-
co is assembled elsewhere in the cells
and then incorporated into a FeMo-co
deficient apo-NifDK, indicating that
FeMo-co biosynthesis and apo-NifDK
biosynthesis occur independently. These
two separate processes will be over-
viewed in this chapter.

Figure 1. FeMo-co structure.

The maturation of apo-NifDK involves a series of nitrogen fixation (*nif*) and non-*nif*
gene products. In a strain lacking NifH, immature apo-NifDK carries two pairs of [4Fe-
4S] clusters that will serve as substrates for the formation of the P-clusters. It is believed
that the cysteine desulfurase, NifS, and the [Fe-S] cluster scaffold protein, NifU, are
involved in the synthesis of the [4Fe-4S] clusters of apo-NifDK, but direct experimental

F. D. Dakora et al. (eds.), *Biological Nitrogen Fixation: Towards Poverty Alleviation
through Sustainable Agriculture.*
© Springer Science + Business Media B.V. 2008

evidence is lacking. P-cluster synthesis occurs "*in situ*" within the NifDK polypeptides, and is dependent on NifH activity. The exact mechanism by which NifH helps catalyze the formation of the P-clusters is unknown. The formation of the P-clusters correlates with the ability of apo-NifDK to accept and insert FeMo-co (Figure 2). The non-*nif* chaperone NafY stabilizes apo-NifDK in a conformation that is fully competent for FeMo-co insertion.

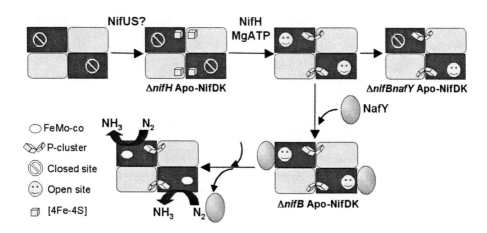

Figure 2. Model for maturation of apo-NifDK (Adapted from Rubio and Ludden, 2005).

A number of *nif* genes are required for the biosynthesis of FeMo-co and the maturation of the nitrogenase component proteins (Dos Santos et al., 2004). The proteins involved in FeMo-co biosynthesis can be functionally classified in three classes; scaffold proteins (NifU, NifB, and NifEN), metallocluster carrier proteins (NifX and NafY), and enzymes (NifS and NifV). The exact roles of NifQ and NifH remain unknown. During the last several years, an intensive effort has been made to purify all the proteins involved in FeMo-co synthesis and insertion. This effort is starting to provide the first insights into the structures of the FeMo-co biosynthetic intermediates and into the biochemical details of FeMo-co synthesis.

A schematic summary of FeMo-co biosynthesis is presented as Figure 3. First, NifB-co, an [Fe-S] cluster of unidentified structure that serves as FeMo-co precursor is assembled on NifB. It is likely that NifU and NifS are involved in the mobilization of Fe and S for the biosynthesis of NifB-co, because NifU and NifS mutants fail to accumulate NifB-co. NifB is a SAM radical enzyme that synthesizes NifB-co in a reaction that requires Fe, S, reductant, and S-adenosyl methionine (SAM) (Curatti et al., 2006). Second, NifX binds NifB-co and transfers it to NifEN, a scaffold protein upon which some steps of FeMo-co synthesis occur. Third, within NifEN, NifB-co is transformed into another [Fe-S] cluster, referred to as the VK-cluster, which would be the next intermediate in the biosynthetic pathway (Hernandez et al., 2007). Similar to NifB-co, the VK-cluster contains neither molybdenum nor homocitrate. However, the VK-cluster exhibits EPR signals in both the dithionite-reduced and the thionine-oxidized states and is, thus, electronically different from the EPR-silent NifB-co. Interestingly, NifX is able to extract and bind the VK-cluster from NifEN. The transfer of NifB-co and the VK-cluster between NifX and NifEN occurs as an equilibrium as would be expected for interacting proteins with similar affinities for the same substrates. It has been proposed that NifX would be involved in the storage of FeMo-co precursors to buffer the flux of FeMo-co precursors through NifEN (Hernandez et al., 2007).

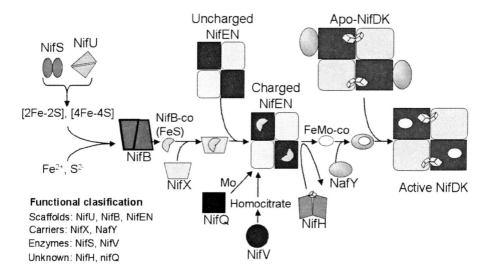

Figure 3. Model for the biosynthesis of FeMo-co.

In addition to the VK-cluster, the NifEN scaffold protein is able to bind molybdenum (Soboh et al., 2006). Presumably, Mo has been previously processed by the activity of NifQ. The accumulation of molybdenum within NifEN is clearly observed in a mutant strain lacking NifH, in which FeMo-co biosynthesis cannot proceed beyond the reactions catalyzed by NifEN. The molybdenum accumulated within NifEN is available to support *in vitro* FeMo-co synthesis and, thus, is relevant to the *in vivo* pathway for the biosynthesis of FeMo-co. The presence of Mo in NifEN purified from a Δ*nifHDK* strain implies that NifH is not essential for the accumulation of Mo within the NifEN complex. Because the addition of molybdenum into the FeMo-co precursor is a NifH-dependent reaction, it is possible that the role of NifH is to catalyze the transfer of Mo from one binding site on NifEN into the VK-cluster-derived FeMo-co precursor. Finally, homocitrate, the metabolic product of NifV, is incorporated into the precursor to generate FeMo-co. The site of homocitrate incorporation into the cofactor is a matter of debate with different reports pointing towards NifX, NifH, or NifEN as putative entry points for homocitrate.

Synthesized FeMo-co then binds to NafY, the product of a non-*nif* gene that also stabilizes the FeMo-co-deficient apo-NifDK protein (see above). Although not essential, NafY is believed to aid in FeMo-co insertion into apo-NifDK converting it into a catalytically active nitrogenase. NafY represents a family of nitrogenase cofactor-binding proteins, that also includes NifX and NifB proteins (Rubio et al., 2002). NifX, the smallest protein in the family, is formed by a single domain involved in binding FeMo-co precursors. NafY contains an N-terminal domain that is involved in apo-NifDK binding and a C-terminal NifX-like domain that binds FeMo-co. The NifX-like domain of NafY consists of five-stranded β-sheets flanked by five α-helices that belongs to the ribonuclease H superfamily and, thus, represents a FeMo-co binding fold that is different from that exhibited by NifDK (Dyer et al., 2003).

Supported by NIH grant 35332 to PWL.

References

Curatti L et al. (2006) Proc. Nat. Acad. Sci. USA 103, 5297–5301.
Dos Santos PC et al. (2004) Chem. Rev. 104, 1159–1174.
Dyer D et al. (2003) J. Biol. Chem. 278, 32150–32156.
Einsle O et al. (2002) Science 297, 1696–1700.
Hernandez JA et al. (2007) Mol. Microbiol. 63, 177–192.
Kim J and Rees DC (1992) Nature 360, 553–560.
Rubio LM and Ludden PW (2005) J. Bacteriol. 187, 405–414.
Rubio LM et al. (2002) J. Biol. Chem. 277, 14299–14305.
Soboh B et al. (2006) J. Biol. Chem. 281, 36701–36709.

NIFB-DEPENDENT "*IN VITRO*" SYNTHESIS OF THE IRON-MOLYBDENUM COFACTOR OF NITROGENASE

L. Curatti, P. W. Ludden and L. M. Rubio

Department of Plant and Microbial Biology, 211 Koshland Hall, University of California-Berkeley, Berkeley, CA 94720, USA

The major part of biological nitrogen fixation is catalyzed by the molybdenum nitrogenase, which carries at its active site the most complex metallocluster known in biology, the iron-molybdenum cofactor (FeMo-co). It is composed of 7 Fe, 9 S, 1 Mo, 1 homocitrate, and 1 unidentified central atom. The genetic and biochemical analysis of nitrogen-fixation (nif) genes and proteins in *Azotobacter vinelandii* and *Klebsiella pneumoniae* during the pre-genomic era led to the identification of at least 11 genes (*nifUS-BQ-ENX-V-H-Y* and *nafY*) proposed to be involved in the biosynthesis and insertion of FeMo-co into apo-dinitrogenase (Rubio and Ludden, 2005; Dos Santos et al., 2004). The importance of the product of the *nifB* gene in this biosynthetic pathway has long been recognized because its metabolic product (NifB-co) is an intermediate of the FeMo-co biosynthetic pathway, and also of the biosynthetic pathways of the cofactors of the alternative nitrogenases. NifB-co is an iron-sulfur cluster of unknown structure that accumulates in cells of mutant strains lacking *nifE*, *nifN* or *nifH* (Shah et al., 1994). The inability to isolate the NifB protein has impaired progress in the understanding of the biochemical aspects of the NifB-catalyzed reaction for more than two decades. Here, we describe the purification and characterization of the *A. vinelandii* NifB protein and the development of a NifB-activity assay. The reaction catalyzed by NifB requires *S*-adenosylmethionine (SAM) and allows, for the first time, the synthesis of FeMo-co from Fe, S, Mo and homocitrate.

To isolate the *A. vinelandii* NifB protein, a mutant strain was created (UW232) in which the *nifB* gene has been shuffled in the *A. vinelandii* chromosome from its original position to form a chimeric operon together with *nifHDK* (encoding the structural genes for nitrogenase). Also, a deca-histidine oligopeptide has been engineered at the N-terminal end of NifB to facilitate its purification. Both the *nifBHDK* chimeric operon and the *his::nifB* allele were functional "*in vivo*". In the strain UW232, the accumulation of the NifB protein was about five-fold higher than that of the wild type strain and the expression of the *his::nifB* gene was normally regulated by ammonium.

F. D. Dakora et al. (eds.), *Biological Nitrogen Fixation: Towards Poverty Alleviation through Sustainable Agriculture.*
© Springer Science + Business Media B.V. 2008

The *A. vinelandii* NifB protein has been purified 2,400-fold from derepressed UW232 cells under rigorous anaerobic conditions, using Co^{2+}-affinity, followed by ionic-exchange, chromatography. Gel filtration analysis show that NifB is a dimeric protein of 108 ± 4 kDa. "*As isolated*" NifB contains 11.8 ± 0.2 iron atoms per dimer and has a UV-visible spectrum with a broad peak around 400 nm and a shoulder at 315 nm, which is characteristic of iron-sulfur proteins. The "*as isolated*" form of the enzyme could be converted into a species containing 18.0 ± 0.4 Fe atoms per dimer, with a corresponding increment in its absorbance at 315 and 400 nm, by chemical reconstitution with Fe^{2+} and S^{2-} under reducing conditions. Interestingly, the addition of SAM to either the "*as isolated*" or the "chemically reconstituted" protein altered the UV-visible spectra, suggesting an interaction between SAM and the [Fe-S] clusters in NifB.

The NifB protein was able to activate the apo-dinitrogenase present in an extract of a NifB-deficient strain (UW45) in a reaction that also required Fe^{2+}, S^{2-}, *S*-adenosylmethionine (SAM), molybdate, homocitrate, ATP and reducing equivalents. The demonstration of the NifB-dependent "*in vitro*" incorporation of ^{55}Fe from ^{55}FeCl$_3$ into apo-dinitrogenase further confirmed that the biosynthetic pathway for FeMo-co, starting from Fe, S, molybdenum and homocitrate, had been successfully reconstituted "*in vitro*". Similar radioactive reactions in which NifB-co substituted for NifB, Fe, S and SAM, showed a dramatic decrease in the amount of ^{55}Fe incorporated into dinitrogenase, indicating that NifB-co (the metabolic product of NifB) contributes most of the Fe atoms present in FeMo-co. The NifB-dependent "*in vitro*" synthesis of FeMo-co was inhibited by the SAM analog S-adenosylhomocysteine, when additional SAM was omitted from the assay, confirming that SAM is involved in the NifB-catalyzed reaction (Curatti et al., 2006).

The primary sequence of *nifB* indicates it is a member of the emerging family of SAM-radical enzymes (Sofia et al., 2001). Proteins from this family display a diversity of biochemical reactions while containing the conserved sequence motif CXXXCXXC. This motif coordinates three Fe atoms of an [4Fe-4S], whereas the fourth Fe atom is coordinated by a SAM molecule. The reduced form of the [4Fe-4S] cluster cleaves the SAM molecule and generates the 5'-deoxyadenosyl radical, which abstracts a hydrogen atom from a substrate during catalysis (Sofia et al., 2001). The requirement of SAM of the NifB-catalyzed reaction strongly suggests that radical chemistry is used early in the FeMo-co biosynthetic pathway.

References

Curatti L et al. (2006) Proc. Natl. Acad. Sci. 103, 5297–5301.
Dos Santos PC et al. (2004) Chem. Rev. 104, 1159–1174.
Hernandez JA et al. (2007) Mol. Microbiol. 63, 177–191.
Rubio LM and Ludden PW (2005) J. Bacteriol. 187, 405–414.
Shah et al. (1994) J. Biol. Chem. 269, 1154–1158.
Sofia HJ et al. (2001) Nucleic Acids Res. 29, 1079–1106.

AZOTOBACTER VINELANDII NITROGENASE MoFe PROTEIN: PRE-STEADY STATE SPECTROSCOPIC STUDIES OF THE METAL COFACTORS

W. E. Newton[1], K. Fisher[1,2], B. H. Huynh[3], D. E. Edmondson[4], P. Tavares[5], A. S. Pereira[5] and D. J. Lowe[2]

[1]Department of Biochemistry, The Virginia Polytechnic Institute and State University, Blacksburg,VA 24061, USA; [2]Department of Biological Chemistry, John Innes Centre, Norwich NR4 7UH, UK;[3] Department of Physics Emory University, Atlanta, GA 30322, USA; [4]Department of Biochemistry, Emory University, Atlanta, GA 30322, USA; [5]Requimte, Centro de Química Fina e Biotecnologia, Faculdade de Ciências e Tecnologia, Universidade Nova de Lisboa, 2829-516 Caparica, Portugal

Rapid freezing of wild-type *A. vinelandii* nitrogenase at a 3:1 molar ratio of Fe protein to MoFe protein, 23°C, and pH 7.4, elicited two transient electron paramagnetic resonance (EPR) signals (signals 1b and 1c) from the FeMo-cofactor within the MoFe protein ca. 500 ms after mixing. The first signal (1b; g = 4.21, 3.76) formed at the expense of the starting S = 3/2 EPR signal (1a; g = 4.32, 3.66), whereas the second signal (1c; g = 4.7, 3.4) formed more slowly and in lower intensity. Both signals occurred independent of substrate (N_2, C_2H_2, H^+). By either increasing the molar ratio of the component proteins or pre-reducing the MoFe protein, both signal 1b and 1c formed more quickly, indicating that they arose from more-reduced MoFe-protein states. The loss of signal-1a intensity was never fully compensated by those of signals 1b plus 1c, suggesting the presence of other EPR-silent species. Simulations of the kinetics of signal-1b formation indicated that it arose from a 3-electron reduced state (E_3) of the MoFe protein and so signal 1c likely arose from an even more reduced state (Fisher et al., 2001). Furthermore, signal 1b closely resembled the EPR signal reported for *Klebsiella pneumoniae* MoFe protein at high pH (Smith et al., 1973), which presumably reflects deprotonation and a changed conformation at the FeMo-cofactor.

To gain insight into whether this putative deprotonation and conformational change is influenced by the FeMo-cofactor's local environment, the variant αH195Q, αH195N, and αQ191K MoFe proteins (where His at position 195 in the α subunit was replaced by Gln/Asn or the Gln at position α-191 was replaced by Lys) were studied similarly. Both the αH195Q and αH195N MoFe proteins developed the 1b and 1c EPR signals, but the αQ191K MoFe protein exhibited neither signal. Why would this be? All three

F. D. Dakora et al. (eds.), *Biological Nitrogen Fixation: Towards Poverty Alleviation through Sustainable Agriculture.*
© Springer Science + Business Media B.V. 2008

variant MoFe proteins can reduce H^+ and C_2H_2, but only the αH195Q MoFe protein catalyzes N_2 reduction (at about 1% of wild-type rate). However, the αH195N MoFe protein can bind N_2 but not catalyze its reduction, whereas the αQ191K MoFe protein cannot even bind N_2 (Fisher et al., 2000). Therefore, the observation of these signals for both the αH195Q and αH195N MoFe proteins, which indicates attainment of the E_3 redox level, correlates with their ability to bind N_2 (as predicted by Lowe and Thorneley, 1984). Further, the more rapid production and accumulation of these EPR signals compared to wild type suggests that variant MoFe-protein species at the E_3 redox level accumulate because they are either very slow (or unable) to go on to the E_4 redox level, where N_2 is irreversibly committed to reduction to NH_3.

The shortest distance between the P cluster and the FeMo-cofactor is about 0.14 Å, indicating possible direct electron transfer between them. Previous stopped-flow spectrophotometry (at 8:1 molar ratio of proteins, 23°C, and pH 7.4) showed putative P-cluster oxidation during nitrogenase turnover (Lowe et al., 1993). Using the αH195Q and αH195N variants, P-cluster oxidation occurred similarly and did so within the same time course as for EPR signal 1b's appearance. These results suggest that P-cluster oxidation is involved in producing the E_3 redox level required for N_2 binding and are supported by the lack of P-cluster oxidation with the αQ191K MoFe protein.

To investigate this possibility further, rapid-freeze ^{57}Fe Mössbauer experiments (at a 3:1 protein molar ratio) were performed. The wild type showed no change in its P-cluster spectrum. However, as found for the generation of the 1b and 1c EPR signals, a higher component-protein molar ratio is likely needed for P-cluster oxidation with wild type. These results suggest that P-cluster oxidation only occurs during N_2 binding/reduction (Fisher et al., 2007). In contrast, the variant β-Y98H [^{57}Fe]-MoFe protein, which is turnover (but not structurally) compromised (Peters et al., 1995) and has the β-Tyr-98 residue substituted by histidine on one of four possible electron-transfer-pathway helices between the P cluster and the FeMo-cofactor, does not require a high component-protein molar ratio to show P-cluster oxidation (ca. 35%); it occurs at a 3:1 Fe protein:MoFe protein ratio. As was found with stopped-flow spectrophotometry (at a 4:1 protein molar ratio; Peters et al., 1995), this initial P-cluster oxidation is followed by re-reduction, which is consistent with the hindered internal electron-transfer properties of the β-Y98H MoFe protein. Thus, compromised (structurally or functionally) P clusters may oxidize differently to wild type and so alter substrate-reduction properties.

Supported by NIH DK37255 (WEN) and GM47295 (BHH) and the BBSRC (DJL).

References

Fisher K et al. (2000) Biochemistry 39, 15570–15577.
Fisher K et al. (2001) Biochemistry 40, 3333–3339.
Fisher K et al. (2007) J. Inorg. Biochem. 101, 1649–1656.
Lowe DJ and Thorneley RNF (1984) Biochem. J. 224, 877–909.

Lowe DJ et al. (1993) Biochem. J. 292, 93–98.
Peters JW et al. (1995) J. Biol. Chem. 270, 27007–27013.
Smith BE et al. (1973) Biochem. J. 135, 331–341.

THE ROLE OF IRON PROTEIN-NUCLEOTIDE INTERACTION AND CONFORMATIONAL CHANGE IN NITROGENASE CATALYSIS

J. W. Peters[1], R. Sarma[1], D. W. Mulder[1], H. Tsuruta[2], R. K. Szilagyi[1] and L. C. Seefeldt[3]

[1]Department of Chemistry and Biochemistry and the Thermal Biology Institutes, Montana State University, Bozeman, MT 59717, USA; [2]Stanford Synchrotron Radiation Laboratory MS 69, 2575 Sand Hill Road, Menlo Park, CA 94025, USA; [3]Department of Chemistry and Biochemistry, Utah State University, Logan, UT 84322, USA

There are a number of key questions in nitrogenase biochemistry pertaining to the biosynthesis of the metal clusters, the mechanism of reduction at the active site, and the role of MgATP in catalysis (Peters and Szilagyi, 2006). We have, for several years, been focusing on the last question, using a variety of biochemical and physical methods to probe Fe-protein structure in defined conformations. These include Fe proteins with or without bound nucleotides and variants resulting from site-specific amino-acid substitution. During catalysis, the Fe protein binds two MgATP molecules and couples their hydrolysis to the transfer of electrons from the Fe protein to the MoFe protein within a complex (Peters and Szilagyi, 2006; Howard and Rees, 1994). Multiple steps of complex formation, MgATP hydrolysis, and electron transfer are required to accumulate enough reducing equivalents to reduce N_2 to ammonia. MgATP-dependent conformational changes in the Fe protein represent a potential mechanism to effectively gate the flow of electrons toward substrate reduction. The structure of the Fe protein confirmed it as an approx. 60 kD homodimer with a single [4Fe-4S] cluster located at the subunit interface. The Fe protein is structurally similar to a large number of proteins in which nucleotide binding and hydrolysis are coupled to conformational changes typically transduced within a macromolecular complex (Howard and Rees, 1994).

The Fe protein has been structural characterized in a number of different forms including the nucleotide-free state, the MgADP-bound state, and as a part of stabilized Fe protein–MoFe protein complexes (recently reviewed in Peters and Szilagyi, 2006). The last complexes established the basis for Fe-protein conformational changes. Within the complex, it was first observed that the Fe protein can exist in an alternative conformation that can be described as the rigid-body reorientation of the Fe protein subunits relative toward one another. These conformational changes result in the formation of new inter-subunit interactions that were proposed to be important in triggering and facilitating nucleotide hydrolysis (Schindelin et al., 1997).

Several years ago, we structurally characterized a variant of the Fe protein with a single deletion (at 127-Leu) in Switch II, which provides a link between the nucleotide-binding site and the [4Fe-4S] clusters (Sen et al., 2004). This Fe protein variant (L127Δ Fe protein) exists in a structural conformation not previously observed for the Fe protein, with the salient differences being a dramatic rigid-body reorientation of subunits away from one another, resulting in a structure that appears more open with limited subunit interactions. Previous biochemical/physical characterization revealed that the altered properties of the nucleotide-free form of the L127Δ Fe protein resembled in part the properties of the native Fe protein with bound MgATP (Ryle and Seefeldt, 1996). In addition, in the presence of the native MoFe protein, the L127Δ Fe protein forms a stable complex and the characterization of this complex revealed a structure that closely resembled the nitrogenase complex stabilized in the presence of MgADP and tetrafluoro-aluminate (Chiu et al., 2001). These features led to the hypothesis that the structure of the L127Δ Fe protein may be related to the structure of the native MgATP-bound state of the Fe protein (Figure 1). The data that link the structure of the L127Δ Fe protein with that of the MgATP bound state are related to properties of the [4Fe-4S] cluster through the assumption that similarities in these properties are a result of the cluster being in a similar structural environment and, thus, a similar conformation may be anticipated. To date, no data have been presented that establish a clear relationship between the L127Δ Fe protein and the MgATP-bound conformation of the Fe protein at the level of protein structure and, thus, an imperative has been placed in our studies on examining the relationship between these Fe-protein conformations.

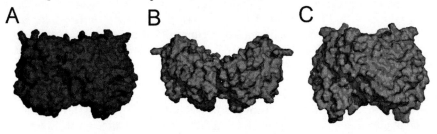

Figure 1. Surface renderings and maximum dimensions of the nitrogenase Fe protein in different structural conformations including (A) the native nucleotide-free (2NIP.pdb) state, (B) L127Δ variant nucleotide-free (1RW4.pdb), and (C) the conformation of the native Fe protein observed in the 2:1 Fe protein:MoFe protein complex stabilized in MgADP-tetrafluoroaluminate (1N2C.pdb).

It is often very difficult to examine the nucleoside triphosphate-bound states of proteins by x-ray crystallography due to the susceptibility of these compounds to hydrolysis in aqueous solution and the length of time that it takes to obtain protein crystals.

Non-hydrolysable nucleoside triphosphate analogs are sometimes used but it is difficult to insure that they are true mimics of the actual nucleoside triphosphate substrates.

Our recent work has focused on using the existing structurally characterized nitrogenase Fe-protein structures as a basis for establishing solution methods to probe states of the Fe protein that are difficult to examine crystallographically. Specifically, we are using small angle solution x-ray scattering (SAXS), limited proteolysis and mass spectrometry, and site-specific spin-labeling to gain insights into the structure of the nitrogenase Fe protein in solution.

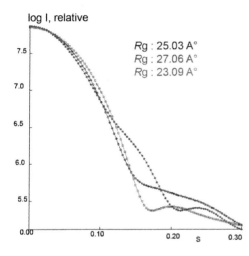

log I, relative

Rg : 25.03 A°
Rg : 27.06 A°
Rg : 23.09 A°

Figure 2. Theoretical scattering curves obtained from crysol for the crystal structures of Fe protein in the conformations shown in Figure 1 with the nucleotide-free native state (2NIP.pdb) in open circles, nucleotide-free L127Δ variant in solid circles (1RW4.pdb), and from the MgADP-tetrafluoroaluminate stabilized complex (1N2C.pdb) in open squares.

We have recently made significant progress, using SAXS, to answer the following outstanding questions. First, is the observed unanticipated structure of the L127Δ Fe protein the same in solution as observed in the crystal structure or does the crystal represent a local minimum of an Fe protein that is highly flexible? Second, is the L127Δ Fe protein a reasonable mimic of the MgATP bound Fe protein conformation? We first simulated SAXS scattering curves for existing Fe protein structures to examine whether the structural differences observed by crystallography could be clearly distinguished by SAXS. The simulated scattering curves for the nucleotide-free Fe protein, the L127Δ Fe protein, and the Fe protein from the nitrogenase complex with MgADP-tetrafluoroaluminate (Figure 2) indicate that Fe proteins existing in these structural conformations have distinct scattering profiles with different values for their respective radius of gyrations (Rg), indicating that SAXS should be effective at differentiating these conformations in solution. In addition, key differences are recognized in the 0.1S-0.2S region in the simulated scattering plots that are indicative of the distinct molecular shape of the Fe protein in the different conformations.

We have been working on optimizing the conditions for the collection of SAXS data on the Fe protein at the Stanford Synchrotron Radiation (SSRL) for several years now.

Previous investigations involving SAXS on the nitrogenase Fe protein have allowed the estimation of the radius of gyration in different nucleotide-bound states (Chen et al., 1994) and, thus, our objective is to obtain SAXS data that will extend the resolution of this analysis such that we can obtain information concerning the molecular shape of the Fe protein in solution in different states. In addition, we are attempting to critically examine the relationship between the nucleotide-bound states of the Fe protein in solution with the structure of the L127Δ Fe protein, which we have suggested may be held in a similar conformation to the MgATP-bound state of the native Fe protein. There are several features of the Fe protein that make SAXS data collection a challenge; these include its O_2 sensitivity and sensitivity to x-rays. We have overcome these obstacles by using a sealed flow-cell sample holder. This system allows continuous flow of the sample in a system that is sealed from the outside atmosphere, thereby limiting the time of x-ray exposure on a given sample of protein over the course of data collection by distributing the x-ray exposure over a larger sample size. In addition, samples can be pumped from sealed serum vials to a closed system that can be kept anaerobic.

Our most recent data collections at SSRL have yielded excellent quality data and our preliminary analysis of scattering curves for the native nucleotide-free Fe protein and the L127Δ Fe protein suggest that the L127Δ Fe protein structure observed by crystallography predominates in solution. In addition, scattering curves of the L127Δ Fe protein incubated with 5–10 mM MgATP or MgADP are indistinguishable from the curves generated for the nucleotide-free form of the L127Δ Fe protein, suggesting that nucleotides do not promote conformational changes in the variant. In contrast, scattering curves for the native Fe protein incubated with 5–10 mM MgATP or MgADP are clearly distinct from that of the native nucleotide-free Fe protein and from each other, indicating that conformational differences are observed upon nucleotide interaction as anticipated. Although MgATP-dependent conformational changes are implicated for the native Fe protein, the preliminary analysis of our results suggests that the conformation of the MgATP-bound state of the native Fe protein is distinct from the structure of the L127Δ Fe protein. These results are clearly promising in providing insights into nucleotide-dependent conformational change in the Fe protein, a key element in defining the role of MgATP binding and hydrolysis in nitrogenase catalysis.

References

Chen L et al. (1994) J. Biol. Chem. 269, 3290–3294.
Chiu HJ et al. (2001) Biochemistry 40, 641–650.
Howard JB and Rees DC (1994) Annu. Rev. Biochem. 63, 235–264.
Peters JW and Szilagyi RK (2006) Curr. Opin. Chem. Biol. 10, 101–108.
Ryle MJ and Seefeldt LC (1996) Biochemistry 35, 4766–4775.
Schindelin H et al. (1997) Nature 387, 370–376.
Sen S et al. (2004) Biochemistry 43, 1787–1797.

THE MECHANISTIC BASIS OF *nif* GENE ACTIVATION

S. Wigneshweraraj[1], P. Burrows[1], D. Bose[2], W. Cannon[1], N. Joly[1],
M. Rappas[2], J. Schumacher[1], X. Zhang[2] and M. Buck[1]

[1]Division of Biology, Sir Alexander Fleming Building, Imperial College
London, London SW7 2AZ, UK; [2]Centre for Structural Biology, Imperial
College London, London SW7 2AZ, UK
(Email: m.buck@imperial.ac.uk)

Many *nif* genes rely upon the major variant RNA polymerase (E) containing the sigma-54 promoter-specificity factor ($E\sigma^{54}$) for their expression (Wigneshweraraj et al., 2005). Specialised bacterial enhancer-binding proteins (EBPs), which belong to the "ATPases Associated with various cellular Activities (AAA+) P-loop ATPase mechano-chemical protein family" and include both the NifA- and NtrC-type proteins, are required to activate the $E\sigma^{54}$ bound at promoters of *nif* genes (Schumacher et al., 2006). EBPs often bind to enhancer sequences which are located ca. 100–150 base-pairs either upstream (usually) or downstream (rarely) from the transcription start site and interact with the initial $E\sigma^{54}$–promoter complex (the closed complex) by a DNA looping event. The ATPase activity of EBPs helps regulate the activity of $E\sigma^{54}$ at the DNA opening step: EBPs couple the energy from nucleotide hydrolysis to remodel the $E\sigma^{54}$ closed complex and trigger a cascade of protein and DNA isomerization events that result in formation of a transcriptionally proficient open complex. Here, DNA strands separate and the template strand occupies the catalytic cleft of $E\sigma^{54}$. Using purified transcription components from enteric bacteria, we have integrated structural biology and functional studies to elaborate a detailed description of the obligatory EBP-driven DNA-opening step in *nif* gene expression and show how ATP hydrolysis changes the functional state of the $E\sigma^{54}$ closed complex to trigger DNA opening and transcription initiation.

A molecular switch within our model EBP (PspF; Phage shock protein F) controls exposure of surface exposed flexible loop L1, that engages directly the σ^{54} in the closed complex (Rappas et al., 2005, 2006). Different functional states of the AAA+ domain of PspF ($PspF_{1-275}$; sufficient for transcription *in vitro* and *in vivo* by $E\sigma^{54}$) form as ATP hydrolysis occurs. Interactions with different functional states of PspF induce different sets of conformational changes in σ^{54}, which in turn control the conformational changes in the RNA polymerase required to form the open complex. Our recent study suggests that different nucleotide-bound states (ATP, ATP-hydrolysis transition state, ADP + Pi, and ADP) during ATP hydrolysis are sensed by an atomic switch pair (N64-E108) in

F. D. Dakora et al. (eds.), *Biological Nitrogen Fixation: Towards Poverty Alleviation through Sustainable Agriculture.*

PspF and that this information is relayed through a conformational signalling pathway within PspF to σ^{54}-interacting L1 loop. ATP binding should change the conformation of N64 to release L1 and L2, allowing PspF to productively interact with the $E\sigma^{54}$ closed complex (Rappas et al., 2006). Alterations in the structure of ATP (approaching the ATP transition state) allow a tight interaction between PspF and σ^{54}. Pi release apparently alters the conformation of N64 and results in L1 returning to a locked state, unable to interact with $E\sigma^{54}$. The threonine residue (T148) of the nucleotide-sensing motif, Sensor I, helps control L1 via use of L2 (J. Schumacher et al., unpublished data, 2006). Mutational analysis of E108 allows the creation of forms of $PspF_{1-275}$ in which ATP-dependent stable binding of $PspF_{1-275}$ to $E\sigma^{54}$ and σ^{54} occurs, supporting the view that E108 functions in switching functional states of PspF in a nucleotide-dependent manner (N. Joly et al., unpublished data, 2006).

How ATP binding and hydrolysis are coordinated between subunits of PspF to enable substrate remodelling is unknown. ADP stimulates the intrinsic ATPase activity of $PspF_{1-275}$ suggesting heterogeneous nucleotide occupancy in a $PspF_{1-275}$ hexamer. Binding of ADP and ATP triggers formation of functional $PspF_{1-275}$ hexamers as shown by a gain of specific activity. Also, ATP concentrations congruent with stoichiometric ATP binding to $PspF_{1-275}$ inhibit ATP hydrolysis and open-complex formation. Demonstration of a heterogeneous nucleotide bound state of a functional $PspF_{1-275}$-$E\sigma^{54}$-closed complex provides clear biochemical evidence for asymmetric nucleotide occupancy in this AAA+ protein. Thus, we propose a stochastic nucleotide-binding and a coordinated hydrolysis mechanism in $PspF_{1-275}$ hexamers (Joly et al., 2006).

A site-directed photo-crosslinking method, which was used to characterise the protein-DNA interactions that govern transcription initiation during *nif* gene expression (P. Burrows et al., unpublished data, 2006), indicate that DNA interactions in the closed complex are solely mediated via the σ^{54} factor. The interaction between the closed complex and $PspF_{1-275}$ with the added ATP hydrolysis transition state analogue (ADP-AlFx) leads to a clear change in the crosslinking pattern, suggesting that ADP-AlFx bound $PspF_{1-275}$ induces conformational changes in the closed complex. Strikingly, no DNA proximity to the RNA polymerase catalytic subunits is observed, suggesting that the DNA has not engaged the catalytic cleft of the RNAP in the presence of $PspF_{1-275}$:ADP-AlFx. Only upon full ATP-hydrolysis-dependent remodelling of the closed complex are interactions between the catalytic RNA polymerase subunits and the DNA strands established. Overall, our site-directed crosslinking results further underscore the level of regulatory precision that operate during *nif* gene expression.

References

Joly N et al. (2006) J. Biol. Chem. 281, 34997–35007.
Rappas M et al. (2005) Science 307, 1972–1975.
Rappas M et al. (2006) J. Mol. Biol. 357, 481–492.
Schumacher J et al. (2006) J. Struct. Biol. 156, 190–199.
Wigneshweraraj SR et al. (2005) Prog. Nucleic Acid Res. Mol. Biol. 79, 339–369.

ARCHITECTURE OF σ^{54}-DEPENDENT PROMOTERS: INTERPLAY BETWEEN CRP-cAMP AND P$_{II}$-NTR SYSTEMS FORMS A NOVEL REGULATORY NETWORK BETWEEN CARBON METABOLISM AND NITROGEN ASSIMILATION IN *ESCHERICHIA COLI*

Y.-P. Wang

College of Life Sciences, Peking University, Beijing 100871, P. R. China

The σ^{54}-RNA polymerase (Eσ^{54}) predominantly contacts one face of the DNA helix in the closed promoter complex, and interacts with the upstream enhancer-bound activator via DNA looping. By introducing protein-induced DNA bends at precise locations between upstream enhancer sequences and the core promoter of the σ^{54}-dependent *glnA*p2 promoter without changing the distance in-between, we observed a strong enhanced or decreased promoter activity. The relative positioning and orientations of Eσ^{54}, the DNA-bending protein, and the enhancer-bound activator on linear DNA were determined by *in vitro* footprinting analysis. Results provide evidence that the activator must approach the Eσ^{54} closed complexes from the unbound face of the promoter DNA helix to catalyze open complex formation. In this case, the activator and Eσ^{54} interact face-to-face and form a complex in which promoter DNA is in between Eσ^{54} and the activator like a 'sandwich'. This is further supported by the modeling of activator-promoter DNA-Eσ^{54} complex (see Figure 1).

In *E. coli*, utilization of carbon sources is regulated by the PTS system (phosphoenolpyruvate-dependent phosphotransferase system), which modulates the intracellular levels of cAMP. The cAMP receptor protein (CRP) controls the transcription of many catabolic genes. The availability of nitrogen is sensed by the P$_{II}$ protein at the level of intracellular glutamine. Glutamine is transported mainly by GlnHPQ and synthesized by glutamine synthetase (GS) encoded by *glnA*. Previous studies suggest that these systems are regulated by CRP by an unknown mechanism (Tian et al., 2001; Maheswaran and Forchhammer, 2003).

Here, we show that at least two mechanisms are involved. First, CRP activates *glnH*p1 via synergistic binding with Eσ^{70} and represses *glnH*p2. As a consequence, in the presence of glutamine, the overall enhancement of *glnHPQ* expression alters GlnB signaling and de-activates *glnA*p2. Second, *in vitro* studies show that CRP can be recruited by Eσ^{54} to a site centered at -51.5 upstream of *glnA*p2. CRP-induced DNA

F. D. Dakora et al. (eds.), *Biological Nitrogen Fixation: Towards Poverty Alleviation through Sustainable Agriculture*.
© Springer Science + Business Media B.V. 2008

bending prevents the nitrogen-regulation protein C (NtrC) activator from approaching the activator-accessible face of the promoter-bound $E\sigma^{54}$ closed complex, and inhibits *glnA*p2. Therefore, as the major transcriptional effector of the 'glucose effect', CRP affects both the signal transduction pathway and the overall geometry of the transcriptional machinery of components of the nitrogen regulon in *E. coli*.

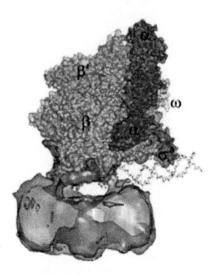

Figure 1. A model of the activator-promoter DNA-$E\sigma^{54}$ complex.

References

Maheswaran M and Forchhammer K (2003) Microbiology 149, 2163–2172.
Tian ZX et al. (2001) Mol. Microbiol. 41, 911–924.

PART 3D

PHOTOSYNTHETIC NITROGEN FIXERS

SIGNALS IN THE REGULATION OF NITROGEN FIXATION AND AMMONIUM ASSIMILATION IN THE PHOTOSYNTHETIC BACTERIUM *RHODOSPIRILLUM RUBRUM*

A. Jonsson, P. Teixeira and S. Nordlund

Department of Biochemistry and Biophysics, Stockholm University, SE-106 91 Stockholm, Sweden

Nitrogen fixation and ammonium assimilation in the photosynthetic bacterium *Rhodospirillum rubrum* are, like in most other diazotrophs, tightly regulated at multiple levels. The key enzyme glutamine synthetase (GS) is subjected to regulation primarily by reversible adenylylation in response to the availability of fixed nitrogen in reactions catalyzed by the bifunctional enzyme adenylyltransferase, GlnE. The expression of *glnA*, which encodes GS, is also regulated, although this is not the primary means of GS regulation in *R. rubrum* (Cheng et al., 1999; Nordlund et al., 1985).

As generally occurs in diazotrophs, *nif* gene expression in *R. rubrum* is strictly regulated in response to N-status. However, in this bacterium, nitrogenase activity is also regulated in response to changes in N-availability as well as to changes in the energy status, i.e., light/darkness (Nordlund and Ludden, 2004). At the molecular level, this regulation is exerted through reversible ADP-ribosylation of the Fe protein, inhibiting its interaction with the MoFe protein. DRAT and DRAG catalyze the modification and de-modification, respectively.

R. rubrum harbors three P_{II} paralogs. They have roles in the regulation of *nif* gene expression, the DRAT/DRAG regulation, and ammonium assimilation to some extent with overlapping specificities (Zhang et al., 2001). GlnB is the P_{II} protein involved in *nif* gene expression and GlnJ seems to be the most important effector of DRAG regulation (Wang et al., 2005), although GlnB can substitute. The function of GlnK, the third P_{II} paralog, is still unknown. The P_{II} proteins are also involved in regulating GlnE activity and, thus, the activity of GS.

P_{II} proteins exert their regulatory role by interacting with the target proteins and this interaction is dependent not only on the uridylylation status of the P_{II} protein, but also on the binding of ATP and 2-ketoglutarate to the P_{II} protein. The reversible modification of P_{II} proteins is catalyzed by the *glnD*-gene product. In *Escherichia coli* the

F. D. Dakora et al. (eds.), *Biological Nitrogen Fixation: Towards Poverty Alleviation through Sustainable Agriculture.*
© Springer Science + Business Media B.V. 2008

activity of this bifunctional enzyme is regulated by glutamine, which stimulates the UMP-removing activity and inhibits the uridylyltransferase activity (Jiang et al., 1998).

We have studied the activities and regulation of GlnD as well as of GlnE and GS in order to determine in greater detail the functions of the P_{II} paralogs in *R. rubrum*. In addition, the roles of 2-ketoglutarate and glutamine as effectors in P_{II} and GS modification were studied.

GlnE, GlnD and the three *R. rubrum* P_{II} paralogs were purified after overexpression in *E. coli*. Furthermore using the purified GlnD, the P_{II} proteins were uridylylated and purified for use in the experiments. GS was purified from *R. rubrum* as described previously. Purified GlnD catalyzed the uridylylation of all three P_{II} proteins, but the efficiency was dependent on the divalent cation, Mg^{2+} or Mn^{2+}, present. For GlnB, Mg^{2+} was more stimulatory, whereas for GlnJ and GlnK, uridylylation was more stimulated by Mn^{2+}. In contrast, Mn^{2+} was required for the deuridylylation of all three P_{II} paralogs. The effect of glutamine on the reactions catalyzed by GlnD was also investigated. In contrast to the *E. coli* system, glutamine did not inhibit uridylylation but had a clear stimulatory effect on the deurididylylation reaction like in *E. coli*.

The concentration of glutamine in *R. rubrum* does not vary greatly with different N-sources, indicating that the activity of GS is tightly regulated. In order to understand the effectors of this regulation, we studied the effect of P_{II} proteins on the activities of *R. rubrum* GlnE. The results clearly show that the adenylylation reaction is dependent on the presence of P_{II}, whereas both P_{II}-UMP and 2-ketoglutarate inhibited the reaction. We suggest that the effect of the latter is on the P_{II} protein and that P_{II}-UMP competes with P_{II} in its interaction with GlnE. We have not found any effectors of the deadenylylation reaction, where addition of phosphate, the substrate, is sufficient.

In conclusion, we propose that the regulatory role of glutamine in nitrogen assimilation is on GlnD, thus determining regulation by P_{II} proteins. 2-ketoglutarate, on the other hand, has an overriding inhibitory function on GlnE regulation and thereby on GS activity. The detailed role of P_{II} proteins in the regulation of DRAT/DRAG remains to be established as does whether or not glutamine and/or 2-ketoglutarate also affects these activities.

References

Cheng J et al. (1999) J. Bacteriol. 181, 6530–6534.
Jiang P et al. (1998) Biochemistry 37, 12782–12794.
Nordlund S and Ludden PW (2004) Genetics and Regulation of Nitrogen-fixing Bacteria (Klipp W et al., eds.), Kluwer, Dordrecht, The Netherlands, pp. 175-196.
Nordlund S et al. (1985) J. Bacteriol. 161, 13–17.
Wang H et al. (2005) FEMS Microbiol. Lett. 253, 273–279.
Zhang Y et al. (2001) J. Bacteriol. 183, 6159–6168.

DEFINING THE ROLES P_{II} AND AmtB PROTEINS IN THE NH_4^+-INDUCED POST-TRANSLATIONAL CONTROL OF NITROGENASE ACTIVITY

L. F. Huergo[1], M. Merrick[2], L. S. Chubatsu[1], F. O. Pedrosa[1] M. B. Steffens[1] and E. M. Souza[1]

[1]Departamento de Bioquímica e Biologia Molecular, Universidade Federal do Paraná, CP 19046, 81531-990, Curitiba, PR, Brazil; [2]Department of Molecular Microbiology, John Innes Centre, Norwich, NR4 7UH, UK

In some diazothophs, such as *Azospirillum brasilense* and *Rhodospirillum rubrum*, nitrogenase is post-translationally controlled by ADP-ribosylation of nitrogenase reductase (NifH). In response to addition of NH_4^+ to the medium, an ADP-ribose is linked to NifH resulting in nitrogenase inactivation (switch-off). This process is catalyzed by the DraT enzyme. When the added NH_4^+ is exhausted by cellular metabolism, the ADP-ribose group is removed by the DraG enzyme leading to nitrogenase activation (switch-on). The activities of both DraT and DraG are regulated by the cellular nitrogen status, although the signaling pathway is not fully understood.

A Model for NH_4^+-Induced Switch-Off

Proteins from the signal-transduction family P_{II} and the ammonium channel AmtB have been implicated in regulation of DraT and DraG in several bacteria, although the signaling pathway has not been elucidated. Our recent studies in *A. brasilense* led us to propose a model for the regulation of DraT and DraG activities in response to NH_4^+ addition (Figure 1). When *A. brasilense* is subjected to an ammonium shock, the P_{II} proteins (GlnB and GlnZ) are de-uridylylated and become membrane associated. Simultaneously, DraG is targeted to the cell membrane and NifH is ADP-ribosylated (Figure 1). Both NifH modification and membrane sequestration of DraG and P_{II} are defective in an *amtB* mutant (Huergo et al., 2006a). These findings indicate that AmtB-dependent DraG membrane association causes its inactivation. We observed that DraG interacts with GlnZ *in vivo* independently of cellular nitrogen levels (Huergo et al., 2006b) and that AmtB interacts with de-uridylylated GlnZ *in vitro* (L.F. Huergo et al., unpublished data, 2006). In our working model, the DraG-GlnZ complex is targeted to the cell membrane through formation of a ternary protein complex between DraG, GlnZ and AmtB, lead ing to the inactivation of DraG (Figure 1). Using pull-down experiments, we have

F. D. Dakora et al. (eds.), *Biological Nitrogen Fixation: Towards Poverty Alleviation through Sustainable Agriculture.*
© Springer Science + Business Media B.V. 2008

also demonstrated that DraT forms a specific complex with de-uridylylated GlnB in *A. brasilense in vivo* (Huergo et al., 2006b). This complex is only formed after an ammonium shock, i.e., in conditions where DraT is active, suggesting that DraT-GlnB complex formation activates DraT (Figure 1). This hypothesis is supported by genetic data showing that NifH modification is defective in a *glnB* mutant (Huergo et al., 2006a).

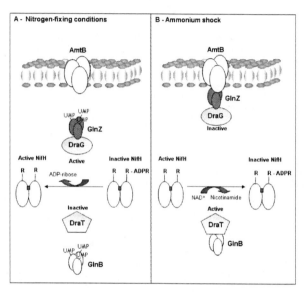

Figure 1. Model for the Ammonium-Induced Switch Off.

Is the Model Applicable to Other Diazotrophs?

As very similar components, e.g., AmtB and one or more P_{II} proteins, are also present in other bacteria that regulate nitrogenase by ADP-ribosylation, it is very likely that this model might be generally applicable. Indeed, recent studies in *R. rubrum* have shown that DraG is membrane-associated under switch-off conditions, and both NifH modification and DraG localization is dependent on AmtB and P_{II} proteins (Wang et al., 2005). The AmtB protein is also required for nitrogenase post-translational control in *Rhodobacter capsulatus* (Yakunin and Hallenbeck, 2002) and *Azoarcus* sp. (Martin and Reinhold-Hurek, 2002). Studies in *R. rubrum* (Zhu et al., 2006) and *R. capsulatus* (Pawlowski et al., 2003) have also shown DraT interaction with de-uridylylated P_{II} proteins, corroborating our model where such an interaction would activate DraT.

References

Huergo LF et al. (2006a) Mol. Microbiol. 59, 326–337.
Huergo LF et al. (2006b) FEBS Lett. 580, 5232–5236.
Martin DE and Reinhold-Hurek B (2002) J. Bacteriol. 184, 2251–2259.
Pawlowski A et al. (2003) J. Bacteriol. 185, 5240–5247.

Wang H et al. (2005) FEMS Microbiol. Lett. 253, 273–279.
Yakunin AF and Hallenbeck PC (2002) J. Bacteriol. 184, 4081–4088.
Zhu Y et al. (2006) J. Bacteriol. 182, 1866–1874.

METABOLIC REGULATION OF NITROGEN FIXATION IN *RHODOSPIRILLUM RUBRUM* AND EVIDENCE FOR AN ENERGY-DEPENDENT AMMONIUM TRANSPORT SYSTEM

H. Wang, A. Norén and S. Nordlund

Department of Biochemistry and Biophysics, Stockholm University, SE-10691 Stockholm, Sweden

In photosynthetic *R. rubrum*, nitrogenase is tightly controlled at both the transcriptional and metabolic level. Nitrogenase activity is regulated by reversible ADP-ribosylation in response to external "switch-off " effectors. In response to a sudden fixed-N increase or energy depletion, the Fe protein is covalently modified by DRAT to inhibit nitrogenase activity. When the fixed-N is metabolized or the energy status of the cell is restored, DRAG removes the ADP-ribose moiety, activating nitrogenase (Nordlund and Ludden, 2004). Two different signal-transduction pathways appear to lead to nitrogen and energy "switch-off" with the pathways merging at or before affecting DRAT/DRAG activities.

P_{II} proteins constitute a family of highly conserved proteins that are involved in signal transduction of the cell's N and C status. In *R. rubrum*, there are three P_{II} paralogues, GlnB, GlnJ and GlnK (Zhang et al., 2001); *glnB* gene is upstream of *glnA* (encoding glutamine synthetase, GS), whereas *glnJ* and *glnK* are upstream of *amtB1* and *amtB2*, respectively, which encode NH_4^+-transport proteins. In *R. rubrum*, the GlnB protein is required for *nif* expression and essential for NifA activation, but GlnJ and GlnK are not involved in *nif* regulation (Zhang et al., 2004). However, GlnB or GlnJ is necessary for nitrogenase "switch-off" in response to both NH_4^+ and darkness (Zhang Y et al., 2001).

DRAG is always associated with chromatophore membranes on cell disruption and is released by 0.5 M NaCl. This reversible membrane association is a likely part of the regulating mechanism that controls DRAG activity (Norén and Nordlund, 1997). We found that DRAG was predominantly in the cytosolic fraction in a *glnB/glnJ* double mutant and a triple P_{II} mutant (Wang et al., 2005), in agreement with the complete loss of metabolic regulation of nitrogenase in these mutants (Zhang et al., 2001).

F. D. Dakora et al. (eds.), *Biological Nitrogen Fixation: Towards Poverty Alleviation through Sustainable Agriculture.*
© Springer Science + Business Media B.V. 2008

The Amt family consists of high affinity ammonium/ammonia transport proteins and is found in all domains of life. In prokaryotes, the Amt structural gene (*amtB*) is always downstream of *glnK*. Crystal structures of AmtB from *E. coli* and Amt-1 from *Archaeoglobus fulgidus* reveal a trimeric trans-membrane protein that forms a central channel to conduct ammonia (Zheng et al., 2004; Andrade et al., 2005). In *E. coli*, *Azotobacter vinelandii*, *Azospirillum brasilense* and *R. rubrum*, one P_{II} homologue (GlnK, GlnZ or GlnJ) is sequestered by the membrane in an AmtB-dependent fashion and the uridylylation state of P_{II} proteins is modulated in response to the intracellular nitrogen status (Coutts et al., 2002; Huergo et al., 2006; Zhang et al., 2006).

We showed that NH_4^+-transport rates are not affected in *R. rubrum* mutants lacking either one or both AmtB proteins under normal photoheterotrophic conditions, therefore, the loss of NH_4^+ "switch-off" in the mutants lacking AmtB1 is due to disruption of the signaling pathway. Similar results occur in *Rhodobacter capsulatus*, suggesting a role of AmtB as an NH_4^+ sensor in regulating nitrogenase activity (Yakunin and Hallenbeck, 2002). Interestingly, NH_4^+-transport studies under different conditions indicate the existence of an energy-dependent NH_4^+-transport system in *R. rubrum*. In wild-type *R. rubrum* cells, the NH_4^+-uptake rate in the dark is about 25% of the rate in the light. In the *amtB1/amtB2* mutant, however, NH_4^+ uptake was completely abolished in the dark, whereas the uptake rate in the light was similar to wild-type. The lower NH_4^+-uptake rate in the dark can be restored by O_2 addition, indicating energy-dependent uptake in *R. rubrum*. In conclusion, NH_4^+ uptake is not affected as long as there is active electron transport, either by aerobic respiration in the dark or by photosynthesis in the light.

References

Andrade SLA et al. (2005) Proc. Natl. Acad. Sci. USA 102, 14994–14999.

Coutts G et al. (2002) EMBO J. 21, 536–545.

Huergo LF et al. (2006) Mol. Microbiol. 59, 326–337.

Nordlund S and Ludden PW (2004) In: Genetics and Regulation of Nitrogen Fixation in Free-living Bacteria (W. Klipp et al., eds.), Kluwer, Dordrecht, The Netherlands, pp. 175–196.

Norén A and Nordlund S (1997) J. Bacteriol. 179, 7872–7874.

Wang H et al. (2005) FEMS Microbiol. Lett. 253, 273–279.

Yakunin AF and Hallenbeck PC (2002) J. Bacteriol. 184, 4081–4088.

Zhang Y et al. (2001) J. Bacteriol. 183, 6159–6168.

Zhang Y et al. (2004) Proc. Natl. Acad. Sci. USA 101, 2782–2787.

Zhang Y et al. (2006) Microbiology 152, 2075–2089.

Zheng L et al. (2004) Proc. Natl. Acad. Sci. USA 101, 17090–17095.

MEMBRANE SEQUESTRATION OF P_{II} PROTEINS AND NITROGENASE REGULATION IN *RHODOBACTER CAPSULATUS*

P.-L. Tremblay[1], T. Drepper[2], B. Masepohl[3] and P. C. Hallenbeck[1]

[1]Département de Microbiologie et Immunologie, Université de Montréal, Montréal, Québec, Canada; [2]Institut für Molekulare Enzymtechnologie, Heinrich-Heine-Univertität Düsseldorf, Forschungszentrum Jülich, Jülich, Germany; [3]Fakultät für Biologie der Ruhr-Universität Bochum, Bochum, Germany

In the photosynthetic bacterium *Rhodobacter capsulatus*, the post-translational regulation of the MoFe nitrogenase is deficient in a strain with a kanamycin insertion in *amtB*, a gene coding for an ammonium channel (Yakunin and Hallenbeck, 2002). GlnB and GlnK, two homologues of the P_{II} protein, are also involved in this regulatory process since there is no nitrogenase switch-off in a *glnB-/glnK-* double mutant strain (Drepper et al., 2003). Here, we have examined the effect of single mutations in either GlnB or GlnK on nitrogenase regulation. There was no nitrogenase switch-off in the *glnB* strain when subjected to the addition of 200 μM NH_4^+. When ADP-ribosylation of the Fe protein in response to a 50 mM NH_4^+ addition was examined, only a moderate response after 80 min was noted for the *glnB* strain, whereas the wild-type strain showed appreciable Fe-protein modification after 40 min, and even more substantial modification after 80 min. Thus, GlnB appears necessary for nitrogenase switch-off and Fe-protein modification in *R. capsulatus*.

Surprisingly, like GlnB, GlnK was also found to be necessary for the NH_4^+-induced switch-off of Mo-nitrogenase activity and ADP-ribosylation of NifH. To explore the potential need for GlnK modification in the regulation of these processes, a mutant strain carrying the GlnK-Y51F allele was examined. In this strain, which is lacking the wild-type GlnK, the mutant GlnK cannot be modified by uridylylation due to the replacement of the amino acid at the site of modification, tyrosine-51, with phenylalanine. However, the regulation of nitrogenase activity (switch-off) and nitrogenase modification were found to be normal in this strain. This result strongly argues that the modification status of GlnK is irrelevant to its function in regulating these processes.

In *E. coli*, GlnB and GlnK are sequestered by the membrane after addition of ammonium in an AmtB-dependant process. Cell fractionation experiments were carried out

F. D. Dakora et al. (eds.), *Biological Nitrogen Fixation: Towards Poverty Alleviation through Sustainable Agriculture*.
© Springer Science + Business Media B.V. 2008

353

to study the possible sequestration of the two *R. capsulatus* P_{II} homologs, GlnB and GlnK, by the two Amt family members of *R. capsulatus*, AmtB and AmtY. Western blot analysis was performed on the cytoplasmic and membrane fractions obtained before, and 15 min after, the addition of 1 mM ammonium. Before the addition of NH_4^+, GlnB was found primarily in the cytoplasmic fraction, whereas after the ammonium addition, GlnB appeared to be equally divided between the cytoplasmic and membrane fractions. This analysis also showed that addition of ammonium causes GlnK to switch from the cytoplasm to the membrane. Only a very small fraction of the GlnK pool was detectable in the membrane fraction before the ammonium shock, and in the cytoplasmic fraction after it. In a *R. capsulatus amtB* mutant, there is no capture of either of the two P_{II} homologs by the membrane, indicating that membrane sequestration is necessarily dependent upon AmtB. AmtY, a homologue of AmtB, was found to be incompetent for the capture of GlnB/GlnK by the membrane.

Various *amtB* mutants were obtained and studied for their transport, regulation, and sequestration capacities. AmtB with a hexa-histidine tag added to its C-terminal end is functional for transport but not for nitrogenase regulation and P_{II} homologue capture. This result suggests that the transport of ammonium by AmtB alone is not sufficient for nitrogenase switch-off. A similar observation was made with a deleted C-terminal tail *amtB* strain. In a *glnB* strain, there appears to be no change of the localization of GlnK upon ammonium shock. On the other hand, contrary to what has been reported for the *A. brasilense glnB* mutant, the *R. capsulatus glnB* mutant, 15 min after exposure to 1mM ammonium, displays an almost complete deuridylylation of GlnK. In fact, as revealed by native PAGE, GlnB appears to be necessary for the efficient uridylylation of GlnK because intermediary forms of the protein are observed in absence of NH_4^+. In contrast, in the wild-type, only GlnK-3UMP, the fully uridylylated form, is present in the absence of ammonium. Thus, GlnB is directly or indirectly involved in the membrane sequestration of GlnK, and it appears that this process has little to do with GlnK deuridylylation status.

To investigate this further, we examined the localization of GlnK-Y51F before and after the addition of ammonium. Indeed, although traces of the P_{II} homolog variant were present in the membrane before the addition of 1mM ammonium, the majority of GlnK-Y51F is found in the cytoplasmic fraction. After the addition of ammonium, the GlnK-Y51F pool is separated almost equally between the two cellular fractions. This is in contrast to the wild-type, where the majority of GlnK is found in the membrane fraction after an ammonium shock. The ammonium-induced change in the membrane localization of GlnK-Y51F strongly argues that deuridylylation of GlnK is not the main signal that provokes the AmtB-dependent membrane sequestration of GlnK in *R. capsulatus*.

References

Yakunin AF and Hallenbeck PC (2002) J. Bacteriol. 182, 4081–4088.
Drepper TS et al. (2003) Microbiology 149, 2203–2212.

MOLYBDENUM REGULATION OF NITROGEN FIXATION AND MO-METABOLISM IN *RHODOBACTER CAPSULATUS*

J. Wiethaus and B. Masepohl

Ruhr-Universität Bochum, Lehrstuhl fur Biologie der Mikroorganismen, Bochum, Germany

The phototrophic purple bacterium *R. capsulatus* can synthesize both a molybdenum-containing nitrogenase (Mo-nitrogenase) and an alternative non-Mo nitrogenase (Fe-nitrogenase). Mo-nitrogenase exhibits higher specific activity for N_2 reduction as compared to the Fe-nitrogenase (Schneider et al., 1997). Accordingly, Mo-nitrogenase is the preferred enzyme as long as sufficient Mo is available to the organism. In the presence of Mo, transcription of *anfA* (coding for the transcriptional activator of Fe-nitrogenase genes) is repressed by either MopA or MopB, thus limiting the amount of Fe-nitrogenase at high Mo concentrations (Kutsche et al., 1996; Wiethaus et al., 2006).

The Mo-responsive regulators MopA and MopB exhibit 52% identity to each other. The *mopA* gene forms part of the *mopA-modABCD* operon, whereas *mopB* is located immediately upstream of *mopA* and transcribed in the opposite direction. The *modABC* genes code for a high-affinity Mo-transport system required for Mo-nitrogenase activity at low Mo concentrations (Wang et al., 1993; Wiethaus et al., 2006).

MopA and MopB repress the *anfA*, *mopA-modABCD*, *morAB*, and *morC* genes in the presence of Mo (Wiethaus et al., 2006). Hence, expression of one of the Mo-responsive regulators, MopA, and the Mo-uptake system, ModABC, is down-regulated as soon as intracellular Mo concentration increases. In contrast to *mopA*, the *mopB* gene is constitutively transcribed. Consequently, the intracellular ratio between MopA and MopB is expected to vary strongly in response to Mo availability. The products of the *morAB* and *morC* genes exhibit clear similarity to ModA, ModB, and ModC, respectively, but no role has yet been assigned to the putative MorABC transport system. Deletion of the *morABC* genes does not affect nitrogen fixation.

In addition to its role as a repressor of diverse target genes, MopA acts as an activator of *mop*-gene expression (Wiethaus et al., 2006). The *mop* gene codes for a putative Mo-binding protein (Mop). MopA activates *mop* gene transcription only when Mo is available. As described above, however, *mopA* itself is Mo-repressed. Therefore, *mop* expression is expected to occur only at intermediate Mo concentrations.

F. D. Dakora et al. (eds.), *Biological Nitrogen Fixation: Towards Poverty Alleviation through Sustainable Agriculture.*
© Springer Science + Business Media B.V. 2008

Binding of MopA and MopB to their respective target promoters is clearly enhanced by molybdenum as shown by DNA mobility shift assays (Wiethaus et al., 2006). Mo-regulated promoters contain conserved DNA sequences of dyad symmetry, so-called Mo-boxes, thought to serve as binding sites for the Mo-responsive regulators. As typical repressor binding sites, Mo-boxes in the promoter regions of *anfA* and *mopA* overlap the transcription start sites. In contrast, the Mo-box of the *mop* gene is located upstream of the transcription start site which is in agreement with *mop* gene activation via MopA.

To reconstruct the Mo-pathway through the cell, interactions of Mo-binding proteins were studied by yeast two-hybrid analysis (Figure 1). MopB, in addition to its role as a transcriptional regulator, might play a central role in Mo homeostasis as it interacts with a variety of proteins. In particular, MopB might donate Mo to proteins involved in biosynthesis of iron-molybdenum (NifQ) and molybdopterin cofactors (MogA, MoeA).

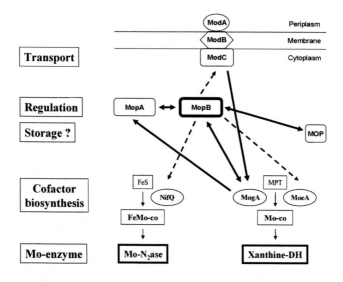

Figure 1. Model of the Mo-pathway through the cell. Strong and weak protein-protein interactions as deduced from yeast two-hybrid studies are indicated by solid and broken arrows, respectively.

References

Kutsche M et al. (1996) J. Bacteriol. 178, 2010–2017.
Schneider K et al. (1997) Eur. J. Biochem. 244, 789–800.
Wang G et al. (1993) J. Bacteriol. 175, 3031–3042.
Wiethaus J et al. (2006) J. Bacteriol. 188, 8441–8451.

HYDROGEN PRODUCTION BY CYANOBACTERIA AS A POTENTIAL TOOL IN ENERGY-CONVERSION PROGRAMMES

H. Bothe, G. Boison and S. Junkermann

Botanisches Institut, Universität Köln, Gyrhofstr. 15, D-50923 Köln, Germany

Among living things, many cyanobacteria have the simplest nutrient requirements. These photoautotrophic organisms thrive on simple inorganic media and can meet their fixed-nitrogen demands by N_2 fixation. They can even offer perspectives for converting solar radiation into combustible energy by coupling photosynthetic electron transport to H_2 production. Cyanobacteria possess two classes of enzyme complexes that catalyze the formation of the hydrogen gas: nitrogenases and hydrogenases. As to the latter enzymes, cyanobacteria can express both an uptake and a bidirectional (reversible) hydrogenase. The uptake hydrogenase, encoded by *hupL* and *hupS*, occurs in fila-mentous forms mainly (or exclusively) in heterocysts, where it recycles the H_2 produced in parallel with ammonia formation during N_2 fixation. The uptake hydrogenase feeds electrons into Complex III of respiration and might serve to protect nitrogenase from damage by oxygen by means of the oxyhydrogen (Knallgas) reaction.

The bidirectional hydrogenase is a Ni-containing enzyme, is encoded by the *hoxEFUYH* genes, is $NAD(P)^+$-dependent, and also occurs outside of heterocysts. The enzyme catalyzes both the uptake and the evolution of the gas. However, since the generation of H_2 from $NAD(P)H$ is thermodynamically unfavourable, H_2 formation catalyzed by the bidirectional enzyme is generally transitory in the cells, e.g., as bursts of excess reductant under high light intensities. Sustained H_2 production by cyano-bacterial hydrogenase as a prerequisite for commercial applications has not yet been reported to our knowledge. Attempts to engineer a foreign membrane-bound hydro-genase from *Ralstonia eutropha* H16 to the acceptor side of photosystem I (to a modi-fied *psaE* gene product) of the cyanobacterium *Synechocystis* PCC 6803 resulted in only marginal rates of H_2 formation (Ihara et al., 2006).

An alternative for H_2 production is provided by nitrogenases because, concomitantly with ammonia formation, nitrogenases catalyze the production of H_2. Cyanobacteria possess two types of nitrogenase, a Mo- and a V-containing enzyme complex, which are encoded by different gene sets (Kentemich et al., 1988; Thiel, 1993). The V-nitrogenase has been reported for only a few cyanobacteria, such as *Anabaena variabilis*, an

F. D. Dakora et al. (eds.), *Biological Nitrogen Fixation: Towards Poverty Alleviation through Sustainable Agriculture.*
© Springer Science + Business Media B.V. 2008

An-baena isolate from the fern *Azolla, Anabaena azotica,* and *Anabaena* sp. CH1 (Boison et al., 2006). H_2 production by the alternative V-containing nitrogenase has not thoroughly been examined in the cyanobacteria, but the data so far available indicate that both N_2- and C_2H_2-reduction rates are lower with V- than with Mo-nitrogenase, whereas net H_2 production is higher in V-grown cells of either *Anabaena variabilis* or *A. azotica*. This occurs despite the fact that most of the H_2 produced is immediately recycled by hydrogenase.

When *A. variabilis* cells were incubated with H_2 during the C_2H_2-reduction assay, it was noted for many years that the amount of H_2 present was significantly higher at the end of the experiment than initially supplied to the vessels. This was the case for both V- and Mo-grown cells. The effect is more distinct in *A. azotica* assayed under lower light intensities and with dilute cell suspensions (Figure 1).

In this experiment, the H_2 added results in smaller amounts of C_2H_4 formed (Figure 2), but causes an increase in H_2 formation depending on the amount of H_2 added (Figure

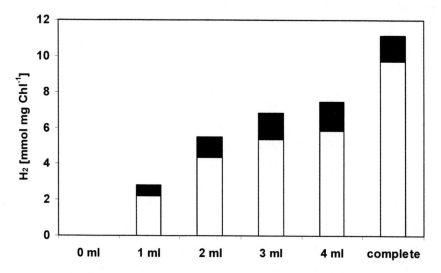

Figure 1. Hydrogen production by V-grown *Anabaena azotica* after 4 h of incubation under anaerobic conditions.White part of the columns indicates the amount of H_2 injected into the vessel (against the pressure in the vessels, therefore, the increase is not linear, but the amount per vessel was determined by gas chromatography at the start of the experiment); complete has a gas phase of H_2 only; black part of column indicates the amount of H_2 formed after 4 h per mg chlorophyll. The experiments were performed in 7.0 mL septa-sealed Fernbach flasks into which C_2H_2 (1 mL) and H_2 (as indicated on the abscissa) were injected. The experiments were performed with 3-mL *A. azotica* suspensions (total chlorophyll content 0.035 mg) with the Fernbach flasks rotating on a horizontal shaker at 20°C and a light intensity at the vessels of about 300 $\mu E\ m^{-2}\ s^{-1}$.

1). High concentrations of H_2 are known to inhibit nitrogen-fixation activity (Dixon, 1972). When large amounts of H_2 are present in the assay vessels (~15% of the gas phase consisting of H_2), the reducing power generated by photosynthetic electron transport and allocated to nitrogenase is mainly directed to H_2 production. Under these conditions, rates are 50–100 µmoles H_2 formed/h x mg chlorophyll a (Figure 1). If these rates were not so high, they would not have been noticed against the high amount of H_2 injected into the vessels at the start of the experiment. This H_2 production in the presence of C_2H_2 and H_2 is light dependent and sensitive to uncouplers, like FCCP or CCP, and therefore is not due to bacterial contaminants in the cyanobacterial extra-cellular polysaccharides. Such H_2 production in the presence of C_2H_2 and H_2 is not observed when N_2 replaces C_2H_2 as the nitrogenase substrate.

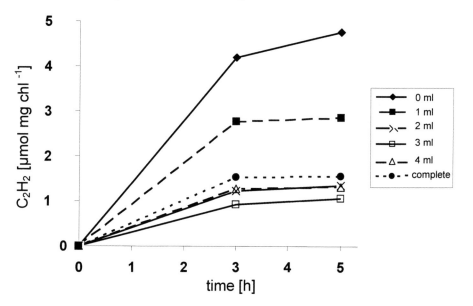

Figure 2. Decreasing C_2H_4 production by increasing H_2 concentrations. Experimental conditions were as described in Figure 1.

If one aims to exploit cyanobacteria for solar energy conversion programmes, H_2 formation by the alternative V-containing nitrogenase offers potentially the best prospect. However, recycling of H_2 by hydrogenase needs to be prevented. This could be achieved by constructing a mutant defective in *both* uptake and bidirectional hydrogenase. As described here, incubating such a mutant with C_2H_2 and H_2 could augment H_2 production significantly. The current experimental situation does not allow us to state whether this H_2 formation is sufficient for potential application. To be so, it must significantly exceed the maximal value for the solar energy conversions factors ($\leq 7\%$ of the irradiation converted to H_2; Mitsui and Kumazawa, 1977; Tamagnini et al.,

2002). In addition, conditions have to be worked out for sustained H_2 production by V-nitrogenase over weeks or months. Over the years, many attempts by the late David Hall from London (Tsygankov et al., 1997) and others have been based on using cyano-bacteria in solar energy conversion programmes. The current finding of enhanced H_2 production in the presence of H_2 and C_2H_2 could well be of significance in achieving this goal.

References

Boison G et al. (2006) Arch. Microbiol. 186, 367–376.

Dixon ROD (1972) Arch. Microbiol. 85,193–201.

Ihara M et al. (2006) Photochem. Photobiol. 82, 676–682.

Kentemich T et al. (1988) FEMS Microbiol. Lett. 51, 19–24.

Mitsui A and Kumazawa S (1977) In: Biological Solar Energy Conversion (Mitsui A et al., eds.), Academic, New York/London, pp. 23–43.

Tamagnini P et al. (2002) Microbiol. Mol. Biol. Rev. 66, 1–20.

Thiel T (1993) J. Bacteriol. 175, 6276–6286.

Tsygankov AS et al. (1997) Int. J. Hydrogen Energ. 22, 859–867.

REGULATORY INTERACTIONS INVOLVED IN NITROGEN SIGNALLING IN CYANOBACTERIA: PARTNER SWAPPING AND ROLE OF PIPX-PII COMPLEXES IN *SYNECHOCOCCUS* SP. PCC 7942

J. Espinosa[1], M. A. Castells[1], K. Forchhammer[2] and A. Contreras[1]

[1]División de Genética, Universidad de Alicante, Apartado 99, E-03080 Alicante, Spain; [2]Institut für Mikrobiologie und Molekularbiologie, Universität Giessen, Heinrich-Buff-Ring 26-32, D-35392 Giessen, Germany

Cyanobacteria are phototrophic organisms that perform oxygenic photosynthesis. Due to lack of 2-oxoglutarate (2-OG) dehydrogenase in cyanobacteria, the cellular 2-OG concentration is an excellent indicator of the cell carbon-to-nitrogen balance. 2-OG modulates the activity and/or binding properties of three key nitrogen regulators in *Synechococcus*; the signal transduction protein P_{II} (Forchhammer, 2004), the transcriptional activator NtcA, and the regulatory factor PipX (Espinosa et al., 2006). The transcriptional activator NtcA is required for the expression of multiple genes subjected to NH_4^+ repression. 2-OG stimulates binding of NtcA to both PipX and DNA-target sites as well as transcription activation *in vitro*. Consistent with a role for PipX as an NtcA co-activator, the *Synechococcus* PipX[-] mutant shows a pleiotropic phenotype reminiscent of NtcA[-] strains. PipX[-] cells showed reduced activity of nitrogen-assimilation enzymes, retarded induction and a slower rate of nitrate consumption, and, when subjected to nitrogen starvation, retarded phyco-bilisome degradation and a faster reduction of the chlorophyll content (Espinosa et al., 2007). In addition, PipX activates the NtcA-dependent promoters of *glnB*, *glnN* and *nblA* under conditions of nitrogen-deficiency, corresponding to high intracellular levels of 2-OG. In *Synechococcus*, P_{II} has been implicated in nitrogen regulation of both nitrate and nitrite uptake and the activities of nitrate reductase, NtcA, and N-Acetylglutamate kinase (NAGK) (Burillo et al., 2004).

We have detected PipX-NtcA-DNA complexes and studied the requirements for their formation by Surface Plasmon Resonance (SPR) analysis. P_{II} inhibits the 2-OG-dependent binding of PipX to NtcA-DNA complexes, an effect counteracted by ATP (Figure 1). Therefore, partner swapping (from P_{II} to NtcA) by PipX provides a mechanism by which signals regulating P_{II}-PipX interactions can be transmitted to the PipX-NtcA pair to fine regulate the transcriptional activity of NtcA according to metabolic signals.

F. D. Dakora et al. (eds.), *Biological Nitrogen Fixation: Towards Poverty Alleviation through Sustainable Agriculture*.
© Springer Science + Business Media B.V. 2008

Stimulation of NAGK activity is the only function of cyanobacterial P_{II} proteins for which the molecular basis is understood. This stimulation requires formation of a P_{II}-NAGK complex that is inhibited by 2-OG. Because both P_{II}-NAGK and P_{II}-PipX complexes seem to form under conditions of nitrogen sufficiency, we wondered whether PipX affects regulation of P_{II}-NAGK complexes and whether the three proteins could form a macromolecular complex compatible with NAGK activation. *In vitro*, the maximal P_{II}-stimulated NAGK activity was not affected by an excess of PipX. However, the time required to achieve maximal NAGK activity was specifically retarded by PipX (Figure 2), suggesting that NAGK activation by P_{II} requires disruption of preformed P_{II}-PipX complexes. NAGK, but not PipX, seems to be able to bind to P_{II} proteins from preformed complexes because, once formed, P_{II}-NAGK complexes were not affected by the presence of PipX. Inactivation or production of constitutively higher levels of the *pipX* gene had only a minor impact on the NAGK activity of *Synechococcus* cells. Other nitrogen-regulated functions, such as P_{II} phosphorylation and inhibition of nitrate uptake, also remained functional in the PipX⁻ strain.

In summary, PipX-P_{II} complexes appear to affect PipX-related functions more than P_{II}-related ones. P_{II} can affect NtcA functions *in vivo* and *in vitro*. In contrast, PipX does not seem to bind to P_{II}-NAGK complexes or interfere with its activity. Considering that *Synechococcus* P_{II} proteins are very abundant, able to sense and adopt different conformations and phosphorylation status according to the nitrogen/carbon ratio, and are predicted to interact with different partners, we propose that PipX-P_{II} complexes serve mainly to modulate PipX availability and/or activity. In other words, PipX would be a P_{II} target rather than a P_{II} regulator.

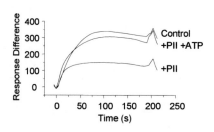

Figure 1. Effect of P_{II} and ATP on PipX binding to immobilized NtcA-P_{glnA} complex (SPR analysis).

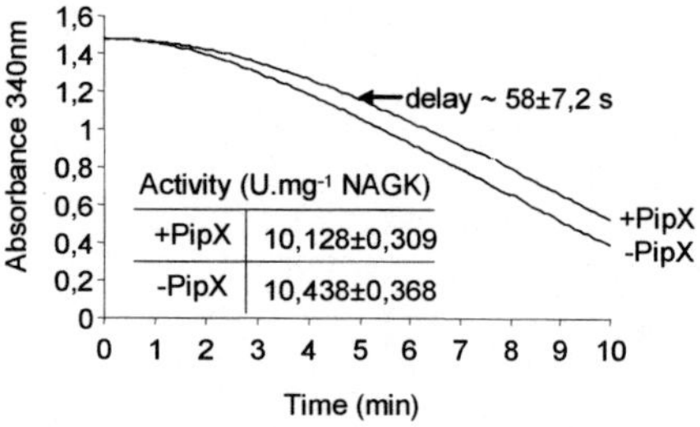

Figure 2. Effect of PipX on PII-stimulated NAGK activity (biosynthetic assay).

References

Burillo S et al. (2004) J. Bacteriol. 186, 3346–3354.
Espinosa J et al. (2006) Mol. Microbiol. 61, 457–469.
Espinosa J et al. (2007) Microbiology 153, 711–718.
Forchhammer K (2004) FEMS Microbiol. Rev. 28, 319–333.

INTERCELLULAR TRANSFER OF NITROGEN
IN HETEROCYST- FORMING CYANOBACTERIA

E. Flores, R. Pernil, V. Mariscal, S. Picossi and A. Herrero

Instituto de Bioquímica Vegetal y Fotosíntesis, Consejo Superior de Investigaciones Científicas y Universidad de Sevilla, Avda. Américo Vespucio 49, E-41092 Seville, Spain

Cyanobacteria are Gram-negative bacteria that perform oxygenic photosynthesis. In some filamentous species, N_2 fixation is carried out in specialized cells called heterocysts (Herrero et al., 2004). Heterocysts do not perform the photosynthetic fixation of CO_2, but receive reduced carbon compounds from the vegetative cells. Sucrose is likely involved as an intermediate in carbon transfer (Curatti et al., 2002). Heterocysts, in turn, transfer fixed nitrogen to their neighboring vegetative cells. Glutamine and some other amino acids might be the transferred N vehicles (Wolk et al., 1976; Thomas et al., 1977). This metabolite exchange is essential for the performance of the heterocyst-forming cyanobacteria as true multicellular organisms, but the exchange mechanism is unknown.

The Periplasm is Continuous Along the Cyanobacterial Filament

It has generally been assumed that transfer of fixed-N from the heterocysts to the vegetative cells in the filament takes place by diffusion through intercellular cytoplasmic connections. However, such connections, which would be analogous to plant plasmodesmata, have not been observed in electron-microscopy studies of filamentous cyanobacteria (Flores et al., 2006). These studies have shown instead that, whereas a cytoplasmic membrane surrounds each cell in the filament, the outer membrane is continuous along the filament, which implies a continuous periplasm. Using a green fluorescent protein exported to the periplasm as a probe, we have observed GFP diffusion in the periplasm, thus confirming its functional continuity. If there are no cytoplasmic bridges between cells in the filament, the intercellular transfer of metabolites could take place through an extra-cytoplasmic route, with permeases being required for the movement of metabolites across the cytoplasmic membranes.

F. D. Dakora et al. (eds.), *Biological Nitrogen Fixation: Towards Poverty Alleviation through Sustainable Agriculture.*
© Springer Science + Business Media B.V. 2008

Amino-acid Transporters Required for Optimal Diazotrophic Growth.

A functional genomics approach has permitted the identification of amino-acid transporters in this cyanobacterium. Three ABC-type transporters account for about 95% of the total amino acid-transport activity: (i) transport system Nat1 for neutral (preferentially hydrophobic) amino acids; (ii) system Nat2 for neutral and acidic amino acids; and (iii) system Bgt for basic amino acids. The Nat2 and Bgt systems have the uncommon characteristic of sharing an ATPase component, BgtA. Whereas inactivation of Bgt has no effect on growth on N_2, mutation of genes encoding system Nat1 (Picossi et al., 2005) or Nat2 impairs diazotrophic growth, while permitting heterocyst development, and the impairment is enhanced in a *nat1/nat2* double mutant.

Inactivation of a TRAP-type transporter, which can transport pyruvate, and of a membrane protein, SepJ, which bears a domain showing similarity to some amino-acid exporters, also impairs diazotrophic growth. The *sepJ* mutant shows filament fragmentation and does not differentiate mature heterocysts, suggesting a developmental role for SepJ.

Expression of Permease Genes and Proteins

Expression of *nat1* genes has been found to take place in vegetative cells but not in heterocysts (Picossi et al., 2005), consistent with a vectorial role of Nat1 in the diazotrophic filament. The subcellular location of BgtA and SepJ could be determined with GFP fusions. Whereas BgtA-GFP localizes to the periphery of the cells, consistent with a localization of the Nat2 and/or Bgt permeases at the cytoplasmic membrane, SepJ-GFP localizes to the cell poles at the intercellular septa in the filament.

Nat1 and Nat2 might be involved in the uptake by the vegetative cells of amino acids exported by the heterocysts and transferred through the periplasm; available data point to Ala and Gln as putative N vehicles (Picossi et al., 2005). On the other hand, the discovery of SepJ opens a scenario, where proteins localized at the intercellular septa might play a specific role in the intercellular relationships that take place in the cyanobacterial filament.

References

Curatti L et al. (2002) FEBS Lett. 513, 175–178.
Flores E et al. (2006) Trends Microbiol. 14, 439–443.
Herrero A et al. (2004) FEMS Microbiol. Rev. 28, 469–487.
Picossi S et al. (2005) Mol. Microbiol. 57, 1582–1592.
Thomas J et al. (1977) J. Bacteriol. 129, 1545–1555.
Wolk CP et al. (1976) J. Biol. Chem. 251, 5027–5034.

GENOMICS AND SYSTEMS BIOLOGY ANALYSIS OF *CYANOTHECE* SP. ATCC 51142, A UNICELLULAR, DIAZOTROPHIC CYANOBACTERIUM WITH ROBUST METABOLIC RHYTHMS

E. A. Welsh[1], M. Liberton[1], J. Stockel[1], H. B. Pakrasi[1], H. Min[2], J. Toepel[2] and L. A. Sherman[2]

[1]Department of Biology, Washington University, St Louis, MO 63130, USA; [2]Department of Biological Sciences, Purdue University, West Lafayette, IN 47904, USA

Cyanothece sp. ATCC 51142, is a unicellular, diazotrophic cyanobacterium with a versatile metabolism and pronounced diurnal rhythms. When grown under 12 h light-dark (LD) periods, it performs photosynthesis during the day and N_2 fixation and respiration at night, with carbohydrate and amino acids compartmentalized in granules in the light and dark, respectively. This strain also grows very well heterotrophically (on glycerol) and can fix N_2 when grown under continuous light conditions. The excellent synchrony of a culture under LD diazotrophic conditions permits analysis of cellular morphology, mRNA levels, proteomics and metabolomics as a function of time. For this reason, *Cyanothece* is the target organism for a Membrane Biology Grand Challenge through DOE and the Pacific Northwest National Labs (PNNL).

As a first step, the *Cyanothece* genome was sequenced at the Washington University Genome Sequencing Center and resulted in a major surprise with two parts to the genome, a 4.9-Mb circle with about 4,750 genes and a 428-kb linear segment with about 438 genes. The linear chromosome is not of phage or bacterial origin, but does have genes for glycolysis, fermentation and one cytochrome oxidase operon. The notable features of the circular chromosome include a large contiguous region of some 40 genes that contains the *nifHDK* genes and all of the genes needed to assemble the nitrogenase complex. At the start of this cluster, there are two regulatory genes, including one with three PAS domains. There are two sets of *adh* and *pdc* genes. The first set encodes the prototypical proteins that are extremely similar to the *Zymomonas* proteins and the second set is on the linear chromosome. There are three cytochrome oxidase operons, one of which seems to be involved with the respiration that increases dramatically during nitrogenase activity. Four plasmids (P1-P4) have been sequenced as well and some details of all six DNA elements are presented in Table 1.

F. D. Dakora et al. (eds.), *Biological Nitrogen Fixation: Towards Poverty Alleviation through Sustainable Agriculture.*
© Springer Science + Business Media B.V. 2008

These data permit us to determine the regulation of cellular components as a function of time and to correlate with respect to physiology and ultrastructure. Another important feature of *Cyanothece* is the assembly/disassembly of metabolic storage granules as a function of time and physiology and we have explored the ultrastructure of *Cyanothece* via the EM, using high pressure freezing and freeze substitution methods (Figure 1). We are most interested in the relationship of the photosynthetic membrane to the cytoplasmic membrane as a function of time. We are also interested in the interaction between the glycogen granules and the photosynthetic membrane and the way in which the granules are assembled. As shown in Figure 1, the glycogen granules are closely appressed to the photosynthetic membrane and appear to be encased by a protein coat. The excellent preservation provided by the cryo EM methods also permits us to use EELS to identify the composition of various intracellular inclusion bodies and this adds substantially to the information that we previously obtained with immunocytochemistry.

The availability of the sequence also facilitates microarray analysis of transcription. A microarray experiment on cultures grown for 2 days in 12 h light-12 h dark N_2-fixing conditions indicated robust rhythms and transcriptional patterns nearly identical over the 2-day light or dark periods. Figure 2 represents the transcript levels of the 35 genes in the nitrogenase cluster that center on *nifHDK* (4.156–4.158). These genes had peak transcript levels at D1 and troughs at L9 and were similar to results from Northern blots (Sherman et al., 1998). Not surprisingly, all photosynthesis genes have a regulatory pattern 12 h out of phase with those of nitrogenase. We are using a similar protocol to determine the protein composition of cells throughout the diurnal cycle to enable us to identify the nature of the genes and gene products that are transcribed at various times and under different physiological conditions.

Table 1. Details of all six DNA elements of *Cyanothece*.

	CIRCLE	LINEAR	P1	P2	P3	P4
SIZE (bp)	4.93 Mb	428 kb	39.6 kb	31.8 kb	14.7 kb	10.2 kb
G + C content (%)	37.9	38.6	36.8	41.5	38.1	37.0
Genes (total = 5,269)						
Protein-coding genes	4,738	438	37	23	20	13
Probable function	1,719	72	3	5	3	1
Possible function	1,403	112	10	7	2	3
Conserved unknown	229	1	–	–	–	–
Conserved hypothetica l	239	10	1	1	1	–
Unknown	200	16	3	4	4	–
Hypothetical	948	227	20	6	10	9
Average ORF size (bp)	899	822	966	860	538	651
Max. ORF size (kbp)	7.7	5.9	5.5	3.8	2.5	2.4
Coding (%)	86.3	84	90	62	73	83
Regulation	188	12	–	1	–	–
Transport and binding	234	14	1	–	–	–
Transposon-related	80	9	2	1	2	1
Regulation (%)	4.0	2.7	–	4.3	–	–
Transport/binding (%)	4.9	3.2	2.7	–	–	–
Transposon-related (%)	1.7	2.0	5.4	4.3	10.0	7.7

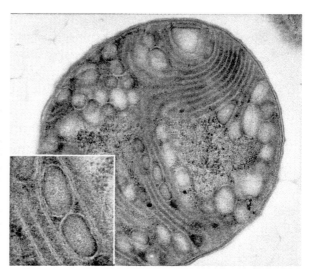

Figure 1. Electron micrograph of *Cyanothece* sp. ATCC 51142 at L5 grown under 12 h light-12 h dark conditions. Cultures were high pressure frozen and freeze substituted with OsO_4. The cell diameter is about 5 μm (micrograph courtesy of Debra Sherman, Purdue University Life Sciences Microscopy). Facility.

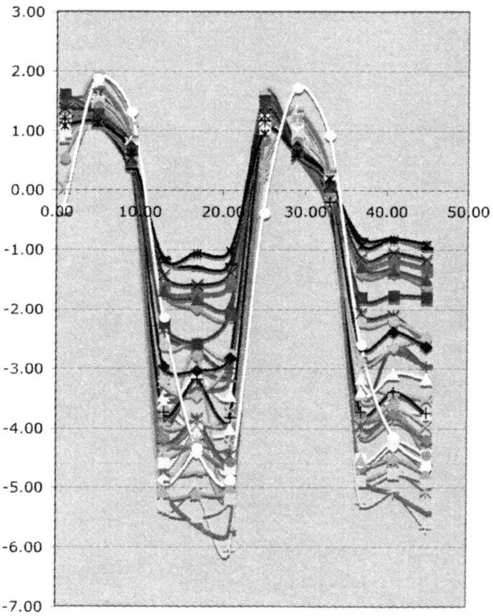

Figure 2. Microarray (Agilent) analysis of *Cyanothece* sp. ATCC 51142 grown under 12 h light-12 h dark, N_2 fixing conditions for 2 days. Samples were taken every 4 h from D1 on day 1 to D1 on day 3 (49 h). The transcript levels of genes in the *nif* cluster (4.142–4.176) are virtually identical during this period.

Funded by the Membrane Biology EMSL Scientific Grand Challenge project at the W. R. Wiley Environmental Molecular Sciences Laboratory, a national user facility sponsored by the U.S. DoE's BER program located at PNNL.

Reference

Sherman LA et al. (1998) Photosyn. Res. 58, 25–42.

PART 3E

SUMMARY PRESENTATIONS

SINORHIZOBIUM ARBORIS GENES ARE DIFFERENTIALLY EXPRESSED UNDER SALT STRESS

P. Penttinen[1], Z. Terefework[2], L. Paulin[3], P. Auvinen[3], D. Greco[3] and K. Lindström[1]

[1]Department of Applied Chemistry and Microbiology, University of Helsinki, Finland; [2]Department of Molecular Cell Physiology, Vrije Universiteit Amsterdam, The Netherlands; [3]Institute of Biotechnology, University of Helsinki, Finland

Sinorhizobium arboris nodulates the tropical trees *Acacia senegal* and *Prosopis chilensis*. The type strain *S. arboris* HAMBI 1552 has good nodulation potential and tolerates harsh environmental conditions. The genetic background of stress tolerance in *S. arboris* was studied with a DNA macro-array method. An *S. arboris* plasmid library was printed on nitrocellulose membranes. Total RNA from cells grown in minimal medium without or with 0.38 M NaCl (salt stress) was transcribed to ^{33}P-labelled cDNA and hybridized on to the membranes. Five plasmid clones that showed up-regulation under salt stress were end-sequenced. The sequences were translated to protein sequences in all six reading frames and a protein blast was performed. Most of the genes found were close matches to the genes of the model rhizobium *Sinorhizobium meliloti*. Close matches against *Rhizobium etli*, *Rhodopseudomonas palustris* and *Frankia alni* were also found. Interestingly, the arrangement of genes differed from that of *S. meliloti*. For example, on the clone carrying matches to *smc00861-smc00865*, there were matches to sequences flanking the *S. meliloti* gene *smc00864*, but not to the gene itself.

In preliminary quantitative RT-PCR analysis, seven genes were up-regulated under salt stress. One of the up-regulated genes coded for an isopentenyl monophosphate kinase, one for a methyl-accepting chemotaxis protein (Mcp), and five for hypothetical proteins. The up-regulation of the isopentenyl monophosphate kinase could result from the enhanced metabolic activity of the rhizobia under stress. Isopentenyl monophosphate kinases take part in the synthesis of isoprenoids, which are involved in respiration and cell-wall biosynthesis. The match to the Mcp was to the nitrate- and nitrite-sensing domain of the Mcp. Mcps, like two component histidine kinases, couple environmental to cellular responses. The genome segment carrying the *mcp*-like sequence will serve as a starting point for sequencing the genome further to find whether the up-regulated gene codes for a chemotactic or other response-linked gene.

F. D. Dakora et al. (eds.), *Biological Nitrogen Fixation: Towards Poverty Alleviation through Sustainable Agriculture*.
© Springer Science + Business Media B.V. 2008

THIOL-REDOX POTENTIAL IN MELANIN PRODUCTION AND NITROGEN FIXATION BY *SINORHIZOBIUM MELILOTI*

S. Castro-Sowinski, O. Matan, L. Star and Y. Okon

The Hebrew University of Jerusalem, Faculty of Agricultural, Food and Environmental Quality Sciences, Rehovot Campus, 76100 Israel (Email: okon@agri.huji.ac.il)

Melanin is a dark brown pigment produced by many bacteria, fungi, plants and animals. It is synthesised by the oxidative polymerisation of polyphenolic compounds involving enzymes, such as tyrosinase and/or laccase. Its production has been linked to resistance to UV and visible light-irradiation and enhanced survival and competitive abilities under environmental stresses. In addition, melanisation is important for both plant- and animal-pathogenic fungi because it reduces the susceptibility of melanised microbes to host-defence mechanisms.

Castro-Sowinski et al. (2000, 2002) reported that many Uruguayan native sinorhizobia produced melanin and showed for the first time that sinorhizobia possess tyrosinase and laccase activity. Laccases are polyphenol oxidases that are widely distributed in plants and fungi, but their occurrence has also been reported among a few prokaryotes. They belong to the multi-copper oxidase family. The melanin-producing isolate CE52G laccase is a homodimeric protein with an apparent molecular mass of 45 kDa for each subunit (Rosconi et al., 2005). A CE52G mutant that does not produce melanin was mapped to a thioredoxin-like gene. The mutant was impaired in the response to paraquat-induced oxidative stress, symbiotic nitrogen fixation, and laccase and tyrosinase activities (S. Castro-Sowinski et al., unpublished data, 2006). Thioredoxins are responsible for maintaining disulfide bonds in cytoplasmic proteins in a reduced state. Thus, laccase and tyrosinase may be included within the proteins regulated by the redox state of structural or catalytic SH groups via a disulfide oxido-reduction system. Thus, *Sinorhizobium* thioredoxin likely affects nitrogen fixation by the modulation of the redox potential of the bacterium in the nodule.

References

Castro-Sowinski S et al. (2000) J. Microbiol. Meth. 41, 173–177.
Castro-Sowinski S et al. (2002) FEMS Microbiol. Lett. 209, 119–125.
Rosconi F et al. (2005) Enz. Microb. Technol. 36, 800–807.

F. D. Dakora et al. (eds.), *Biological Nitrogen Fixation: Towards Poverty Alleviation through Sustainable Agriculture.*
© Springer Science + Business Media B.V. 2008

USE OF CHIMERIC PROTEINS TO INVESTIGATE THE ROLE OF THE GAF DOMAIN OF *HERBASPIRILLUM SEROPEDICAE* NIFA IN THE RESPONSE TO FIXED NITROGEN

R. A. Monteiro[1], R. Dixon[2], R. Little[2], M. A. S. Oliveira[1], R. Wassem[3], M. B. R. Steffens[1], L. S. Chubatsu[1], L. U. Rigo[1], F. O. Pedrosa[1] and E. M. Souza[1]

[1]Department of Biochemistry and Molecular Biology, CP 19046, Universidade Federal do Parana, Curtiba, Brazil; [2]Department of Molecular Microbiology, John Innes Centre, Norwich, UK; [3]Department of Genetics, C. Postal 19046, Universidade Federal do Paraná, Curitiba, Brazil

Herbaspirillum seropedicae (*Hs*) NifA activity is controlled by external ammonium and O_2 levels. Ammonium control involves the N-terminal GAF domain and P_{II}-like signal-transduction proteins (Souza et al., 1999). When expressed separately the N-terminal GAF domain of NifA inhibits the activity of AAA+ and C-terminal domains in the presence of excess ammonium (Monteiro et al., 1999). To study the response of the N-terminal GAF domain to ammonia, we constructed chimeric proteins between the GAF domain of *Hs* NifA and the AAA+ and C-terminal domains of *K. pneumoniae* (*Kp*) and *A. vinelandii* (*Av*) NifA, which are not regulated by O_2 (Dixon, 1998).

The NifA chimeric proteins activities were tested by monitoring expression of a *nifH-lacZ* reporter fusion. Cultures were grown aerobically, either under N-limiting or N–excess (NH₄Cl, 20 mM) conditions. *In vivo* studies showed that *Hs* NifA is inactive in *E. coli*. In contrast, the chimeric proteins, HsGAF+KpAAA+-C-terminal and HsGAF+AvAAA+-C-terminal, are more active in *E. coli* under N-limiting conditions. Thus, the activity of the *Kp* and *Av* AAA+-C-terminal domains are regulated by the *Hs* GAF domain in response to ammonium levels. Similar levels of NifA activity were observed in the wild type and the *glnB* mutant strain, suggesting that *E. coli* GlnB is not involved in the regulation of the *Hs* NifA GAF domain. However, in the *E. coli glnB-glnK* and *glnK* mutant strains, NifA activity was similar in both N-limiting and N-excess conditions, suggesting that GlnK may be responsible for the increased activity of the chimeric proteins. *E. coli* GlnK seems to play a role in the regulation of the GAF

F. D. Dakora et al. (eds.), *Biological Nitrogen Fixation: Towards Poverty Alleviation through Sustainable Agriculture.*
© Springer Science + Business Media B.V. 2008

domain of *Hs* NifA and may signal the nitrogen status to this domain. Interaction of the uridylylated form of GlnK with GAF may increase the activity of the chimeric proteins.

This research was supported by JIC, Instituto do Milênio/MCT/CNPq, and CAPES.

References

Souza EM et al. (1999) J. Bacteriol. 181, 681–684.
Monteiro RA et al. (1999) FEMS Microbiol. Lett. 180, 157–161.
Dixon R (1998) Arch. Microbiol. 169, 371–380.

MOLECULAR CHARACTERIZATION OF SYMBIOTIC BACTERIA ASSOCIATED WITH THE PASTURE LEGUME *ORNITHOPUS* SP. NATIVE TO PORTUGAL

C. Vicente[1], M. A. Pérez-Fernández[2], R. Costa[1], G. Pereira[1], N. Simões[1] and M. Tavares-de-Sousa[1]

[1]Estação Nacional de Melhoramento de Plantas, Apartado 6, 7350 Elvas, Portugal; [2]Área de Ecologia, Dpto. de Ciencias Ambientales, Universidade Pablo de Olavide, Carretera a Utrera, Km 1, 41013 Sevilla, Spain

Serradela, *Ornithopus* sp., is an annual pasture legume characterized by its ability to tolerate stress conditions, such as drought, and saline and infertile soils (Ovalle et al., 2006), thus being a good candidate to rehabilitate land quality and improve soil structure. This legume establishes symbiotic associations with *Bradyrhizobium* spp. (Safronova et al., 2004), a slow-growing bacterium, that enters dormancy faster and survives desiccation better than other rhizobial species. This study aimed to identify rhizobial diversity associated with leguminous plants from the genus *Ornithopus*, using serradela nodules collected in five different locations in Portugal.

Forty-six isolates were characterized by their phenotype (morphologic traits) and genotype (rep-PCR, using BOX and ERIC primers). Isolates displayed *Bradyrhizobium* characteristics. A total of seven bands were considered for the genetic analysis. The dendrogram (Figure 1) reflects the biodiversity of serradela isolates, suggesting that serradela may be promiscuous with respect to its symbiont and possibly nodulating with other rhizobia. Twenty-three isolates were chosen for cross-inoculation tests on *Biserrula pelecinus*, a pasture legume that behaves like serradela under adverse environments. The specificity of *Bradyrhizobium* isolates towards its host legume and biserrula concerning its microsymbionts will be analysed.

F. D. Dakora et al. (eds.), *Biological Nitrogen Fixation: Towards Poverty Alleviation through Sustainable Agriculture*.
© Springer Science + Business Media B.V. 2008

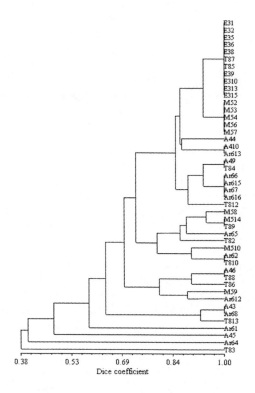

Figure 1. Dendogram UPGMA of serradela isolates (A, Ajuda; Ar, Arronches; E, Elvas; M, Monsaraz; T, Terena).

References

Ovalle C et al. (2006) Agric. Téc. 66, 196–209.
Safronova VI et al. (2004) Ant. van Leewenhoek 85, 115–127.

SIDEROPHORE GENES IN GRAM POSITIVE AND GRAM NEGATIVE NITROGEN-FIXING BACTERIA

A. Giongo, A. Beneduzi, A. Ambrosini, R. Farina, P. B. Costa,
S. B. Campos, M. H. Bodanese-Zanettini and L. M. P. Passaglia

Núcleo de Microbiologia Agrícola, Departamento de Genética,
Universidade Federal do Rio Grande do Sul, RS, Brazil. C. P. 15053,
91501-970, Porto Alegre, RS
(Email: lpassaglia@terra.com.br)

Although iron is one of the most abundant elements, its assimilation depends on the development of effective iron-sequestration and uptake strategies by microorganisms. Its most common states, ionic forms, especially iron(III), are insoluble under physiological conditions (Neilands, 1995). To solubilize iron, many microbes synthesize and utilize siderophores, which are relatively low molecular weight, ferric ion-specific chelating agents. When grown under iron-deficient conditions, microorganisms synthesize and excrete siderophores to sequester and solubilize iron from the environment. Many bacteria and fungi produce more than one type of siderophore or have more than one iron-uptake system to take up multiple siderophores (Neilands 1981). Research focusing on the iron-siderophore transport systems is not only important to understand how cells are able to acquire iron, but is also needed to understand transport mechanisms as a whole. Although Gram-negative and -positive bacteria have differences in their cell structure, they share some genes in common for both specific siderophore transport and iron-binding proteins (Clarke et al., 2000).

Aiming to identify genes related to siderophore transport and the differences between the two bacteria types, we have chosen one nitrogen-fixing bacterium from each group; *Paenibacillus* (Gram positive) and *Bradyrhizobium* (Gram negative). Transport and regulator genes, like *fhuA*, *fegA* and *fur*, were identified and characterized using PCR and Southern-blot hybridization techniques. We confirmed the presence of siderophore uptake systems through the amplification of these three genes both bacteria. The respective gene products differed in structure between the two bacterial groups. Further experiments using mutants will elucidate the siderophore-uptake mechanisms in these two groups of bacteria.

F. D. Dakora et al. (eds.), *Biological Nitrogen Fixation: Towards Poverty Alleviation through Sustainable Agriculture.*   379
© Springer Science + Business Media B.V. 2008

References

Clarke et al. (2000) Nature Struct. Biol. 7, 287–291.
Neilands JB (1981) Ann. Rev. Biochem. 50, 715–731.
Neilands JB (1995) J. Biol. Chem. 270, 26723–26726.

GENOMICS AND GENETIC TRANSFORMATION IN THE UNICELLULAR DIAZOTROPHIC CYANOBACTERIUM *CYANOTHECE* SP. ATCC 51142

H. Min[1], E. A. Welsh[2], H. B. Pakrasi[2] and L. A. Sherman[1]

[1]Department of Biological Sciences, Purdue University, West Lafayette, IN 47907, USA; [2]Department of Biology, Washington University, St. Louis, MO 63130, USA

Cyanothece sp. ATCC 51142, is a unicellular, diazotrophic cyanobacterium with a versatile metabolism and very pronounced diurnal rhythms. When grown under 12 h light-dark (LD) periods, it performs photosynthesis during the day and N_2 fixation and respiration at night. Such interesting metabolism and regulation has led to the sequencing of its genome. The genome includes a large cluster that contains all of the genes required for nitrogenase assembly and function. We have developed a genetic transformation system for *Cyanothece* by using electroporation and $CaCl_2$ treatment. The effectiveness of the treatment and the efficiency of the transformation depend upon metabolic modes of growth.

A broad-host-range plasmid, pRL1383a (kindly provided by Dr. Peter Wolk), was introduced into *Cyanothece* by transformation. The transformants were stable and the plasmid was isolated from the transformed strain and reintroduced into *E. coli* by electroporation. We have used this system to construct mutants in genes such as *ntcA* and *nifHDK* by using single- or double-recombination events. However, unlike other cyanobacteria, *Cyanothece* seems to have a high level of non-homologous recombination.

Genome sequencing data and optical mapping of the *Cyanothece* DNA suggest that there is a 0.43 Mb (about 425 genes) linear chromosome in addition to a 4.9 Mb (about 4,750 genes) circular chromosome. There are phage-like recombinase/integrase genes on the linear DNA. The linear DNA may complicate recombination and may take up exogenous genes by non-homologous recombination. We now are developing a transformation protocol by using single-stranded DNA in order to favor homologous recombination so that we can knock out genes important for nitrogen metabolism.

F. D. Dakora et al. (eds.), *Biological Nitrogen Fixation: Towards Poverty Alleviation through Sustainable Agriculture.*
© Springer Science + Business Media B.V. 2008

SIPA, A NOVEL REGULATOR OF THE PHOTOSYNTHESIS-RELATED SENSOR HISTIDINE KINASE NBLS

P. Salinas, D. Ruiz and A. Contreras

División de Genética, Universidad de Alicante, Apartado 99, E-03080 Alicante, Spain

Cyanobacteria respond to fixed-nitrogen (N) deprivation and other stress conditions by degrading their phycobilisomes (PBS), the light-harvesting complexes for photosynthesis. The small protein, NblA, is a key effector of this process of chlorosis or bleaching. In *Synechococcus* sp. PCC 7942, transcription of the *nblA* gene is regulated in response to a variety of signals by NblS and by the response regulator, NblR. In previous work (Espinosa et al., 2006), we used yeast two-hybrid approaches to identify a protein, named SipA (Nbl<u>S</u> interacting <u>p</u>rotein <u>A</u>), which binds to the ATP-binding domain of NblS. Here, we perform genetic analysis to infer the role of SipA in chlorosis and survival.

Inactivation of the *sipA* gene had a very small effect on pigment content and *nblA* expression. SipA⁻ cultures subjected to N starvation or high light stress degraded their PBS normally. Interestingly, constitutive expression of the *sipA* gene from an ectopic promoter (SipA^C strain) resulted in a strong *non-bleaching* phenotype. The SipA^C mutant had a higher level of phycocyanin than the wild type and was not able to reduce them under high light or N stresses.

The lethality of the NblR⁻ mutant was significantly suppressed by inactivation of *sipA*, suggesting that SipA impairs recovery from chlorosis in the absence of a functional *nblR* gene and that signals from NblS-SipA could be transmitted to the target genes by an additional response regulator.

We propose that, under standard growth conditions, SipA (in concert with NblS) prevents *nblA* expression and stimulates photosynthetic activity. In the absence of a functional NblR protein, this stimulatory effect would result in failure to recover from chlorosis.

Reference

Espinosa J et al. (2006) FEMS Microbiol. Lett. 254, 41–47.

F. D. Dakora et al. (eds.), *Biological Nitrogen Fixation: Towards Poverty Alleviation through Sustainable Agriculture.*
© Springer Science + Business Media B.V. 2008

T